Current Topics in the Theory and Application of Latent Variable Models

This book presents recent developments in the theory and application of latent variable models (LVMs) by some of the most prominent researchers in the field. Topics covered involve a range of LVM frameworks including item response theory, structural equation modeling, factor analysis, and latent curve modeling, as well as various non-standard data structures and innovative applications.

The book is divided into two sections, although several chapters cross these content boundaries. Part one focuses on complexities which involve the adaptation of latent variable models in research problems where real-world conditions do not match conventional assumptions. Chapters in this section cover issues such as analysis of dyadic data and complex survey data, as well as analysis of categorical variables. Part two of the book focuses on drawing real-world meaning from results obtained in LVMs. In this section there are chapters examining issues involving assessment of model fit, the nature of uncertainty in parameter estimates, inferences, and the nature of latent variables and individual differences.

This book appeals to researchers and graduate students interested in the theory and application of latent variable models. As such it serves as a supplementary reading in graduate level courses on latent variable models. Prerequisites include basic knowledge of latent variable models.

Michael C. Edwards is an Associate Professor in the Quantitative Area of the Department of Psychology at The Ohio State University.

Robert C. MacCallum is Professor Emeritus of Psychology at both the University of North Carolina at Chapel Hill and The Ohio State University.

Current Topics in the Theory and Application of Latent Variable Models

Edited by
Michael C. Edwards
The Ohio State University
and
Robert C. MacCallum
The University of North Carolina at Chapel Hill

Routledge
Taylor & Francis Group
NEW YORK AND LONDON

First published 2013
by Routledge
711 Third Avenue, New York, NY 10017

Simultaneously published in the UK
by Routledge
27 Church Road, Hove, East Sussex BN3 2FA

Routledge is an imprint of the Taylor & Francis Group, an informa business

Library of Congress Cataloging-in-Publication Data
Current topics in the theory and application of latent variable models / edited by Michael C. Edwards and Robert C. MacCallum.
p. cm.
1. Latent structure analysis. 2. Latent variables. I. Edwards, Michael C. (Michael Charles), 1977– II. MacCallum, Robert C.
QA278.6.C87 2013
519.5'35–dc23 2012017548

ISBN: 978-1-84872-951-3 (hbk)
ISBN: 978-0-415-63778-7 (pbk)
ISBN: 978-0-203-81340-9 (ebk)

Typeset in Times New Roman
by Cenveo Publisher Services

Contents

Foreword

I have lived long enough that I was privileged to know many of the revered stalwarts of quantitative psychology and, when a graduate student, was fortunate to have taken classes with several of them (R. E. Bargmann, R. B. Cattell, L. J. Cronbach, L. G. Humphreys, H. F. Kaiser, L. R Tucker). Another quantitative psychologist, however, one much nearer my own age, remains one of my favorite QPIHKPs (quantitative psychologists I have known personally). He is Michael W. Browne, the honoree of this volume. I was well aware of Michael Browne's work and reputation before I first met him in 1972 at the University of Pennsylvania, where we were attending a workshop on multidimensional scaling sponsored by Bell Labs and the University of Pennsylvania. We became acquainted both during that workshop's scheduled events, which included a "rip-roaring"' evening talk by Warren G. Torgerson, and at an informal dinner at a middle-Eastern restaurant in Philadelphia featuring an attractive belly-dancer. The food was not memorable; the dancer, however, was.

That Michael Browne was truly "good," to use an adjective that was common then to characterize someone who was clearly rising to the top of his or her field, was obvious from the outset. Had I been more prescient, I would have realized that he was to become one of the very best ever in the quantitative modeling of multivariate data. As readers of this volume will soon learn, Michael's contributions to estimation, model fitting and evaluation, factor analysis, time series modeling, etc., have been extremely influential for the past four decades and will remain so, for the foreseeable future.

But I saw later in other venues, such as meetings of the Society of Multivariate Experimental Psychology, that there was much more to Michael Browne than supreme expertise in statistics and modeling. Michael possessed a precision of thinking, a dignity, an integrity, and a sense of fair play that has been all too rare in my time. His standards of professional ethics and behavior are impeccable—we would all do well to emulate them.

Michael's influence on my own thinking has been tremendous. When he spent a sabbatical quarter at the University of Virginia in 1999 I daresay that I benefitted more from his company than anyone. Michael and I had the opportunity to become much better acquainted through direct intellectual exchange, and did so in numerous discussions and friendly arguments over lunches, dinners, and not a few evening libations of a certain kind of tea that, while approximately the right color, was served over ice cubes and tasted remarkably like Jack Daniels.

During the past decade, Michael Browne and I have worked together on a number of research papers, and that collaboration has been a career highlight for me. I am proud that we have emphasized a focus on the individual as the unit of analysis in behavioral research, and on the study of process and change rather than static attributes. We have enthusiastically promoted multivariate measurement applied on many successive occasions as a path for realizing these emphases. In turn, this emphasis has fostered the role of Cattell's P-technique factor analysis as much more than a covariation technique but, rather, as a way of thinking about empirical research on behavior. This effort has been helped immeasurably by capitalizing on Michael Browne's expertise in time series modeling. In the process, we have stood shoulder-to-shoulder to question some venerated traditions and blindly adhered-to dogma of our field, and have suggested some alternatives to the dominant practices of today that seem to us to warrant reconsideration. After several unsuccessful tussles with leading quantitative psychology journals regarding these matters, we remain "bloody but unbowed" and continue to push for broadening perspectives on key matters having to do with measurement and modeling.

In short, my collaboration with Michael Browne is an honor and a privilege about which I have only two regrets—it began too late and has been much too infrequent! Having an opportunity to enter into the written record these comments acknowledging Michael Browne's singular contributions to his field as well as to my own career at such an appropriate place and time as this festschrift honoring his work affords me the greatest pleasure.

John R. Nesselroade
University of Virginia

Preface

In 1904 Charles Spearman published an article of monumental importance in the history of psychology in which he proposed and sought to provide compelling evidence for his theory of the structure of intelligence. Of equal importance within that paper was Spearman's development of basic concepts and methods of what was in fact a latent variable model; in particular, a factor analysis model positing that correlations among cognitive tests could be accounted for by the effects of the latent variable "general intelligence" on those tests. Spearman didn't use the term "factor analysis" or "latent variable," and didn't call his proposed structure a model. But it was, as we can see now, a latent variable model. And he attempted to evaluate that model by showing that it could accurately account for empirical data. In the ensuing century, latent variable modeling (LVM) has grown into a broad field characterized by intensive methodological development and research. Techniques such as structural equation modeling (SEM), item response theory (IRT), multilevel models, latent class models, and more have been developed and are used extensively across a wide range of scientific disciplines.

In this context of continuing growth in the theory, methods, and application of LVM, the purpose of the current volume is two-fold. Our first objective is to provide a snapshot of current topics and ongoing research in LVM methodology and application. The chapters in this volume were contributed by participants in a conference held in September, 2010 focusing on current developments in theory and application in this field. Importantly, there is considerable diversity among the contributors on several dimensions. Some are senior investigators, some are more junior. There are quantitative psychologists, statisticians, and investigators focusing on applied problems in areas such as social and cognitive psychology. Particular LVM interests of the contributors vary across issues in factor analysis, SEM, IRT, as well as across different types of data structures and applied issues. As a group, the contributors reflect a range of methodological and applied interests in LVM as well as a breadth and depth of expertise.

The second purpose of this volume is to recognize and honor the scientific contributions of a key figure in the history of LVM – Michael W. Browne, Professor Emeritus of Psychology and Statistics at The Ohio State University. Professor Browne has been at the center of development of theory and methods in LVM for more than 40 years. He has made major contributions to topics such as estimation of model parameters, robustness to violation of assumptions, model evaluation, factor rotation, nonlinear latent curve models, and models for multitrait-multimethod data. In recent years he has pursued highly innovative development of dynamic factor analysis models for time series data. Throughout his career, Professor Browne has been known for his deep insights into LVM issues as well as for his thorough analysis and solution of problems and the highest level of scientific rigor. The conference that eventually resulted in the publication of this volume was organized and convened to honor Professor Browne's most distinguished career.

The 15 invited speakers were asked to deliver presentations, and subsequently to provide chapters for this volume, on a topic of their choosing fitting the occasion and the focus of the conference. The result was a collection of high-quality contributions reflecting an array of methodological issues and applied topics within LVM. The resulting chapters came to be organized across two themes. The first is *complexities*, referring to the extension and adaptation of standard LVM in order to manage the various complexities of data and research questions that go beyond the standard frameworks. Some such issues include the modeling of dyadic data, models involving categorical variables, and models for complex survey samples. The second theme is *meaning*, referring to the drawing of real-world meaning from results obtained in LVM. This section of the book includes contributions on issues of model fit and statistical inference, among other topics. While the book is organized into these sections, this thematic structure is not rigid - beyond the quality of the individual contributions, a strength of the book is that many of the chapters cross these boundaries.

The book is intended to be of value and relevance to several audiences. One is researchers studying, developing, or applying latent variable models such as factor analysis, SEM, and IRT. Another is graduate students seeking readings about current issues and ongoing work in this field. Many of the chapters should provide useful material in graduate seminars on these LVM topics. The chapters vary somewhat in technical complexity, depending on the topic and the intention of the author(s), but readers of any of the chapters should already be familiar with basic models and methods (e.g., factor analysis, SEM, etc.) relevant to the chapters of interest.

As organizers of the conference, editors of this volume, and colleagues and friends of the honoree, we hope that this volume contributes in a useful way to the body of literature on LVM, and also that it serves as an appropriate honor to Michael W. Browne and his distinguished place in the history of this field.

Robert C. MacCallum
Michael C. Edwards

Acknowledgements

This volume is based on presentations made by the authors at a conference held at The Ohio State University, September 9-10, 2010. The conference was held to celebrate the lifetime of scientific contributions by Professor Michael W. Browne. The conference was made possible in part by generous contributions from the Departments of Psychology and Statistics at Ohio State, the Society of Multivariate Experimental Psychology, and Scientific Software International, Inc.

The staffs at the Ohio Union and the Blackwell Hotel were tremendously supportive in helping our speakers feel welcome and in making the event enjoyable for all participating. Additional assistance was provided by Angela Mercer, who helped coordinate the entire event and insured its smooth running. We would also like to thank the graduate students and faculty in the Quantitative Psychology program at Ohio State for their efforts in support of the conference.

We would also like to thank those who gave us valuable feedback on the book proposal: Ronald H. Heck, University of Hawaii at Manoa, Melanie M. Wall, Columbia University, Jason T. Newsom, Portland State University, and one anonymous reviewer. Thanks are also due to Debra Riegert and the support staff at Taylor and Francis. Her guidance throughout this process has been invaluable.

Finally, we would like to express our appreciation to all of the speakers and other attendees of the conference. The quality of the talks, the caliber of the conversations, and the sincere appreciation of Michael's work all helped to make this a once in a lifetime experience.

Contributors

William Howard Beasley is interested in graphical data analysis and computationally intensive procedures including resampling methods and Bayesian estimation routines. He has conducted simulation studies of Bayesian inference, bootstrapping, IRT DIF analysis, and multilevel models with multiple dependent variables. He has worked on applied problems involving spatial association, behavior genetics, microbial communities, cardiovascular rehabilitation, adolescent development, and screening test validity. Will is President of Howard Live Oak, LLC (a small statistical consulting and software company) and an Assistant Professor of Research at the University of Oklahoma Health Sciences Center (Department of Pediatrics). In 2010 he received his PhD in Quantitative Psychology from the University of Oklahoma, Norman.

Steven M. Boker is Professor of Quantitative Psychology and directs the Quantitative Psychology PhD program, the Human Dynamics Laboratory, and the LIFE Academy at the University of Virginia. He is an internationally recognized expert in modeling longitudinal data from a dynamical systems perspective. Dr. Boker's novel developments include the Differential Structural Equation Modeling (dSEM) and Latent Differential Equations (LDE) methods for testing and comparing models of dynamical systems in mixed longitudinal and cross-sectional data, and the Windowed Cross-Correlation (WCC) method for determining nonstationary relative phase in multivariate time series. Dr. Boker developed a path analysis method (RAMpath) that is used by modern structural equation modeling programs for calculation of components of covariance and converting between path diagrams and covariance expectation algebra. He is lead Principal Investigator on two project grants: from the National Institutes for Health Roadmap for development of the OpenMx Structural Equation Modeling software (http://openmx.psyc.virginia.edu); and from National Science Foundation for a project using real-time facial tracking avatars to study nonverbal communication of emotion during dyadic conversation (http://faculty.virginia.edu/humandynamicslab). Dr. Boker has edited two books and is author or co-author of over 60 journal articles and book chapters.

Michael W. Browne obtained his BA (1958) and MSc (1965) degrees from the University of the Witwatersrand and his PhD (1969) from the University of South Africa. He was employed at the South African National Institute for Personnel Research from 1959 to 1973, ending as Chief Research Officer, and was Professor of Statistics at the University of South Africa from 1973 to 1990. During this time he spent several years on sabbatical in the USA, being a Visiting Research Fellow in Frederic Lord's group at Educational Testing Service in 1966, 1967, and 1972, Visiting Scholar at UCLA in 1980, and Hill Visiting Professor at the University of Minnesota in 1988. He immigrated to the USA in 2001 and was Professor of Psychology and of Statistics at The Ohio State University (OSU) from January 1991 to June 2011. Currently he is Emeritus Professor at OSU. Dr Browne's research has been primarily concerned with statistical modeling of multivariate psychological data. He has made contributions on asymptotically distribution free estimation in the analysis of covariance structures, asymptotic robustness of multivariate normal theory against violation of assumptions, multiplicative models for multitrait–multimethod data, circumplex models for investigating the personality circle, nonlinear latent curve models, aspects of cross-validation, rotational methodology for factor analysis, and dynamic factor analysis. He is a past president of the South African Statistical Association, the Psychometric Society, and the Society for Multivariate Experimental Psychology. He also is the recipient of Lifetime Career Awards from the Society for Multivariate Experimental Psychology and the Psychometric Society.

Li Cai is an Associate Professor of Advanced Quantitative Methodology in the Graduate School of Education and Information Studies and Associate Professor of Quantitative Psychology (joint appointment) in the Department of Psychology at University of California, Los Angeles (UCLA). He received his PhD in Quantitative Psychology from the L. L. Thurstone Psychometric Lab at the University of North Carolina at Chapel Hill. Broadly speaking, Li's methodological research agenda involves the development, integration, and evaluation of latent variable modeling techniques that have wide-ranging applications in educational, psychological, and health-related domains of study. A key component of this agenda is statistical computing, particularly as related to item response theory (IRT) and multilevel modeling. Li has developed new algorithms that can significantly improve the accuracy and efficiency of parameter estimation for multidimensional IRT models. These methods now form much of the computational basis of the new IRT programs IRTPRO (from Scientific Software International, Inc.) and flexMIRT (from Vector Psychometric Group, LLC). The development of both programs was led by Li. Li collaborates with substantive researchers at UCLA and RAND on National-Institute-on-Drug-Abuse-funded projects examining measurement issues in substance abuse, treatment, and outcomes research. Li is a member of American Educational Research Association, American Statistical Association, National Council on Measurement in Education, and Psychometric Society.

Peter F. Craigmile received a BSc in Mathematics and Statistics from the University of Glasgow, Scotland, in 1996. He received the Diploma in Mathematical Statistics from Cambridge University, England in 1997, and a PhD in Statistics from the University of Washington, Seattle in 2000. He is currently a Reader in Statistics in the School of Mathematics and Statistics at the University of Glasgow, and an Associate Professor in the Department of Statistics at The Ohio State University. His research interests include time series analysis, spatial statistics, space-time modeling, and longitudinal methods. He is interested in the use of spectral and wavelet methods to investigate dependency structures and to analyze periodicities and trends. One application of this is to the study of long memory processes. He also works on methodology for spatial extrema and exceedances. He enjoys application-oriented research in areas such as climatology, environmental sciences, and psychology. He is a member of the American Statistical Association, the Institute of Mathematical Statistics, The Royal Statistical Society, and the International Environmetrics Society.

Robert Cudeck received his PhD from the University of Southern California and is currently a member of the Psychology Department at The Ohio State University. His research activities are concerned with applications of factor analysis, structural equation models, and random coefficient models for repeated measures data in the behavioral sciences.

Kristin A. Duncan completed her graduate studies at The Ohio State University and is Assistant Professor of Statistics at San Diego State University. Her research interests include Bayesian and nonparametric Bayesian models for item response data with applications in education and health.

Stephen du Toit is the Senior Vice President of Scientific Software International, Inc. He received his PhD in Mathematics and Statistics from the University of South Africa. He is a recipient of the University of Pretoria medal for outstanding achievement in research (1990) and the P. M. Robbertse medal for outstanding contribution towards research at the Human Sciences Research Council of South Africa (1985). Stephen has worked on numerous statistical software packages such as LISREL, a program for fitting structural equation models, SuperMix, a program for multilevel modeling of continuous and categorical outcomes, and IRTPRO, a program for fitting unidimensional and multidimensional IRT models. In addition to his involvement in the technology field, he currently holds the positions of Adjunct Professor in the Department of Psychology at the University of North Carolina at Chapel Hill, and Professor Emeritus in the Department of Statistics at the University of Pretoria, South Africa. Stephen is a Fellow of the South African Statistical Association and a member of the American Psychometric Society. He has reviewed technical papers for several journals, including *Psychometrika*, *Multivariate Behavioral Research*, and *Structural Equation Modeling*. His publications cover topics on optimization techniques, structured means and covariances, nonlinear growth models, and vector time series processes.

Michael C. Edwards is an Associate Professor in the Quantitative Area of the Department of Psychology at The Ohio State University. He received his PhD in 2005 from the L. L. Thurstone Psychometric Laboratory at the University of North Carolina at Chapel Hill. His research focuses on measurement issues in the social sciences with specific topics including multidimensional item response theory, computerized adaptive testing, local dependence, and measurement invariance.

Richard Gonzalez received his PhD from Stanford University and is currently Professor at the University of Michigan in the Psychology, Statistics, and Marketing departments. He is also a Research Professor in the Research Center for Group Dynamics and in the Center for Human Growth and Development. His general area of research is in judgment and decision making.

Dale Griffin received his PhD from Stanford University and is currently Professor and Associate Dean in the Sauder School of Business, University of British Columbia. His general area of research is in judgment and decision making and prediction processes in consumers and managers. Gonzalez and Griffin became interested in dyadic processes in the early 1990s when studying social interaction in dryads. The available analytic models made an assumption of independence, which was counter to the key psychological phenomenon of interdependence present in dyadic data.

Jeffrey R. Harring is an Associate Professor in Measurement, Statistics and Evaluation at the University of Maryland, College Park. Dr. Harring teaches quantitative methods courses encompassing topics on experimental design and multiple regression analysis. He also teaches advanced graduate seminars on longitudinal data analysis, simulation design, computational statistics, and finite mixture models. His research focuses on nonlinear models for repeated measures data, nonlinear structural equation models, and finite mixtures of both linear and nonlinear growth models. Dr. Harring has published methodological papers in prominent journals such as *Multivariate Behavioral Research*, *Psychological Assessment*, *Journal of Educational and Behavioral Statistics*, *Structural Equation Modeling*, and co-authored an invited submission in the *Annual Review of Psychology*. Recently, Dr. Harring co-authored a book entitled, *Comparing Groups: Randomization and Bootstrap Methods Using R*, which was published by Wiley in June, 2011. He served as the program chair for AERA: Division D, Section 2—Statistical Theory and Methods from 2008 to 2010. Dr. Harring has also been a reviewer for many publications including *Multivariate Behavioral Research*, *Journal of Educational and Behavioral Statistics*, *Psychological Methods*, *Psychometrika*, *Structural Equation Modeling*, *Statistics in Medicine*, *Computational and Graphical Statistics*, and *The Journal of Applied Statistics*.

Chun-Ting Lee is currently a PhD student in the quantitative psychology program at the University of Notre Dame. His main area of interest is in factor analysis and structural equation modeling with ordinal variables. He is also interested in the bootstrap and robust methods for latent variable models.

Taehun Lee received his PhD in quantitative psychology from the University of North Carolina at Chapel Hill in 2010, where he worked with Dr. Robert MacCallum. He acquired his MS degree in mathematical statistics from the Department of Statistics at the University of North Carolina at Chapel Hill in 2009. His research interests concern methodological/statistical issues surrounding estimation, evaluation, and selection of latent variable models, including factor analysis models, structural equation models, multilevel models, and item response theory models. His current research in this area involves addressing uncertainties in parameter estimation, model-data fit evaluation, and missing data imputation. He is also interested in applications of statistical models for the study of individual differences and change over time. He is currently a post-doctoral researcher in the advanced quantitative methodology program in the School of Education at the University of California at Los Angeles.

Robert C. MacCallum is Professor Emeritus of Psychology at the University of North Carolina (UNC) at Chapel Hill. He received his graduate training at the

University of Illinois and then spent 28 years on the faculty in the Quantitative Psychology program at The Ohio State University, moving to UNC in 2003. His research interests focus on methods for analysis and modeling of correlational and longitudinal data, including factor analysis, structural equation modeling, and latent curve models. Within these areas he has worked on various issues including model estimation and evaluation, power analysis for testing models, and the nature and management of sources of error in modeling. His current interests involve the study of uncertainty inherent in results of statistical models. He is former Director of the L. L. Thurstone Psychometric Laboratory at UNC, and former president of the Society for Multivariate Experimental Psychology. In 2011 he received the Samuel J. Messick Award for distinguished scientific contributions from Division 5 of the American Psychological Association.

Steven N. MacEachern is Professor in the Department of Statistics at The Ohio State University, where he has been since receiving his PhD in Statistics from the University of Minnesota. His undergraduate training was in mathematics at Carleton College, Minnesota. He has been a visiting faculty member at Duke University and Carnegie Mellon University. Steve's research interests range widely, from methodology to applications to computation, including both classical and Bayesian paradigms. He has worked extensively on nonparametric Bayesian methods and is the developer of the dependent Dirichlet process. His interests in psychology include psychometric measurement and the development of choice models.

Mike Martin, PhD, is Professor for the Psychology of Aging, Director of the Center of Gerontology and Co-director of the International Normal Aging and Plasticity Imaging Center (INAPIC) at The University of Zurich, Switzerland. Dr. Martin received his MA in psychology at The University of Georgia, his PhD at The University of Mainz, Germany, and his habilitation at The University of Heidelberg, Germany. Together with Steve Boker, Dr. Martin is part of the faculty of the international LIFE consortium. In his research the focus is on the longitudinal lifespan development of cognition, social relationships, personality, and quality of life, and how individuals actively manage to stabilize quality of life through the orchestration of resources.

Roger E. Millsap is a Professor in the Department of Psychology at Arizona State University (ASU) and a faculty member in the Doctoral Program in Quantitative Psychology at ASU. He received his PhD in psychology in 1983 from the University of California—Berkeley. Dr. Millsap's research interests include psychometrics, latent variable models, and multivariate statistics. He has published more than 60 papers in professional journals and co-edited the

SAGE Handbook of Quantitative Methods in Psychology with Alberto Maydeu-Olivares in 2009. He recently completed a new book, *Statistical Approaches to Measurement Invariance*, published by Routledge. Dr. Millsap is a Past-President of the Psychometric Society, of Division 5 of the American Psychological Association, and of the Society of Multivariate Experimental Psychology. He is a past editor of *Multivariate Behavioral Research* and is the current executive editor of *Psychometrika*.

Irini Moustaki obtained her PhD in 1996 from the London School of Economics and Political Science. She has been a Reader in Social Statistics in the Department of Statistics at the London School of Economics since 2007. Her main research interests are latent variable models, analysis of categorical data, composite likelihood estimation, detection of outliers, and missing values. She has co-authored two books: *Analysis of Multivariate Social Science Data* (2008) with Bartholomew, Steele and Galbraith, and *Latent Variable Models and Factor Analysis: A Unified Approach* (2011) with Bartholomew and Knott. She is associate editor and member of the editorial board for four journals including *Computational Statistics and Data Analysis* and the *Journal of Educational and Behavioural Statistics*. Some of her main publications are: 'Generalised latent trait models,' *Psychometrika*, 2000 with Knott; 'Factor analysis for ordinal variables: a comparison of three approaches,' *Multivariate Behavioural Research*, 2001 with Joreskog; 'The forward search algorithm for detecting aberrant response patterns in factor analysis for binary data,' *Journal of Computational and Graphical Statistics*, 2009 with Mavridis; 'Composite likelihood inference in latent variable models for ordinal longitudinal responses,' *Psychometrika* (2012) with Vasdekis and Cagnone.

W. Alan Nicewander is Chief Psychometrician for the Pacific Metrics Corporation. He received his PhD in quantitative psychology from Purdue University. Dr. Nicewander is an emeritus faculty member at the University of Oklahoma where he held posts as Professor (1970–94) and Chair (1987–91). Since his retirement from academia, he has served as Director of Statistical Research, at ACT, Inc. (1994–97) and Director of the Personnel Testing Division, Department of Defense (1997–2004). Dr. Nicewander's research has been focused on multivariate statistics and item response models. He currently is an associate editor of *Psychometrika*.

Mario Peruggia is a Professor in the Department of Statistics at The Ohio State University. He received a PhD (1990) and an MS (1986) in statistics from Carnegie Mellon University and a BS (1983) in mathematics from the Università degli Studi di Milano, Italy. Peruggia joined Ohio State as an Assistant Professor in 1990.

While at Ohio State, he also served as a part-time biostatistician (2004–8) in the General Clinical Research Center, and was a Visiting Associate Professor (Winter 2004) in the Department of Psychology. While on leave from Ohio State, he was an Associate Professor (1997–8) in the Department of Health Evaluation Sciences at the University of Virginia. Peruggia is a Fellow of the American Statistical Association (elected 2004). Peruggia's research interests include Bayesian modeling and inference, and modeling of human performance experiments in psychology.

Joseph Lee Rodgers is a Quantitative Psychologist in the Department of Psychology at the University of Oklahoma (OU), where he is a George Lynn Cross Research Professor and the Robert Glenn Rapp Foundation Presidential Professor. He received his PhD from the University of North Carolina in 1981, and began his faculty career at OU in that year. He has held visiting teaching/research positions at The Ohio State University, University of Hawaii, University of North Carolina, Duke University, the University of Southern Denmark, and the University of Pennsylvania. He and his research team have been continuously funded by NIH since 1987 to develop mathematical models of adolescent development, young adult fertility, and family/friendship interactions. His methodological interests include mathematical modeling, resampling theory, linear statistical models, quasi-experimental design methods, exploratory data analysis (EDA), and multidimensional scaling. He has been the editor of the applied methods journal *Multivariate Behavioral Research* from 2006 to 2011, and is a past president of the Society of Multivariate Experimental Psychology, the Society for the Study of Social Biology, and the American Psychological Association's Division 34 (Population and Environmental Psychology). His wife Jacci is chair of the Accounting Division at Oklahoma City University. They have two teenage daughters, Rachel and Naomi.

Megan Schiller has worked for over a decade in pharmaceutical market research. Prior to joining Tateneni Consulting, she was a Senior Manager in the Client Service Group at Eidetics. In addition to managing marketing research assignments, Megan is an experienced qualitative moderator, having worked extensively with physicians, payers, patients, and other healthcare providers. Megan completed a Bachelors of Science in Marketing at Boston College.

Alexander Shapiro is a Professor in the School of Industrial and Systems Engineering at Georgia Institute of Technology. His areas of interest include: stochastic programming, simulation-based optimization of stochastic systems, and multivariate statistical analysis. He has published more than 100 research articles in peer reviewed journals and is a coauthor of several books. His research is widely cited and he is listed as an ISI (Institute for Scientific Information) Highly Cited Researcher in 2004 on http://isihighlycited.com/. He is currently an area editor of *Operations Research* and on the editorial board of several professional journals, such as *Mathematics of Operations Research*, *Mathematical Programming*, *ESAIM: Control, Optimization and Calculus of Variations*.

Krishna Tateneni received his PhD in quantitative psychology from The Ohio State University in 1998. Before starting his own consulting practice, Krishna worked as a Director in the Decision Modeling Group at Eidetics, a pharmaceutical marketing research and consulting firm in Boston. Prior to that he worked as the Lead Statistician for the College Board's Advanced Placement Program at Educational Testing Service (ETS).

Trisha Van Zandt is an Associate Professor in the Department of Psychology at The Ohio State University, where she is a faculty member in the Quantitative Psychology Program. She has a courtesy joint appointment in the Department of Statistics. She received her PhD in Psychology in 1992 from Purdue University. Her research interests include judgment and decision making, and Bayesian statistical methods for the analysis of reaction time data. She is a member of the editorial boards for the *Journal of Mathematical Psychology* and the *Psychonomic Bulletin and Review*. She is past president of the Society for Mathematical Psychology.

Guangjian Zhang is an Assistant Professor in the Department of Psychology at the University of Notre Dame. He received his PhD in quantitative psychology from The Ohio State University in 2006, and also received a MSc in statistics at the same university. His graduate training at The Ohio State University was under the direction of Michael W. Browne. His previous training includes clinical medicine, clinical and counseling psychology, and social psychology. He was a licensed medical doctor in China and practiced psychiatry. His research interests are in factor analysis, structural equation modeling, and time series analysis.

Chapter 1
Complexity and Meaning in Latent Variable Modeling

Michael C. Edwards and Robert C. MacCallum

1.1 INTRODUCTION

Latent variable modeling (LVM) has been in existence for over 100 years. Its first appearance in Spearman's 1904 paper, *"General Intelligence," Objectively Determined and Measured*, was in the form of factor analysis. On the occasion of the 100th anniversary of the publication of Spearman's paper, a conference was organized and an edited volume of chapters was subsequently published (Cudeck & MacCallum, 2007) examining historical developments and modern directions and issues in the field of factor analysis. In one sense, the purpose of the present volume is to expand the scope of that endeavor beyond factor analysis to the broader field of LVM. Despite the century of evolution that saw the rudimentary concepts of factor analysis eventually yield a wide array of latent variable models, when we consider the core of modern LVM methods we see the unmistakable traces of Spearman's work.

Factor analysis itself developed as a primary vehicle for representing and evaluating theories of mental abilities, and of course was eventually used in a similar fashion in other domains (e.g., personality). Methodological developments in factor analysis focused for decades on techniques for exploratory factor analysis, with breakthroughs providing for confirmatory factor analysis coming in the 1970s. Around that time, factor analysis and path analysis were essentially merged to yield structural equation modeling (SEM), a general approach for fitting and testing models of patterns of relationships among measured and latent variables. SEM remains one of the most heavily used statistical methods in psychological research. Special cases of SEM, such as latent curve models for the study of change, and extensions of SEM, such as multigroup models to study group differences, including specific issues such as factorial invariance, are used extensively in applied research. Another major line of development in LVM is represented in measurement theory, first by classical test theory and then by item response theory (IRT), which dominates modern development and analysis of tests and scales. In the study of hierarchically nested data structures, multilevel models can often be recognized as latent variable models; in particular, multilevel models that specify and estimate random coefficients are essentially defining and estimating latent variables. And there is much more. There is melding of models, such as that represented by multilevel SEM; there are models involving categorical latent variables, such as latent class models. The field of LVM is far broader than we can convey in this short summary.

The primary reason for this long and intensive development and study of LVM is of course its relevance to, and successful implementation in, a wide range of empirical research. Latent variables are routinely at the core of research questions across a range of disciplines including psychology and other social sciences, education, health sciences, and much more. LVM provides vehicles for attempting to answer many of those questions. While the application of LVM has been highly successful in practice, methodological research in the area of LVM also remains extremely active. There is much continuing research on the development of new models and methods, extensions and merging of existing models, and the study of performance and properties and relationships among these approaches.

To examine some of the key lines of work in this area, a conference was held at The Ohio State University in September of 2010 titled *Current Topics in the Theory and Application of Latent Variable Models: A Conference Honoring the Scientific Contributions of Michael W. Browne*. The 15 invited speakers were given very little direction about the content of their presentations other than to keep in mind the theme of the conference. The resulting presentations and corresponding chapters provide a fascinating overview of the progress made in LVM since its creation. Despite the lack of any strict thematic requests to the authors, it became clear to us upon reading the chapters that two themes emerged naturally from the contributions. The first theme we are labeling with the general term *complexity*, referring to complexities in data structures and research questions that arise in practice that are beyond the conventional framework of standard LVM. The chapters that fell into this theme were all concerned with extending or adapting basic/standard latent variable models and methods so as to accommodate the kinds of data and questions that are encountered when applying LVM in contexts beyond those originally conceived. The second theme we are calling *meaning*, referring to the connection between LVM and the empirical phenomena and issues being studied. The contributions under this heading all deal, in some way, with how we connect our LVM results back to the world. Stated simply, the *complexity* chapters focus on obtaining the right answers (from a statistical standpoint) and the *meaning* chapters focus on what those answers tell us about the world beyond our data.

1.2 COMPLEXITIES

The first theme covered in this volume, *complexity*, contains nine of the 15 chapters. In our view, these works cluster into three strands: dependence in data, categorical measures, and complex samples. The first strand involves developments made to deal with various kinds of dependence. Cudeck and Harring (Chapter 2) consider the problem of estimating the correlation between two variables in the presence of certain design complexities: for each individual, the variables are measured a different number of times. The authors resolve the problem with the insightful use of a patterned covariance matrix approach, and they show how their approach can be viewed in terms of both hierarchical linear models (HLM) and factor analysis. Gonzalez and Griffin (Chapter 3) focus on the dependence that is encountered in modeling interactions between members of dyads. They develop latent variable models and show their relationship to both HLM and SEM, and address the problem of estimating parameters and standard errors in dyadic data structures. The authors also offer interesting suggestions about using modern computational tools to facilitate graduate training in quantitative methods. Craigmile, Peruggia, and Van Zandt (Chapter 4) describe a Bayesian hierarchical model for response time data. These response time data are collected repeatedly for a given individual, which requires a hierarchical structure to account for the within subject dependencies in the data. Their accessible treatment of such a complex topic will provide readers with a better understanding of both the power of Bayesian methods and the difficulties encountered in modeling response times.

The second strand within the general theme of complexities contains chapters that address the necessary changes that must be made to take the latent variable models originally developed for continuous outcomes and use them in instances where the measured variables are categorical in nature. The first work in this strand is by Moustaki (Chapter 5), which transitions nicely from the previous strand by providing an excellent review of estimation methods for cross-sectional and longitudinal data where some (or all) of the measured variables are categorical. Moustaki herself has contributed greatly to this literature, and readers will benefit from this clear perspective by one of the leading researchers in the area. In Chapter 6, Zhang and Lee demonstrate an adaptation of Proposition 2 in Browne's famous paper (1984) on asymptotically distribution free (ADF) estimation to provide standard errors for ordinary least squares (OLS) estimates. They demonstrate the approach using both dynamic factor analysis and factor analysis with categorical data. Their creative expansion of Browne's earlier work to these new modeling contexts coincides with a renewed interest in the utility of OLS estimation (Forero, Maydeu-Olivares, & Gallardo-Pujol, 2009; Yang-Wallentin, Jöreskog, & Luo, 2010), which may be especially useful when normality assumptions are severely violated. Cai (Chapter 7) also draws inspiration from Browne's (1984) ADF paper, using Proposition 4 to generate accurate goodness-of-fit tests for commonly encountered situations with both linear and nonlinear factor analysis models. Although we have placed these two chapters in the categorical strand, the contributions from Zhang and Lee and from Cai also cross into other complexities. The method described by Zhang and Lee is also applied to dynamic factor analysis, which very easily fits within our dependence-in-data strand. Cai touches on the issue of missing data, which is another complexity inevitably encountered when dealing with real data. Duncan and MacEachern (Chapter 8) examine a nonparametric Bayesian analog to the widely used three-parameter logistic model in IRT. In a compelling fashion they demonstrate the flexibility and utility of their approach with an empirical example, and also make an interesting argument that the idea of nesting is fundamentally Bayesian. The last contribution to the categorical strand is Nicewander's chapter (Chapter 9) which presents methods to estimate correlations among latent variables measured by separate test items as well as correlations of IRT latent variables with observed covariates. His results could be applied in the process of selecting items in adaptive testing.

The final strand in the complexity theme is represented by du Toit (Chapter 10), who focuses on expanding the functional capacity of SEM to include design weights appropriate for complex survey data. The method proposed can accommodate continuous or ordinal data, missingness, and stratum/cluster variables, while providing corrected standard errors and goodness-of-fit statistics. Although this work clearly includes elements of dependence and categorical data, it also stands apart in its treatment of sampling. This kind of expansion is crucial as SEM makes its way into fields where complex survey designs are the norm.

It is interesting to note that one aspect of complexity that crosses the three strands just described involves the treatment of complex data structures. Within this first major section of the book, there are contributions that focus on, or are relevant to the analysis of, repeated measures data (Chapters 2, 5, 6), dyadic data (Chapter 3), response time data (Chapter 4), item level data (Chapters 7, 8, 9) and data from complex samples (Chapter 10). These contributions reflect the fact that many new developments in LVM in recent years have been driven by the desire to model the structure in a wide array of data beyond the usual one of a simple random sample of individuals, measured on a set of variables that are assumed to follow a multivariate normal distribution in the population. That general theme is reflected in many chapters in the first section of this book and we have no doubt that development will continue along these lines.

1.3 MEANING

The second theme, *meaning*, contains the remaining six chapters. These chapters also group together into three strands: assessment of model fit, inferential approaches to model evaluation, and epistemology. Regarding model fit, Millsap (Chapter 11) proposes a simulation-based approach to evaluate model misfit in terms of particular parametric misspecifications. In the author's own words the question of "If the RMSEA = 0.03 in the population for this model, what types of misspecification are possible" conveys the dilemma facing an individual trying to relate evidence of model fit (or misfit) to something beyond the various indices themselves. While Millsap focuses on how perturbations of a model affect root mean square error of approximation (RMSEA), MacCallum, Lee, and Browne (Chapter 12) focus on how small perturbations of the RMSEA can affect parameter estimates. Their examination of fungible parameter estimates offers a fascinating insight into instances where sets of parameter values that provide a miniscule decrement to model fit may say something vastly different about the world than the "optimal" parameter estimates.

Turning to the issue of inference, Shapiro (Chapter 13) provides a thorough presentation of large-sample statistical inference in covariance structure models. The author uses the general framework of minimum discrepancy function (MDF) analysis and presents different estimators, lays out the implication of a model being incorrect in the population, and also examines constrained estimation. Shapiro's chapter is an excellent technical reference on these issues. Rodgers and Beasley (Chapter 14) speculate about the fate of the null hypothesis significance test had Fisher and Gossett had access to modern computation resources. They suggest an alternative approach wherein the hypothesis to be tested is not the null, but rather the alternative, meaning the investigator's scientific hypothesis. They show how this approach could be implemented for the multiple correlation coefficient within the multiple linear regression framework.

The final strand within the meaning theme is epistemology. The last two chapters by Boker and Martin (Chapter 15) and Tateneni and Schiller (Chapter 16) ask more fundamental questions about how we know what we know about latent variables. Boker and Martin eloquently ponder the ways in which we assign meaning to the latent variables that lie at the heart of the LVMs that are the focus of this volume. They urge a broader view of the notion of a latent variable, describing how the meaning of a latent variable can vary across individuals and can shift with the context in which it is measured. Tateneni and Schiller argue that in some cases more useful knowledge is gained by relying on components rather than factors, and they highlight the usefulness of rotation in creating interpretable components in the context of consumer segmentation in market research.

1.4 CONCLUSION

Readers of this collection will recognize quickly that many of the chapters touch on more than one of the strands we have identified. This is fitting, as most data sets contain multiple complexities and multiple challenges regarding inference and interpretation. But how did the humble factor analysis first proposed by Spearman well over 100 years ago give rise to the LVM machinery we see before it today? The contents of this volume offer some insight. First, LVM is able to address questions researchers find important. This is not a trivial thing. Many other methods no longer exist or exist in limited form because, over time, researchers have lost interest in the specific questions those methods and models were designed to answer. Second, LVM has proven to be incredibly expandable. The researchers in the complexity theme of this volume provide an excellent overview of some of the ways in which the first factor analysis model proposed all those years ago has been expanded to meet the changing needs of researchers. Finally, almost from its inception, LVM researchers have been willing to step back and critically examine the fundamental meaning of the models and methods. This sort of self-criticism and self-reflection is incredibly important as the models expand in complexity, but is also important for the most basic knowledge regarding LVM. As methodologists refine their understanding of these fundamental issues, researchers will be better able to use the methods to answer their own research questions. If we are to be good scientists we must occasionally revisit old assumptions and make sure they still seem adequate in the light of new facts. As the authors of chapters in the meaning theme demonstrate, there are still many new angles from which to examine LVM and many new insights left to discover.

In as much as this story is about the history and current state of LVM, it is also about the contributions of Professor Michael W. Browne, which have largely paralleled this story in their importance, innovativeness, and fundamental curiosity about the methods themselves. We have both thoroughly enjoyed our association with Professor Browne, have been elevated by his research, and have been delighted to interact with the world-class researchers who have contributed to this volume. We hope that readers will benefit as much from reading this volume as we have from helping guide it to completion.

REFERENCES

Browne, M. W. (1984). Asymptotically distribution-free methods for the analysis of covariance structures. *British Journal of Mathematical and Statistical Psychology, 37*, 62–83.

Cudeck, R., & MacCallum, R. C. (2007). *Factor analysis at 100: Historical developments and future directions*. Mahwah, NJ: Erlbaum.

Forero, C. G., Maydeu-Olivares, A., & Gallardo-Pujol, D. (2009). Factor analysis with ordinal indicators: A Monte Carlo study comparing DWLS and ULS estimation. *Structural Equation Modeling, 16*, 625–641.

Spearman, C. (1904). "General Intelligence," objectively determined and measured. *American Journal of Psychology, 15*, 201–293.

Yang-Wallentin, F., Jöreskog, K. G., & Luo, H. (2010). Confirmatory factor analysis of ordinal variables with misspecified models. *Structural Equation Modeling, 17*, 392–423.

PART 1
Complexities in Latent Variable Modeling

Chapter 2

Estimating the Correlation between Two Variables when Individuals are Measured Repeatedly

Robert Cudeck and Jeffrey R. Harring

2.1 INTRODUCTION

In many research projects, individuals are measured on two or more variables under different conditions. Often the main question concerns the possibility of mean differences between conditions on the variables. In the research that motivated this work, the conditions were several treatments that were evaluated for a skin ailment. Although the treatments were not the same, they did not differ markedly from each other in their effect on the disorder. The two variables were (x) the patient's rating of his or her general comfort level under the treatment, and (y) a clinician's rating of the effectiveness of the treatment in reducing symptoms. The principle issue was the correlation between the variables. There were two complicating problems however. The first was that each individual was assessed under several different conditions. Consequently, the observations were not independent but rather were nested within individuals. The second complication was that the conditions administered to each subject varied from person to person, and also the number of conditions was unequal. To illustrate, the first three participants had $n_1 = 3$, $n_2 = 2$, and $n_3 = 4$ pairs of measurements, giving the data shown in Table 2.1. This kind of study also arises in

Table 2.1 Measurements for the first three participants on Comfort (x) and Effectiveness (y) for several treatments

Individuals					
1		*2*		*3*	
x	y	x	y	x	y
28	90	12	95	30	45
36	80	0	80	48	80
24	95			16	93
				12	70

educational problems where different teachers evaluate several children, and also for rater–ratee problems that are common in personnel psychology, at least for those in which one class is assumed to be hierarchically nested within the other. With this method, there is no requirement that all raters evaluate all ratees.

Because measurements are nested within persons, a standard product–moment correlation between x and y is not appropriate. To estimate the correlation accurately, the hierarchical structure of the measurements has to be taken into account. There are several ways in which that can be accomplished. The approach reviewed here is flexible and intuitive. It can be generalized if the design is more complicated, and specialized to include the usual product–moment correlation when observations actually are independent. A statistically satisfactory estimator of the correlation in this design can be viewed both as a hierarchical model and also as a structural equation model, in particular, as a patterned covariance matrix. Patterned covariance structures for balanced designs where each subject is observed under the same conditions have been investigated by a number of authors and have a long history (Bock & Bargmann, 1966; Guttman, 1954; Lawley, 1963; McDonald, 1974; Mukherjee, 1970; Olkin & Press, 1969; Steiger, 1980; Votaw, 1948; Wilks, 1946). Browne (1977) reviewed a class of patterned covariance matrices, outlined a closed form solution, and proved the asymptotic equivalence between normal maximum likelihood and generalized least squares. He also included several realistic and interesting examples.

Repeated, multiple measurements, in which each subject is assessed on two or more variables on several occasions, also has a long history, both in the experimental design literature and in hierarchical linear models. Similar to the present problem, hierarchical models for repeated measures do not require balanced data, that is, that participants be observed at the same time points or have the same number of measurements. A recent summary of models for multivariate repeated measures is given by Fitzmaurice, Laird & Ware (2004, Ch. 16). Several writers (Lee, Koh, & Ong, 1989; Snijders & Bosker, 1999, Ch. 12; Weiss, 2005, Sec. 13.4) review simple models for bivariate correlation when individuals are measured repeatedly in a longitudinal study, including designs with varying numbers of observations within subject. An important difference between the present design and a repeated measures study is that in the latter, variables observed at different time points are not necessarily considered to be similar either at the level of means or covariances. In this approach, the different conditions are assumed to be comparable in the manner described below.

The main issues with a measurement protocol of this kind were noted by Bland and Altman (1994; 1995). Hamlett and co-workers (2003; 2004) showed how the main correlation of interest between variables can be estimated as a hierarchical model, and that three other informative correlation indices related to these data can be estimated as well. Roy (2006; 2009) outlined alternative structures and reviewed a more detailed numerical example. In this chapter, we review the model and describe a way to fit it using structural equation modeling (SEM) software. We also suggest alternative structures including latent variable models that place the hierarchical model in a wider context, and present the analysis of the comfort/effectiveness data as an example.

2.2 A STRUCTURE FOR MULTIPLE PAIRS OF MEASUREMENTS

For the ith individual there are $n_i \geq 1$ pairs of measurements on variables x and y,

$$(x_{i1}, y_{i1},\ x_{i2}, y_{i2},\ x_{i3}, y_{i3}, \ldots, x_{in_i}, y_{in_i})$$

It is assumed that the mean and standard deviation of x is the same over conditions and that the mean and standard deviation of y is also the same over conditions

$$\begin{aligned} \mu_x &= E(x_{ij}) = E(x_{ik}) \qquad \mu_y = E(y_{ij}) = E(y_{ik}) \\ \sigma_x &= sd(x_{ij}) = sd(x_{ik}) \qquad \sigma_y = sd(y_{ij}) = sd(y_{ik}) \end{aligned}$$

These assumptions are convenient, although they can be relaxed in various ways. For example, the means and standard deviations can vary across conditions if the level and scale of the variables over treatments differ. The simplicity of the assumption of equal means and standard deviations is attractive when appropriate.

A key idea of the model is distinguishing between one pair of measurements observed concurrently under condition j, (x_{ij}, y_{ij}), and any two pairs of measurements observed under different conditions, j and k, $(x_{ij}, y_{ij}, x_{ik}, y_{ik})$. For an (x_{ij}, y_{ij}) pair obtained concurrently, the covariance matrix has the usual form

$$\Sigma_c = cov(x_{ij}, y_{ij}) = \begin{pmatrix} var(x_{ij}) & \\ cov(x_{ij}, y_{ij}) & var(y_{ij}) \end{pmatrix} = \begin{pmatrix} \sigma_x^2 & \\ \rho_{yx}\sigma_y\sigma_x & \sigma_y^2 \end{pmatrix} \tag{2.1}$$

where ρ_{yx} is the correlation. This often is the main quantity of interest.

Define the 2×2 covariance matrix between elements of any two different pairs to have the following structure

$$\Sigma_a = cov[(x_{ij}, y_{ij}), (x_{ik}, y_{ik})] = \begin{pmatrix} cov(x_{ij}, x_{ik}) & cov(x_{ij}, y_{ik}) \\ cov(y_{ij}, x_{ik}) & cov(y_{ij}, y_{ik}) \end{pmatrix} = \begin{pmatrix} \rho_{xx}\sigma_x^2 & \dot{\rho}_{yx}\sigma_y\sigma_x \\ \dot{\rho}_{yx}\sigma_y\sigma_x & \rho_{yy}\sigma_y^2 \end{pmatrix} \tag{2.2}$$

There are two asynchronous correlations between the same variable under different conditions, one for the correlation between x-measurements in different conditions and a second for the y-variables

$$\rho_{xx} = corr(x_{ij}, x_{ik}) \qquad \rho_{yy} = corr(y_{ij}, y_{ik})$$

It is assumed that the cross correlations between the different variables under conditions j and k are identical

$$\dot{\rho}_{yx} = corr(y_{ij}, x_{ik}) = corr(x_{ij}, y_{ik})$$

For any two pairs of measurements, the covariance matrix among variables $(x_{ij}, y_{ij}, x_{ik}, y_{ik})$ combines (2.1) and (2.2)

$$\Sigma = \begin{pmatrix} \Sigma_c & \\ \Sigma_a & \Sigma_c \end{pmatrix} \tag{2.3}$$

The corresponding correlation matrix is

$$\mathbf{P} = \begin{pmatrix} 1 & & & \\ \rho_{yx} & 1 & & \\ \rho_{xx} & \dot{\rho}_{yx} & 1 & \\ \dot{\rho}_{yx} & \rho_{yy} & \rho_{yx} & 1 \end{pmatrix} \tag{2.4}$$

A path diagram of the pattern is shown in Figure 2.1.

The number of observations for individuals varies, so both the mean vector and covariance matrix of the measurements depend on the person. Let μ_i $(2n_i \times 1)$ and

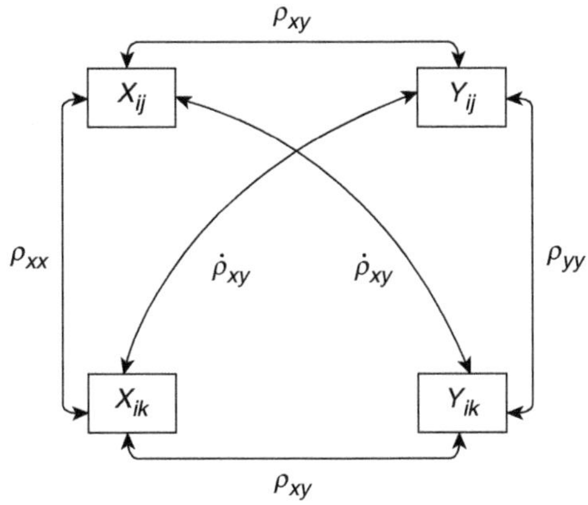

Figure 2.1 Diagram of the correlation structure in model (1) and (2) for two variables measured under two conditions. The correlation between variables measured concurrently is ρ_{xy}. There are three additional correlations among the variables that are measured under different conditions. For x observed under different conditions and y observed under different conditions, the correlations are ρ_{xx} and ρ_{yy}. The correlation between x and y from different conditions is $\dot{\rho}_{xy}$.

Σ_i $(2n_i \times 2n_i)$ be the mean vector and covariance matrix of the ith participant. The mean vector is

$$\mu_i = \mathbf{1}_i \otimes \mu_c$$

where $\mu_c = (\mu_x, \mu_y)'$ are means of the two variables, $\mathbf{1}_i$ is the unit vector of order n_i, and $\otimes$ denotes Kronecker product. For example, with $n_i = 3$ the mean vector is of order 6 but is composed of two distinct elements,

$$\mu_i = (\mu_x, \mu_y, \mu_x, \mu_y, \mu_x, \mu_y)'$$

The covariance matrix, generalizing (2.3), is

$$\Sigma_i = \mathbf{I}_i \otimes \Sigma_c + (\mathbf{J}_i - \mathbf{I}_i) \otimes \Sigma_a \tag{2.5}$$

where $\mathbf{I}_i$ is the identity matrix of order n_i and $\mathbf{J}_i$ the $(n_i \times n_i)$ matrix of unities. For example the individual covariance matrices for the first three individuals in this study are

$$\Sigma_1 = \begin{pmatrix} \Sigma_c & & \\ \Sigma_a & \Sigma_c & \\ \Sigma_a & \Sigma_a & \Sigma_c \end{pmatrix} \quad \Sigma_2 = \begin{pmatrix} \Sigma_c & \\ \Sigma_a & \Sigma_c \end{pmatrix} \quad \Sigma_3 = \begin{pmatrix} \Sigma_c & & & \\ \Sigma_a & \Sigma_c & & \\ \Sigma_a & \Sigma_a & \Sigma_c & \\ \Sigma_a & \Sigma_a & \Sigma_a & \Sigma_c \end{pmatrix}$$

$$n_1 = 3 \qquad\qquad n_2 = 2 \qquad\qquad n_3 = 4$$

It is worth noting that if $n_i = 1$, then $\Sigma_i = \Sigma_c$. Therefore, even though the model is developed assuming that participants are measured under several conditions, the standard bivariate design is a special case. In the general case, regardless of the number of pairs obtained on any individual, the parameters consist of two means, two standard deviations, and four correlations, $(\mu_x, \mu_y, \sigma_x, \sigma_y, \rho_{yx}, \rho_{xx}, \rho_{yy}, \dot{\rho}_{yx})$.

The asynchronous correlations between the same variable in different conditions, ρ_{xx} and ρ_{yy}, summarize the extent to which pairs of x-variables or pairs of y-variables covary. The cross correlations between different variables, ρ_{yx} and $\dot{\rho}_{yx}$, determine the degree of scatter of the within-subject clusters. These relationships can be illustrated graphically. Figure 2.2 shows six groups of $N = 10$ individuals each with $n_i = 4$ pairs of measurements that were constructed according to model (2.5). The four measurements for an individual are connected with a line. In each of the six panels, the variables have $\mu_x = \mu_y = 50, \sigma_x = \sigma_y = 10$. First, consider the examples in the top row. The three samples have the same (x, x) and (y, y) correlations, $\rho_{xx} = \rho_{yy} = 0.95$, but the two kinds of (x, y) correlations across the left-, center, and right-hand panels differ: they are $\rho_{xy} = \dot{\rho}_{yx} = 0.95$, $\rho_{xy} = \dot{\rho}_{yx} = 0$, $\rho_{xy} = \dot{\rho}_{yx} = -0.95$, respectively. Because the same-variable correlations ρ_{xx} and ρ_{yy} are high, the four measurements for a person are clustered closely together, which is apparent in each panel. When the cross correlations ρ_{yx} and $\dot{\rho}_{yx}$ are high and positive, the individual clusters are scattered tightly in an upward sloping ellipse. When the cross correlations are strongly negative, the overall shape is tightly clustered in a downward sloping pattern. And when the cross correlations are zero, the individual clusters form a circular pattern.

In each panel of the bottom row the two same-variable correlations are $\rho_{xx} = \rho_{yy} = 0.10$. These low values of the correlation mean that the four measurements for an individual are generally quite dissimilar. This can be seen in the relatively large distances between measurements for a subject. The cross correlations in the bottom left, center, and right panels, are $\rho_{xy} = \dot{\rho}_{yx} = 0.32$, $\rho_{xy} = \dot{\rho}_{yx} = 0$, $\rho_{xy} = \dot{\rho}_{yx} = -0.32$, respectively. (The range covered by ρ_{xy} and $\dot{\rho}_{yx}$ in the examples of the bottom row is not as wide as in those of the upper row. The possible values of these correlations must be chosen in a way that ensures that the covariance matrix is positive definite.) When the cross correlations for different variables either from the same or different conditions, ρ_{xy} and $\dot{\rho}_{yx}$, are positive but moderate size, the shape of the individual clusters again is upward sloping but in this case is not scattered tightly. When these correlations are negative, the overall shape is downward sloping but not tightly compact. And when the cross correlations are zero, the shape is essentially circular.

This illustration shows that the model specifies two correlations that describe the degree of similarity of individual measurements, and also two correlations that represent the extent to which the individual clusters are tightly or loosely aligned as well as their overall orientation. As a way of representing the within- and between-subject design and for accommodating the unequal number of measurements, the model is simple, effective, and understandable.

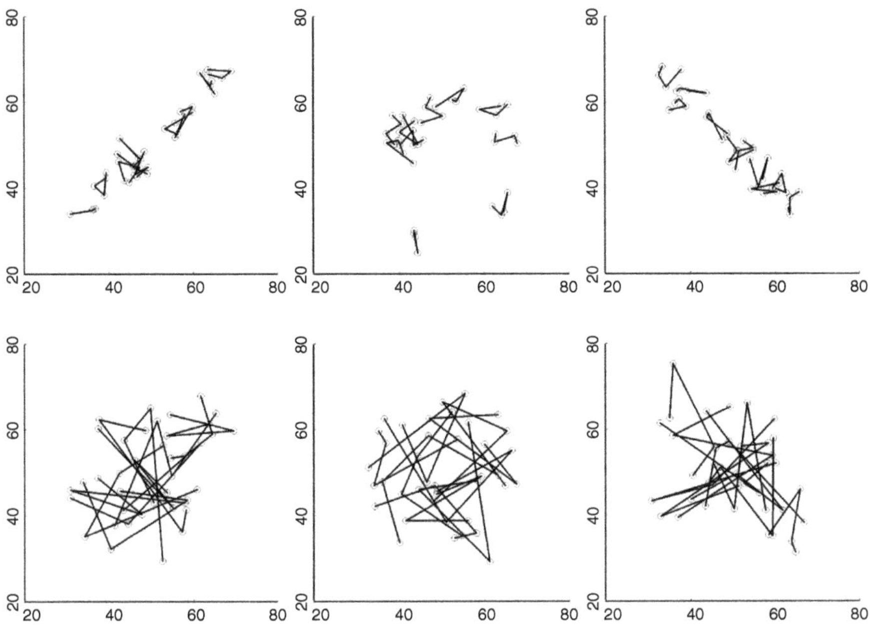

Figure 2.2 Graphs of six samples of size $N = 10$ with $\mu_x = \mu_y = 50$, $\sigma_x = \sigma_y = 10$. In the upper row of the figure $\rho_{xx} = \rho_{yy} = 0.95$, but cross correlations are $\rho_{xy} = \dot{\rho}_{yx} = 0.95$, $\rho_{xy} = \dot{\rho}_{yx} = 0$, $\rho_{xy} = \dot{\rho}_{yx} = -0.95$ in the left-, center, and right-hand panels. In the lower row, $\rho_{xx} = \rho_{yy} = 0.10$, while the two cross correlations are $\rho_{xy} = \dot{\rho}_{yx} = 0.32$, $\rho_{xy} = \dot{\rho}_{yx} = 0$, $\rho_{xy} = \dot{\rho}_{yx} = -0.32$.

2.3 RELATED LATENT VARIABLE MODELS

Latent variable models imply patterned structures for a covariance matrix (Browne, 1982). Consequently, apart from the fact that in this situation individuals are measured repeatedly, the basic covariance structure of (2.3) often is equivalent to other well-known models. In particular, a one-factor, and more frequently a two-factor model, fits many versions of this pattern. A latent variable framework can be an informative model for this situation because it provides a broader theoretical context for the data. It can be useful to know that the manifest variables covary because of their association with latent variables.

Define the factor loading matrix for a typical set $(x_{ij}, y_{ij}, x_{ik}, y_{ik})$ to be

$$\Lambda = \begin{pmatrix} \lambda_1 & 0 \\ 0 & \lambda_2 \\ \lambda_3 & 0 \\ 0 & \lambda_4 \end{pmatrix}$$

where the factors for the respective columns of coefficients are f_x and f_y. Assume $var(f_x) = var(f_y) = 1$, $corr(f_x, f_y) = \varphi$, and a diagonal matrix of unique variable variances, Ψ. The covariance structure for two factors is $\Sigma = \Lambda\Phi\Lambda' + \Psi$ so that

$$\Sigma = \begin{pmatrix} \lambda_1^2 + \psi_1 & & & \\ \lambda_2\lambda_1\varphi & \lambda_2^2 + \psi_2 & & \\ \lambda_3\lambda_1 & \lambda_3\lambda_2\varphi & \lambda_3^2 + \psi_3 & \\ \lambda_4\lambda_1\varphi & \lambda_4\lambda_2 & \lambda_4\lambda_3\varphi & \lambda_4^2 + \psi_4 \end{pmatrix}$$

If the x-variables and y-variables are parallel, then $\lambda_1 = \lambda_3$, $\lambda_2 = \lambda_4$, $\psi_1 = \psi_3$, and $\psi_2 = \psi_4$. In this case the pattern of (2.1) and (2.2) is equivalent to a two-factor model for parallel tests. The reliability of the x variables, for example, is $\rho_{xx}^2 = \lambda_1^2/(\lambda_1^2 + \psi_1)$.

The pattern in (2.3) with elements (2.1) and (2.2) also holds approximately for other models. For example, if the two x-variable loadings are equal and their unique variances not greatly dissimilar, and if the same is true for the y-variable loadings and unique variables, then the covariance pattern approximates the model for tau equivalent tests. If the loadings in each set are of comparable size, then model (2.1) and (2.2) approximates the model for congeneric tests also. Finally, the model for two correlated restricted factors is equivalent to a model for simple latent variable regression

$$f_y = \gamma f_x + d$$

where γ is the regression coefficient for f_y on f_x, d is the regression residual, and where the measurement model for the pair of latent variables is again the pattern for two sets of parallel variables. Consequently the covariance structure of (2.3) is also equivalent or approximately equal to this basic latent variable regression model. Which of these various alternatives is preferable of course depends on the goal of the study, and the pattern of the observed variances and covariances. Although the covariance structure of model (2.3) is interesting and informative in its own right, it may be useful to exploit the correspondence between the patterned covariance structure and a latent variable model in some situations. One important consideration with latent variable models for this design is that the model must make sense in light of the measurement plan in which no pair of variables is clearly antecedent to others.

2.4 ESTIMATION WITH SEM SOFTWARE

Estimation of this structure is straightforward. It is assumed that observations are independent but not identically distributed as $\mathbf{y}_i \sim N(\mu_i, \Sigma_i)$. The log-likelihood is

$$\ln L = k - \frac{1}{2}\sum_i \left[\ln|\Sigma_i|^{-1/2} + (\mathbf{y}_i - \mu_i)'\Sigma_i^{-1}(\mathbf{y}_i - \mu_i)\right]$$

Because the measurement model is hierarchical, HLM programs can fit it. Hamlett et al. (2004) illustrated an approach using a setup for SAS PROC MIXED. Because the model is a structured covariance pattern, most SEM programs can fit it too. One way to accomplish this is to define Σ for the individual with the largest number of measurements, and then estimate the structure for other participants using the available data under the assumption that the incomplete observations satisfy a missing at random process. For example, if the largest number of measurements is $n_i = 3$, then the six unique elements of the covariance structure are arranged as

$$\Sigma_i = \begin{pmatrix} \theta_1 & & & & & \\ \theta_2 & \theta_3 & & & & \\ \theta_4 & \theta_5 & \theta_1 & & & \\ \theta_5 & \theta_6 & \theta_2 & \theta_3 & & \\ \theta_4 & \theta_5 & \theta_4 & \theta_5 & \theta_1 & \\ \theta_5 & \theta_6 & \theta_5 & \theta_6 & \theta_2 & \theta_3 \end{pmatrix} \tag{2.6}$$

where from (2.1) and (2.2)

$$\begin{array}{lll} \theta_1 = \sigma_x^2 & \theta_2 = \rho_{yx}\sigma_y\sigma_x & \theta_3 = \sigma_y^2 \\ \theta_4 = \rho_{xx}\sigma_x^2 & \theta_5 = \dot{\rho}_{yx}\sigma_y\sigma_x & \theta_6 = \rho_{yy}\sigma_y^2 \end{array}$$

The general two-level hierarchical model is a convenient framework for this problem. The model for the variables of the ith case is

$$\mathbf{y}_i = \mathbf{X}_i\beta + \mathbf{Z}_i\mathbf{b}_i + \mathbf{e}_i$$

There are no residuals in the covariance structure of (2.5), so $\mathbf{e}_i = \mathbf{0}$. Define $\Phi = cov(\mathbf{b}_i)$. The mean vector and covariance matrix are then

$$E(\mathbf{y}_i) = \mathbf{X}_i\beta \qquad cov(\mathbf{y}_i) = \mathbf{Z}_i\Phi\mathbf{Z}_i'$$

In terms of the covariance structure reviewed above, the HLM model parameters are $\mu_c = \beta$ and $\Sigma = \Phi$, in which Φ has the pattern of (2.5), or more concretely of (2.6).

To arrange the problem so it can be estimated either with an HLM program or an SEM program with a flexible system for incomplete data, the observed variables, $\mathbf{y}_i$, and design matrices $\mathbf{X}_i$ and $\mathbf{Z}_i$ are needed for each individual. Suppose that $\mathbf{y}_1$ ($2n_1 \times 1$), $\mathbf{X}_1$ ($2n_1 \times 2$) and $\mathbf{Z}_1$ ($2n_1 \times 2n_1$) are the matrices for the individual with the largest number of measurements. For all other cases, $\mathbf{X}_i$ is set equal to the first $2n_i$ rows of $\mathbf{X}_1$, and $\mathbf{Z}_i$ is set equal to the first $2n_i$ rows of $\mathbf{Z}_1$. For example, if the first two individuals have $n_1 = 3$ and $n_2 = 2$, then the coding scheme gives

$$\underset{\mathbf{Y}_1}{\begin{pmatrix} x_{11} \\ y_{11} \\ x_{12} \\ y_{12} \\ x_{13} \\ y_{13} \end{pmatrix}} \quad \underset{\mathbf{X}_1}{\begin{pmatrix} 1 & 0 \\ 0 & 1 \\ 1 & 0 \\ 0 & 1 \\ 1 & 0 \\ 0 & 1 \end{pmatrix}} \quad \underset{\mathbf{Z}_1}{\begin{pmatrix} 1 & 0 & 0 & 0 & 0 & 0 \\ 0 & 1 & 0 & 0 & 0 & 0 \\ 0 & 0 & 1 & 0 & 0 & 0 \\ 0 & 0 & 0 & 1 & 0 & 0 \\ 0 & 0 & 0 & 0 & 1 & 0 \\ 0 & 0 & 0 & 0 & 0 & 1 \end{pmatrix}}$$

$$\begin{pmatrix} x_{21} \\ y_{21} \\ x_{22} \\ y_{22} \end{pmatrix} \quad \begin{pmatrix} 1 & 0 \\ 0 & 1 \\ 1 & 0 \\ 0 & 1 \end{pmatrix} \quad \begin{pmatrix} 1 & 0 & 0 & 0 & 0 & 0 \\ 0 & 1 & 0 & 0 & 0 & 0 \\ 0 & 0 & 1 & 0 & 0 & 0 \\ 0 & 0 & 0 & 1 & 0 & 0 \end{pmatrix}$$
$$\mathbf{Y}_2 \qquad\qquad \mathbf{X}_2 \qquad\qquad\qquad \mathbf{Z}_2$$

Setting up these design matrices and the pattern in (2.6) are the only real issues in estimating the model. So although it is non-standard as a latent variable model, it is not difficult to estimate the structure in one way or another.

2.5 MULTIPLE RATINGS OF COMFORT AND EFFECTIVENESS

The purpose of the study reviewed in the Introduction was to investigate the judged level of comfort and effectiveness of several treatments for a chronic skin disorder. At two hours after administration, each lotion was rated by the individual patient for the level of comfort it provided. Each lotion was also judged by a clinician for the condition of the affected skin area. For both judgements the rating scale was 0 to 100. The main research issue was the correspondence between the two ratings, x, Comfort, and y, Effectiveness. This was complicated by the fact that patients could choose whether to use a lotion, and also by the fact that patients were free to try more or fewer of the alternatives. Consequently, the number of ratings obtained for an individual varied. After initial analyses it was decided that the mean ratings did not vary significantly over conditions and that the covariance pattern was a reasonable way to handle the multiple measurements.

The sample consisted of $N = 65$ individuals with a total of 114 (x, y) pairs. Forty-six percent of the sample had two or more measurements, as is shown in the distribution of n_i below

n_i:	1	2	3	4	5	7
f:	35	19	7	2	1	1

The standard product–moment correlation, which incorrectly assumes independence of the observations, was $r = -0.49$. The usual standard error of the correlation is $se(r) = [(1 - r^2)/df]^{\frac{1}{2}}$. Under the assumption of normality and independence, the degrees of freedom is $df = n_p - 2$, where $n_p = 114$ is the total number of pairs. This is clearly incorrect here because the observations are nested. Another approach is to take $df = N - 2$, where $N = 65$ is the number of individuals. This is also incorrect because nearly half the sample have $n_i \geq 2$ measurements, and this option for degrees of freedom ignores the within-subject replications. Both approaches to computing $se(r)$ are incorrect because the estimate r is incorrect with nested observations.

A more complete analysis of the correlation between Comfort and Effectiveness is given by model (2.1) and (2.2). Maximum likelihood estimates of the parameters follow, with standard errors in parentheses

$\hat{\mu}_x = 75.7(2.1)$	$\hat{\sigma}_x = 17.9(2.7)$	$\hat{\mu}_y = 24.7(1.4)$	$\hat{\sigma}_y = 25.5(1.7)$
$\hat{\rho}_{yx} = -0.49(.08)$	$\hat{\rho}_{xx} = 0.59(.10)$	$\hat{\rho}_{yy} = 0.20(.13)$	$\hat{\rho}^{\bullet}_{yx} = -0.24(.11)$

The estimate of the concurrent correlation is moderately negative, $\hat{\rho}_{yx} = -0.49$, with a small standard error of 0.08. Thus there is a tendency across conditions for lotions to seem comfortable to the patients yet to be judged as somewhat ineffective by the clinicians. Similarly, the estimated correlation of Comfort in one condition with Effectiveness in a different condition is also slightly negative, $\hat{\rho}^{\bullet}_{yx} = -0.24$, a finding that again implies an inverse although weaker relationship. On the other hand, there is a fairly strong trend across conditions for the Comfort measurements to be moderately positively correlated, $\hat{\rho}_{xx} = 0.59$, which suggests that patients who find one lotion to be comfortable tend to judge other lotions similarly. The other same-variable correlation is estimated to be $\hat{\rho}_{yy} = 0.20$, with $se(\hat{\rho}_{yy}) = 0.13$. Consequently, the parameter may be zero. Thus the clinicians' ratings of Effectiveness are not systematically associated over conditions. Those who judge one lotion to be highly effective are as likely to judge another as high as they are likely to judge it low.

2.6 DISCUSSION

Although the product moment estimate of correlation and the mixed model estimate are the same with these data, $r = \hat{\rho}_{xy} = -0.49$, the two indices have quite different derivations, and domains of application. The former requires independent observations, and from this assumption the standard error of estimate is derived. In the design of the example, both the traditional estimator and standard error, $se(r)$, are not correct. In contrast, both the estimator ρ_{xy} and its standard error of estimate are valid for this hierarchical design. Furthermore, the covariance structure of (2.1) and (2.2) gives a more complete representation of the correlation between x and y in this design than does the simple product moment correlation over observations, ignoring the fact that multiple measurements are obtained within subject. The model includes the concurrent correlation, ρ_{yx}, the correlation between x and y when variables are observed in the same condition. It also provides information about the correlation of the same variable in different conditions, ρ_{xx} and ρ_{yy}. And it defines a correlation, $\dot{\rho}_{yx}$, for x measured in one condition and y in another. These indices reflect very different but well-defined aspects of the different types of covariance that exist in this design.

This structure can be extended in several ways. One obvious elaboration is to incorporate means for the various conditions, and instead of assuming $\mu_x = E(x_{ij}) = E(x_{ik})$ over all conditions, to introduce location parameters for variables in each condition, such as $\mu_{x_j} = E(x_{ij})$, $\mu_{x_k} = E(x_{ik})$. Instead of two parameters for the means, one for all x-variables and a second for all y-variables, there would be two for each condition. This would allow the variables to differ in level over conditions, while the covariance structure remains the same.

A second extension is to specify different standard deviations for each variable under each condition as well as different means, while keeping the correlation structure the same. A practical consideration in this extension is that the number of parameters may be large if the number of conditions is large. This model still may be of interest as a way to investigate a basic correlation structure for variables that differ in level and scale.

As illustrated in the example, the model easily handles multiple measurements on x and y, with any number of pairs $n_i \geq 1$. Individuals with $n_i = 1$ do in fact contribute information to the asynchronous covariance matrix Σ_a in (2.2) under maximum likelihood estimation. It is also possible to accommodate individuals who are not measured on both x and y in every condition but instead on only one variable of the pair. Although in general the number of measurements is $2n_i$, if the assumption of missing at random is maintained then it is possible for participants to contribute just a single score.

Patterned covariance structures are an interesting and useful class of statistical model, with many possible applications. They are parsimonious and adapt easily to specific contexts. In this domain, Browne's (1977) article is an important contribution, as much for the convincing empirical examples as the statistical methodology.

REFERENCES

Bland, J. M., & Altman, D. G. (1994). Correlation, regression and repeated data. *British Medical Journal, 308*, 896.

Bland, J. M., & Altman, D. G. (1995). Calculating correlation coefficients with repeated observations. Part 1 – correlation within subjects. *British Medical Journal, 310*, 446.

Bock, R. D., & Bargmann, R. E. (1966). Analysis of covariance structures. *Psychometrika, 31*, 507–534.

Browne, M. W. (1977). The analysis of patterned correlation matrices by generalized least squares. *British Journal of Mathematical and Statistical Psychology, 30*, 113–124.

Browne, M. W. (1982). Covariance structures. In D.M. Hawkins (Ed.), *Applied multivariate analysis* (pp. 72–141). Cambridge: Cambridge University Press.

Fitzmaurice, G. M., Laird, N. M., & Ware, J. H. (2004). *Applied longitudinal analysis*. New York: Wiley.

Guttman, L. (1954), A new approach to factor analysis: The radex. In P. F. Lazarsfeld (Ed.), *Mathematical thinking in the social sciences* (pp. 258–348). Glencoe, IL: Free Press

Hamlett, A., Ryan, L., Serrano-Trespalacios, P., & Wolfinger, R. (2003). Mixed models for assessing correlation in the presence of replication. *Air and Waste Management Association, 53*, 442–450.

Hamlett, A., Ryan, L., & Wolfinger, R. (2004). On the use of PROC MIXED to estimate correlation in the presence of repeated measures. SAS Users Group International, Proceedings of the Statistics and Data Analysis Section. Paper 198v29, 1v7.

Lawley, D. N. (1963). On testing a set of correlation coefficients for equality. *Annals of Mathematical Statistics, 34*, 149–151.

Lee, J., Koh, D., & Ong, C. N. (1989). Statistical evaluation of agreement between two methods for measuring a quantitative variable. *Computers in Biology and Medicine, 19*, 61–70.

McDonald, R. P. (1974). Testing pattern hypotheses for covariance matrices. *Psychometrika, 39*, 189–201.

Mukherjee, B. N. (1970). Likelihood ratio tests of statistical hypotheses associated with patterned covariance matrices in psychology. *British Journal of Mathematical and Statistical Psychology, 23*, 89–120.

Olkin, I., & Press, S. (1969), Testing and estimation for a circular stationary model, *Annals of Mathematical Statistics, 40*, 1358–1373.

Roy, A. (2006). Estimating correlation coefficient between two variables with repeated measures observations using mixed effects model. *Biometrical Journal, 48*, 286–301.

Roy, A. (2009). An application of linear mixed effect model to assess the agreement between two methods with replicated observations. *Journal of Biopharmaceutical Statistics, 19*, 150–173.

Snijders, T., & Bosker, R. (1999). *Multilevel analysis: An introduction to basic and advanced multilevel modeling*. Thousand Oaks, CA: Sage.

Steiger, J. H. (1980). Testing pattern hypotheses on correlation matrices: Alternative statistics and some empirical results. *Multivariate Behavioral Research, 15*, 335–352.

Votaw, D. F. (1948). Testing compound symmetry in a normal multivariate distribution. *The Annals of Mathematical Statistics 19*, 447–473.

Weiss, R. E. (2005) *Modeling longitudinal data*. New York: Springer.

Wilks, S. S. (1946). Sample criteria for testing equality of means, equality of variances, and equality of covariances in a normal multivariate distribution. *Annals of Mathematical Statistics, 17*, 257–281.

Chapter 3

Deriving Estimators and Their Standard Errors in Dyadic Data Analysis: Examples Using a Symbolic Computation Program

Richard Gonzalez and Dale Griffin

Michael Browne's foundational papers on the analysis of covariance structures (e.g., Browne, 1974, 1982) were influential in our conceptualization of models for dyadic data analysis. He provided a framework for working with covariance structures that makes it relatively easy to derive new results. We began working on dyadic models in the early 1990s, before the explosion of research on multilevel models and random effect models. Motivated by the ubiquity of dyadic data in social psychological research, we recognized that a structural equation modeling (SEM) approach with latent variables representing dyadic similarity would be helpful in modeling multivariate dyadic data. Multilevel models, let alone multivariate multilevel models, were relatively new at that time and we explored their usefulness for dyadic data analysis. But it was the SEM approach and the Browne (1974, 1982) papers that we found most helpful in our initial model formulation.

Our interest was sparked by Kenny and LaVoie (1985), who presented a dyadic latent variable model based on an analysis of variance (ANOVA) approach to the intraclass correlation. They provided an estimator of a latent dyadic correlation but did not provide significance tests. We wanted to extend the Kenny and LaVoie approach to allow researchers of dyadic and group processes to use a more flexible modeling approach, including both observed and latent variables, covariates, plus tests of mediation, and other standard analytical tools. We wanted closed form estimators based on formal statistical approaches that allowed clear interpretation of the parameters, so we could understand why estimators were sometimes out of bounds, develop better intuition about the parameters, and provide a formal linkage between the latent variable dyadic approach and the well-studied (but often misapplied) aggregate approaches such as the correlation of group means. We wanted our dyadic data analysis contributions to illuminate the analytic issues in a simple manner so the tools could be used to answer research questions in a meaningful manner. We also wanted to provide more intuition about models that partition variance in a multilevel manner into individual-level and group-level sources, so there was a pedagogical aspect to our work as well.

It was in this context of exploration that we came across Michael Browne's very clear theoretical formulation of covariance structures. Both the 1974 article and the 1982 chapter were ahead of their time, and illuminated the deep and magnificent structure that characterizes covariance structures. We studied these papers carefully. We remember the excitement each page provided, and how we spent hours working through each page of the papers. Not only did we learn about covariance structures at a much deeper level, but we also developed an appreciation for the underlying statistical theory and the importance of clear, well-justified recommendations for practicing researchers. This influence continues to run deep in our research and in our thinking, even outside the scope of dyadic data analysis.

The problem we were working on when we learned about the 1974 and 1982 papers was that we had noticed a discrepancy between the standard errors we derived analytically for one of our dyadic models and the numerical standard errors that emerged from structural equation modeling programs. We noticed that the way in which an analyst identifies a latent factor in a structural equation model (i.e., how the scale of a latent variable is set) can influence the standard error in a nonlinear way, thus influencing the Wald Z test for a parameter. We noticed that statistical programs were giving different Wald tests for identical parameters based on how latent variables were identified, that is, whether the latent variance was set to 1 or which of the indicators was given a path weight of 1. These different Wald tests occur even though the goodness-of-fit indices are invariant to where the scale identifying "one" is placed. This confused us because we had derived the standard errors for a special case model by hand and our standard errors were different from those that emerged from the SEM programs. At first we thought we had made a mistake in our derivation, but we eventually discovered that the issue had to do with how most SEM programs approximate the standard errors. Michael Browne's 1974 and 1982 papers were instrumental in giving us the background and understanding to see this relation. This work led to the Gonzalez and Griffin (2001a) paper that illustrated the problem, and provided a solution using the likelihood ratio test to avoid the curvature approximation (Hessian) that is used in the estimation of the standard error in the Wald test.

It is wonderful to honor Michael Browne's contribution to statistical methodology in psychology. In this chapter we illustrate how we used some of the results in Browne's 1974 and 1982 papers to derive standard errors for particular parameters in standard models of dyadic data analysis. Our goal in this chapter is to illustrate a little of the underlying statistical theory, and to show how one can use a symbolic software package like Maple (http://www.maplesoft.com/products/maple/) or Mathematica (http://www.wolfram.com/mathematica/) to derive useful results. One of the many things we learned from Michael Browne's scholarly contributions is the importance of analytic derivation to facilitate understanding of the underlying statistical properties. Not all psychology students have the background to perform some of the analytic computations by hand. Computer packages like Maple can go a long way toward empowering students to play with analytic results, provided the students know a few fundamental ideas from calculus and linear algebra. We believe that it is possible to incorporate more theoretical results and properties from statistics into advanced graduate training in psychology, and tools like Maple and Mathematica can provide helpful bridges for students who have not had much quantitative background. If taught in the right way, the student can achieve not only a deeper understanding of the underlying statistical toolbox but also know how to adapt the tools (or know when the tools do not apply) to solve new problems. Given the rapid pace of advances in quantitative psychology, graduate methods training might serve students well by a teaching style that facilitates adaptability to new methods throughout one's career. Textbooks teaching more theoretical statistical concepts using these tools have already begun to emerge (e.g. Rose & Smith, 2002).

3.1 DYADIC DATA ANALYSIS

3.1.1 Intuition, Psychology, and Models

Psychology studies the mind and its related processes, so it becomes especially interesting when we consider two individuals in a social relationship, such as a married couple, a parent and child, or two team members. We can study not just an individual in isolation but also an individual in the context of another person. We can frame this problem in several ways. One framing could be in terms of two levels of analysis: an individual level and a dyadic level. We could explore psychological processes in terms of the dyad's shared variance and refer to what is left over after removing that shared variance as individual-level variance (i.e., a type of residual variance after the shared dyad variance has been removed). This framing of the dyadic analysis problem involves a decomposition of individual- and group-level variance.

A second framing involves a different conceptualization of dyadic interaction, not in terms of shared variance but instead focusing on processes involving "influence" of one person on another. What is the magnitude of one person's influence on the other individual in the dyad? Is the cross-person influence different than a person's influence over his or her own behavior (e.g., a spouse's anxiety influencing her frequency of risky choices versus her own anxiety influencing her own risky behavior)? This second framing of dyadic processes focuses more on the social influence part of the interaction. It pits actor- and partner-influence processes against each other rather than dyadic- and individual-level processes as in the first framing.

The first framing above is called the latent variable interdependence model and the second framing is called the actor-partner interaction model, or APIM (e.g., Gonzalez & Griffin, 2001b; Kenny, Kashy, & Cook, 2006). Both models can be used in the context of modeling a covariance matrix with interdependence. Both models can be expressed either as a structural equation model or as a multilevel model. The latent variable and actor-partner models lead to slightly different generalizations, although they can lead to identical goodness-of-fit under some special cases.

Consider a heterosexual married couple. The two individuals in the couple produce data on the key dependent variable, say, risky behavior. There may also be a predictor variable collected from each dyad member, say, anxiety. The goal is to understand the relation between anxiety and risky behavior in the context of interdependence. The latent variable and actor-partner models approach this goal in different ways. A statistical model can be estimated for each individual but because the two individuals are interdependent, the residuals resulting from such models are correlated. This kind of correlated error is well-known in the econometrics literature, such as seemingly unrelated regression (SUR) models. The goal in those models is to correct the estimation procedure for the presence of such correlated residuals. But the interesting psychological properties may be in the degree of interdependence between the individuals in the dyad. So, rather than treat interdependence as a nuisance that needs to be eliminated from the data to achieve unbiased parameter estimates and tests of significance, as is typically done in SUR models, proponents of dyadic data analysis models conceptualize the correlated error as something of potential or even central psychological interest. The degree of interdependence could inform us about dyadic processes, which could be a theoretically interesting attribute of the data and the model.

The dyadic data structure for this example on two variables (anxiety and risky behavior) for each dyad member has four observations per dyad. The APIM views these as four variables in a two regression framework with correlated residuals and correlated predictor variables. In symbols,

$$Y_w = \beta_{0w} + \beta_{ww}\text{anxiety}_w + \beta_{wh}\text{anxiety}_h + \epsilon_w$$

$$Y_h = \beta_{0h} + \beta_{hw}\text{anxiety}_w + \beta_{hh}\text{anxiety}_h + \epsilon_h$$

where the intraclass correlation $r_{XX'}$ handles the interdependence on the predictor variable anxiety and the residuals are correlated. This model is represented

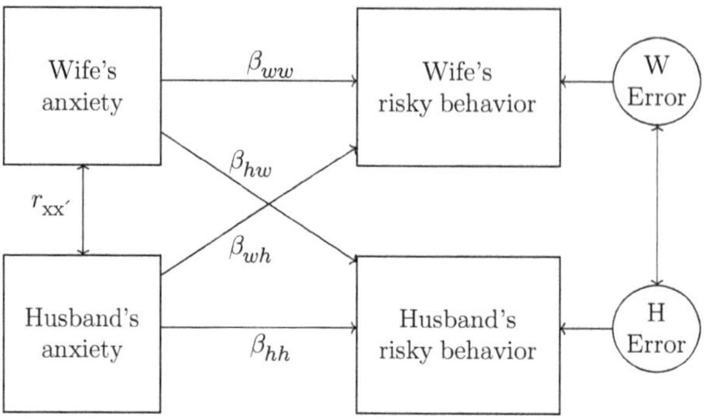

Figure 3.1 Actor-partner model for anxiety and risky behavior. The figure depicts the intraclass correlation $r_{XX'}$ and the correlation between the residuals.

in Figure 3.1. Of course, in the present correlational structure it is not possible to establish causality; it is possible that risky behavior causes anxiety, or that a third variable causes both. Additional information or different designs would be necessary to gain some purchase on the causality question for these variables.

The latent variable model also uses two regression equations, but the usual parameterization involves a partitioning of shared dyadic-level variance and unique individual-level variance. This is accomplished by using two sets of latent variables, or as we will see below, an equivalent formulation in terms of a multi-level model using random effects. One set of latent variables involves individual-level (i.e., residual) variance and the second set of latent variables involves dyadic-level (i.e., factor) variance. In symbols, the variance of each observed variable Y (each square in Figure 3.2, using the convention that observed variables are depicted as squares and latent variables as circles) is modeled as

$$\text{var(Y)} = r_{XX'}\text{var(shared)} + \text{var(unique)} \tag{3.1}$$

and the covariances among the observed variables follow standard covariance algebra rules through the r_i and r_d terms. The implementation of the model requires identification constraints, such as setting the dyad variance to one or the indicator path to one. The fitting of this model can be conducted in either a traditional SEM approach, where the observed covariance matrix is compared to the model-implied covariance matrix, or in a multilevel modeling (MLM) approach. The SEM and MLM approaches can give identical results, such as when data are balanced and the same estimation procedure is used (e.g., maximum likelihood estimation). The SEM representation is presented in Figure 3.2.

Models of dyadic processes provide interesting possibilities for psychologists interested in the study of social interaction. The models help quantify a somewhat fuzzy

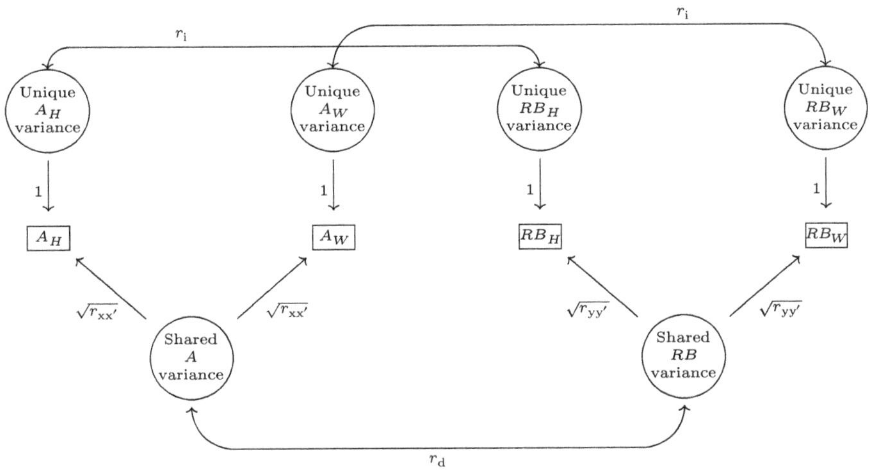

Figure 3.2 Latent variable approach to estimating the intraclass correlation between husband and wife on Anxiety (A) and Risky Behavior (RB), the dyad level correlation r_d, the individual level correlation r_i, and the intraclass correlations $r_{XX'}$ and $r_{YY'}$.

notion about what it means to be interdependent. The models assess interdependence, treating it as a parameter that can be used in more complicated models with additional predictors and outcome variables. The models can be extended to longitudinal data so that temporal processes in dyads can be examined.

In the remainder of the chapter we show how to estimate the parameters of these models in terms of observable covariances and correlations. Interesting relations between statistical parameters emerge, with the intraclass correlation playing a key role. To simplify our analysis we focus on the case of distinguishable dyads, where dyad members can be distinguished on theoretically relevant variables, such as gender in heterosexual couples and age in same-sex sibling pairs. We refer the reader interested in the issue of exchangeable dyads, such as homosexual dating couples and same-sex twins, to Griffin and Gonzalez (1995) and Kenny et al. (2006).

3.2 INTRACLASS CORRELATION: UNIVARIATE MODELS

We begin with an essential ingredient to dyadic models: the intraclass correlation. Interdependence between interval scaled data in the context of linear models can be assessed through the intraclass correlation. The basic intuition for the intraclass correlation is that it is the percentage of total variance that can be attributable to between dyad variance. The intraclass correlation appears in both the APIM and the latent variable model, and we will show that it plays an important role in parameter and interval (i.e., standard error) estimation. In the APIM (Figure 3.1) the intraclass correlation appears as the correlation between the predictor variables and, as we develop later in this chapter, appears in the formula for the βs and their standard errors. In the latent variable model (Figure 3.2) the square root of the intraclass correlation is equal to the respective indicator paths and also appears in the formula for the individual and dyad-level correlations as well as their standard errors.

One standard formulation for the intraclass correlation takes the variance associated with a dyad-level variable and normalizes it by the sum of the dyad-level variance plus the variance of the individual-level error variance. The underlying structural model for an observation Y for the jth person in the ith dyad is

$$Y_{ij} = \mu + \beta_j + \alpha_i + \epsilon_{ij} \tag{3.2}$$

where the μ is the fixed effect constant, β_j is the fixed effect term for the jth dyad member (such as husband and wife), α_i is a random effect term for the dyad that is assumed to be normally distributed with mean 0 and variance σ_α^2, and the usual error term ϵ with variance σ_ϵ^2. There are slightly different estimation formulas for the intraclass correlation depending on whether one uses maximum likelihood or restricted maximum likelihood (the latter accounts for degrees of freedom as in an ANOVA) estimation but the basic logic is similar. The intraclass correlation becomes the ratio

$$\frac{\sigma_\alpha^2}{\sigma_\alpha^2 + \sigma_\epsilon^2}. \tag{3.3}$$

This model (Equation 3.2) can be placed in the context of a multilevel model with two levels, such that the first level represents the raw data and the second level represents the part of the model that is between dyads:

$$Y_{ij} = \gamma_i + \beta_j + \epsilon_{ij} \tag{3.4}$$

$$\gamma_i = \mu + \alpha_i \tag{3.5}$$

where β is a fixed effect term that estimates the difference between the two distinguishable dyad members, γ is a random effect dyad term, and the ϵ is the usual error term. If one substitutes Equation (3.5) into Equation (3.4), then the result is the same as Equation (3.2).

A third way to conceptualize the intraclass correlation is as a structural equation model with two indicators, one latent factor and a specific set of restrictions. If one sets the variance of the latent factor to 1, the two indicator paths to the observed variables equal to each other, and the error variances equal to each other, then the indicator paths are equal to the square root of the intraclass correlation (see Figure 3.3). The two indicator paths are constrained to be equal because the interpretation of the

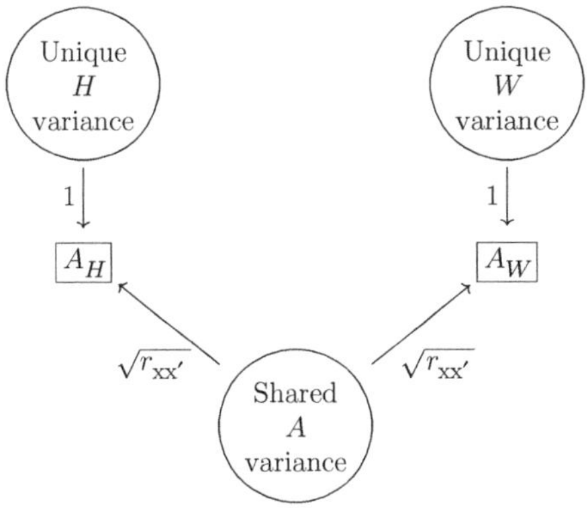

Figure 3.3 Structural equation modeling approach to estimate the intraclass correlation $r_{XX'}$ between husband and wife anxiety scores (A_H and A_W, respectively).

latent variable is one of "shared variance" where both individuals contribute equally. This is a different parameterization and interpretation than the usual latent factor, where indicators can have different paths and can relate differentially to the latent variable.

Thus, there are several ways of conceptualizing the logic of interdependence as indexed by the intraclass correlation, and they lead to the same result. One can model the intraclass correlation as a linear mixed model, as a multilevel model, or as a structural equations model. The results will be the same as long as the same estimation procedure is used (e.g., maximum likelihood) and the proper constraints on the parameters are imposed throughout. All three approaches impose the identical structure on the relevant covariance matrix.

3.2.1 Derivations Using Maple

To illustrate the approach and some basic Maple commands, we present a simple asymptotic Z test for two dependent covariances, such as testing whether the covariance between the wife's anxiety and risky behavior is different than the covariance for the husband for those two variables. This analysis involves two variables (e.g., anxiety and risky behavior) from two dyad members for a total of four observations per dyad.

Browne (1974) made use of the asymptotic property that the covariance of two elements of a covariance matrix is given by

$$\text{cov}(c_{ij}, c_{kl}) = (c_{ik}c_{jl} + c_{il}c_{jk})/N$$

(see also Kendall & Stuart, 1966). This is useful when deriving standard errors for elements of a covariance matrix, and some of the subscripts i, j, k, and l can be the same value, such as in the case of a variance V_{jj}. This asymptotic result forms the building block for one approach to deriving analytic results for dyadic models of interdependence.

We begin by defining the observed covariance matrix for the wife's anxiety (variable 1), the wife's risky behavior (2), the husband's anxiety (3), and the husband's risky behavior (4). In Maple we type the following command

```
covo := matrix(4, 4, [[v1, c12, c13, c14], [c12, v2, c23, c24],
                                  [c13, c23, v3, c34], [c14, c24, c34, v4]])
```

which produces the matrix of symbolic terms

$$\begin{bmatrix} v1 & c12 & c13 & c14 \\ c12 & v2 & c23 & c24 \\ c13 & c23 & v3 & c34 \\ c14 & c24 & c34 & v4 \end{bmatrix}$$

The observed covariance matrix has 10 unique terms: there are 4 observed variances and 6 observed covariances. We apply the asymptotic result in Equation 3.6 repeatedly to create the 10×10 covariance matrix V of the 10 observed elements in the observed covariance matrix. In Maple one can execute the following commands to initialize the observed covariance matrix:

```
#lines beginning with pound sign are comments
#define procedure to compute asymptotic covariance

#initialize observed covariance matrix
covx := covo;
acov := proc(i,j,g,h);
                    covx[i,g]*covx[j,h] + covx[i,h]*covx[j,g];
             end;

k := 10;
ind := matrix(k,2);
counter := 0;
for c1 from 1 to rowdim(covx) do
    for c2 from 1 to c1 do
            counter := counter + 1;
            ind[counter,1] := c1;
            ind[counter,2] := c2;
    od;
od;

#initialize variance matrix
V := matrix(k,k);
```

```
#apply asymptotic covariance formula to compute every term in V
for c1 from 1 to k do
      for c2 from 1 to k do
            V[c1,c2] := acov(ind[c1,1],ind[c1,2],ind[c2,1],ind[c2,2])
      od
od;
```

We illustrate the first 4 rows and columns of the output V

$$\begin{bmatrix} 2\,v1^2 & 2\,c12\,v1 & 2\,c12^2 & 2\,c13\,v1 \\ 2\,c12\,v1 & v2\,v1 + c12^2 & 2\,v2\,c12 & c23\,v1 + c13\,c12 \\ 2\,c12^2 & 2\,v2\,c12 & 2\,v2^2 & 2\,c23\,c12 \\ 2\,c13\,v1 & c23\,v1 + c13\,c12 & 2\,c23\,c12 & v3\,v1 + c13^2 \end{bmatrix}$$

Another approach to estimate matrix V is to compute the Hessian matrix directly from the maximum likelihood function, which can also be done in Maple. We opted to illustrate the asymptotic result mostly for simplicity in exposition. In our experience the Hessian approach can be relatively slow in Maple compared to the asymptotic approach we outline here, and additional manipulation is usually necessary to simplify the resulting expressions into forms that are more instructive.

Now that we have the covariances of the observed covariance matrix, the next step in the process of getting a Wald Z test for two dependent covariances is to apply the delta rule to matrix V for the desired function on the observed covariances. The function takes the difference between the covariance of the two variables for the wife C12 and the covariance for the two variables for the husband C34. We need the standard error of that difference in order to compute the Z test. To compute the standard error of a difference of two covariances, we take the Jacobian of the difference of the two covariances with respect to all 10 observed covariances (with the terms listed in the same order as the matrix V). The Maple command is

```
da := jacobian([c12-c34], [v1, c12, v2, c13, c23, v33, c14, c24, c34, v4])
```

with output

$$\begin{bmatrix} 0 & 1 & 0 & 0 & 0 & 0 & 0 & 0 & -1 & 0 \end{bmatrix}$$

The final step in computing the standard error for the difference C12–C34 involves applying the delta rule by pre- and post-multiplying the matrix V with the Jacobian we just computed (following standard rules of linear algebra to achieve the multivariate extension of multiplying V by the square of the first derivative).

```
new := multiply(da, V, transpose(da))
```

with output

$$\begin{bmatrix} v2\,v1 + c12^2 - 2\,c24\,c13 - 2\,c23\,c14 + v4\,v3 + c34^2 \end{bmatrix} \tag{3.6}$$

We now have all the ingredients to write the equation symbolically for the Z test between wife's C12 and husband's C34 (i.e., the test for equality between the wife's covariance between risk and anxiety and the husband's covariance between risk and anxiety). This Z test is needed to test one of the equality constraints necessary to justify the pooling across dyad members such as equivalent individual-level correlations r_{i} for each dyad member. The Z test for two dependent covariances (e.g., C12 and C34) is given by

$$Z = \frac{(c_{12} - c_{34})\sqrt{N}}{\sqrt{v_1 v_2 + v_3 v_4 + c_{12}^2 + c_{34}^2 - 2(c_{13}c_{24} + c_{14}c_{23})}}$$

The denominator is the square root of the variance of the difference between two covariances that we computed using the asymptotic result and the delta rule (Equation 3.6). This Z can be compared to the critical value of 1.96 for a two-tailed test at $\alpha = .05$.

The computation we just developed shows that it is relatively straightforward to compute the standard error of a function of elements of a covariance matrix. The same logic extends to dyadic models (and SEM more generally) where we estimate parameters that are more complicated functions of elements from the observed covariance matrix and need to estimate the standard errors of those parameters. We illustrate with the computation of the intraclass correlation for the dyad (as depicted in Figure 3.3). The one key addition to the setup is that we first estimate the parameters directly from the likelihood function following Browne's (1974, 1982) papers.

We begin with the observed covariance matrix for two observed variables, say anxiety measured in each of the husband and wife. The covariance matrix is

$$\begin{bmatrix} v1 & c13 \\ c13 & v3 \end{bmatrix}$$

We implement the model in Figure 3.3 by parameterizing it with paths set to 1, though we could have parameterized it by setting the latent variance to one and estimating the constrained paths. We let *vlat* represent the variance of the dyad-level latent variable and *ve* represent the variance of the residual, or individual-level latent variable. The model-based covariance matrix following the SEM approach to the intraclass correlation in Figure 3.3 is

$$\begin{bmatrix} vlat + ve & vlat \\ vlat & vlat + ve \end{bmatrix}$$

The following Maple commands set up the likelihood function (Browne, 1974, 1982), take derivatives and solve for each of the two parameters *vlat* and *ve* in terms of the observed covariance matrix.

```
fitfuncml1 := normal(rationalize((-N/2) * (log(factor(det(covm)))+
                           trace(multiply(covo, inverse(covm)))))):
g1 := grad(fitfuncml1, [vlat, ve]);
mle := solve(convert(g1, set), {ve, vlat});
```

We then run the same asymptotic variance procedure as above, with $k = 3$ since there are three unique elements in the observed covariance matrix, and apply the delta rule to derive the standard error.

```
templ :=
    transpose(jacobian([rhs(mle[1]), rhs(mle[2])], [covo[1, 1], covo[1, 2], covo[2, 2]]))
va := scalarmul(multiply(transpose(templ), scalarmul(V, 1/N), templ), 1)
```

The resulting standard error for the variance of the dyad-level latent variance is

$$\sqrt{\frac{v3\,v1 + c13^2}{N}}$$

If one plugs the three observed values from the covariance matrix into this expression, the result will be identical to the numerical result in SEM programs, as long as the covariances are computed the same way as one's program (i.e., either N or $N - 1$ in the denominator).

These examples illustrate the usefulness of a symbolic system to do some relatively complicated operations that may not be within the reach of many students in the social sciences. Standard errors are placed in the context of observed elements of the covariance matrix, so one can understand the components of the standard error to get a better sense of how the confidence interval and the test of significance behave as a function of the observed covariances. The understanding that can emerge by examining the definitional equations of both the parameter estimates and their standard errors can provide a deeper intuition about the underlying estimates and their tests. We now turn to a discussion of the latent variable model, showing maximum likelihood estimates and their standard errors.

3.3 INTRACLASS CORRELATION: MULTIVARIATE MODELS

The model depicted in Figure 3.2 can be estimated (equivalently) using either SEM or MLM formulations. We begin with the SEM approach to the multivariate latent dyadic model as per Gonzalez and Griffin (1999). The model is a simple generalization of the single variable case (Figure 3.3). Each variable has the "V" shape graphical representation, with a single latent variable representing the dyadic variance, the variables for each dyad member having the unique error variance, and the indicators being the square root of the respective intraclass correlation. The multivariate structure adds correlations across the separate "V" structures in the graph. There is a set of individual-level correlations r_{i} that represents the correlations between a single dyad member's error variance across two variables, and a dyad-level correlation r_{d} that represents the correlation between the two dyad-level latent factors. This model performs a partitioning of individual-level and dyad-level variance separately for each variable and correlates across these partitioned terms to estimate r_{i} and r_{d}, respectively. In the case of distinguishable dyads one can also test the assumption of equality of the two individual-level correlations r_{i}.

The Gonzalez and Griffin (1999) paper presents maximum likelihood estimates of the two intraclass correlations (one for each variable), r_{i} and r_{d} as well as their standard errors and Wald Z tests. The paper also presents SEM syntax. In terms of the present chapter, the derivation of the parameter estimates and their standard error proceeds just like in the previous section, where the maximum likelihood estimate for the intraclass correlation and its standard error was presented. The maximum likelihood estimates for the four key parameters ($r_{\mathrm{xx'}}$, $r_{\mathrm{yy'}}$, r_{i}, and r_{d}) are estimated by taking derivatives of the likelihood function, setting each derivative equal to zero, and solving for the derivatives. Formulas for the asymptotic standard errors follow the procedure discussed in the previous section of using the asymptotic result for a covariance of two covariance terms (Equation 3.6) and the delta method.

Here we highlight a few key aspects of the parameter estimate for r_{d} and its standard error. As showen in Gonzalez and Griffin (1999), r_{d} is related to the intraclasses of both variables. So, in our running example, the intraclass correlation for the anxiety variable and the intraclass correlation for the risky behavior variable appear as reliability corrections in the formula for r_{d} (with small shared variance implying low reliability of measuring the dyadic covariance, and hence an "inflation factor" that increases r_{d}. This closed form estimate for r_{d} provides an explanation for why in some datasets the estimate of r_{d} is out of bounds (e.g., greater than 1). If the product of the intraclass correlations is relatively close to zero, then r_{d} can be overly inflated. One should question the applicability of the latent variable model when either or both intraclass correlations are close to zero—there may not be sufficient shared variance (interdependence) on which to develop a dyadic model. Another interesting point to highlight is the role that the intraclass correlations play in the standard error of r_{d}. Using the method described above we showed (Gonzalez & Griffin, 1999; Griffin & Gonzalez, 1995) that the product of the two intraclass correlations also enters the standard error of r_{d}. All other things being equal, the standard error of r_{d} will be smaller the greater the product of the two intraclasses.

3.3.1 Connections to Multilevel Model

This latent variable model can be equivalently estimated in the context of a multilevel model. We develop the parallel language for multilevel models because it is instructive to show the equivalence of the two approaches, and we encourage students to develop the skills to switch easily between the two approaches. Unfortunately, the description of the multivariate model is not straightforward because multilevel models are typically expressed for a single variable with a single residual term ϵ at level 1. But we need separate error terms for each of the two variables, and in the case of distinguishable dyads, possibly separate error terms for each dyad member. In order to fit the multilevel model to the multivariate example in Figure 3.2 it is necessary to express the multilevel model in a way that allows for different residual variances for each variable and person. We present one method to model multiple variables in the multilevel model using a switching regression technique.

The first level model is

$$\mathbf{Y} = \beta_d D1 + \beta_s D2 \tag{3.7}$$

One dummy code $D1$ assigns a 1 to all anxiety scores and a 0 to all risky behavior scores. The other dummy code $D2$ assigns a 1 to all risky behavior scores and a 0 to all anxiety scores. The first level equation does not include an error variance nor an intercept—those terms appear in the second level separately for each variable. Thus, this switching regression merely partitions the "outcome variable Y" into either anxiety scores or risky behavior scores (in our running example).

The second level of regression equations are:

$$\beta_a = \text{intercept}_a + v_{ag} + u_{ai} \tag{3.8}$$

$$\beta_r = \text{intercept}_r + v_{rg} + u_{ri} \tag{3.9}$$

where v and u are random effects that code group and individual terms (respectively), each equation has its own fixed effect intercept term, the subscripts a, r, g, and i refer to anxiety, risky behavior, group and individual, respectively. The random effect v assesses group-level variance and the random effect u assesses individual-level variance.

Now comes the important part of this particular formulation. We formulate a covariance structure on each of the random effects v (group-level) and u (individual-level). Let the two group-level v's be bivariate normally distributed with covariance matrix

$$\Omega_v = \begin{bmatrix} \sigma^2_{va} & \\ \sigma_{var} & \sigma^2_{vr} \end{bmatrix} \tag{3.10}$$

where a and r denote anxiety and risky behavior, respectively. This means that the random effect v associated with anxiety has variance σ^2_{va}, the random effect v associated with risky behavior has variance σ^2_{vr}, and the two vs have covariance σ_{var}. Similarly, an analogous covariance is formulated on the two individual-level us

$$\Omega_u = \begin{bmatrix} \sigma^2_{ua} & \\ \sigma_{uar} & \sigma^2_{ur} \end{bmatrix} \tag{3.11}$$

This covariance matrix Ω_u gives the variances and covariance between anxiety and risky behavior at the individual level. In this formulation we require equality of all individual-level correlations (i.e., referring back to Figure 3.2, this particular multilevel model implementation forces the two individual-level correlations for husband and wife to be identical).

These two matrices provide all the information necessary to compute the terms r_d and r_i in the latent group model as well as each of the two intraclass correlations. Using the terms in those two covariance matrices we have

intraclass correlation for anxiety:

$$\frac{\sigma_{va}^2}{\sigma_{va}^2 + \sigma_{ua}^2}$$

intraclass correlation for risky behavior:

$$\frac{\sigma_{vr}^2}{\sigma_{vr}^2 + \sigma_{ur}^2}$$

individual level correlation between anxiety and risky behavior:

$$\frac{\sigma_{uar}}{\sqrt{\sigma_{ua}^2 \sigma_{ur}^2}}$$

dyad level correlation between anxiety and risky behavior:

$$\frac{\sigma_{vda}}{\sqrt{\sigma_{va}^2 \sigma_{vr}^2}}$$

This multilevel model formulation leads to identical parameters (i.e., intraclass correlations, r_i and r_d correlations) to the SEM approach as long as one uses similar estimation methods, such as maximum likelihood, across both multilevel and SEM approaches. One can use Maple to estimate these parameters directly from the likelihood function to get closed form solutions. Of course, the closed form tractability of these estimates and their standard error follows because we have equal sized clusters (dyads). It becomes difficult to derive closed form solutions when the clusters have different sizes, such as the case when individuals are nested in families and the sample includes families of different sizes. In that case numerical methods are needed, as implemented in popular multilevel modeling programs.

3.3.2 Actor-partner Model

The actor-partner model (Figure 3.1) provides another opportunity to highlight how the exercise of deriving closed form solutions for parameter estimates and their standard errors provides deeper understanding of how the model works. As with the latent variable model, the intraclass correlations play an important role. Before presenting the regression parameter and its standard error, we discuss the pairwise approach to the actor-partner model that we described in Gonzalez and Griffin (1999) and Griffin and Gonzalez (1995). The pairwise approach organizes the data matrix in "long form" in such a way that each variable is entered twice (e.g., in one column the data are ordered as husband and wife anxiety, in another column the same data are reordered as wife and husband anxiety). We call one variable X or Y and its reordered form X' or Y'. We showed that the maximum likelihood intraclass correlation for variable X is equivalent to the Pearson correlation between X and X' (e.g., Griffin & Gonzalez, 1995). The pairwise organization to the data makes some derivations a little easier, and they follow the procedure described earlier for estimating parameters and their standard errors.

In the case of the actor-partner interaction model described in Figure 3.1 let's define the predictors as X and the dependent variable as Y. The raw score actor effect (i.e., the actor path from X to Y in Figure 3.1) is given by

$$\frac{s_y(r_{xy} - r_{xy'}r_{xx'})}{s_x(1 - r_{xx'}^2)} \tag{3.12}$$

where s_y and s_x are the standard deviations of the criterion variable X and the predictor variable Y, respectively. It is interesting that the intraclass correlation for the predictor variable X appears both in the numerator and the denominator (in squared form) for the regression parameter. One can see that the regression parameter increases with increasing values of the intraclass, and when the intraclass correlation is 1 (perfect agreement between dyad members) the regression parameter is undefined, as expected. The interested reader can refer to our papers for interpretations of the correlations r_{xy} and $r_{xy'}$. The regression coefficient for the partner effect has the same form with the role of r_{xy} and $r_{xy'}$ interchanged.

The standard error of the regression weights is also instructive. Under the null hypothesis that the population $\beta = 0$, the estimate of the variance for the actor regression slope reduces to

$$V(\beta_{\text{actor}}) = \frac{s_y^2(r_{xy'}^2 r_{xx'}^2 - r_{xx'}r_{yy'} + 1 - r_{xy'}^2)}{2Ns_x^2(1 - r_{xx'}^2)} \tag{3.13}$$

where N is the number of dyads. The standard error for the partner regression effect is analogous, except that r_{xy} appears in Equation 3.13 in place of $r_{xy'}$. Clearly, the intraclass correlation of the predictor variable X and the dependent variable Y play an important role in the standard error of the slopes in the APIM.

If the APIM is constrained to have equal actor paths and equal partner paths (e.g., husband to wife is constrained to equal wife to husband), then the APIM yields the identical goodness-of-fit to the latent variable model under the constraint that the two individual-level correlations r_i (one for each dyad member) are constrained to be equal. In other words, they are identical models under a set of constraints. The APIM can be generalized in different ways from the latent variable model, and in this sense the two models can assess different dyadic processes.

3.3.3 Summary of Multilevel Models

The take-home message of this type of modeling is that the observed variances and covariances between variables and between individuals in a dyad can be partitioned into individual-level and dyad-level terms or into an actor-partner regression framework. The approaches can be formulated either as structural equation models or as multilevel models. Once the partition is performed, substantive research questions can be tested at each level. For example, one can test for predictors of the dyad and individual-level terms, and can examine the outcome variables associated with the decomposed variances. Of course, these models can be extended to cases with more than two variables. For example, in the latent variable model one would have a separate "V-structure" for each variable to model each variable's interdependence. One could then correlate the individual-level variances and the dyadic-level variances across variables, or could construct more complicated models with directional paths (e.g., the dyadic variance from one variable predicting the dyadic variance of another variable, mediational models at each of the individual-level and dyadic-levels, etc.). The actor-partner model could also be extended to multiple variable and even longitudinal designs; one needs to keep track of all the actor paths and the partner paths across the different variables and times as one would do in standard cross-lagged designs.

Further, the two modeling approaches highlight different dyadic processes. The latent variable model highlights shared and individual variance processes (e.g., conformity and social norms), whereas the actor-partner model highlights social influence. It is in this way that the analysis of dyadic data goes beyond "correcting" data for violations of independence by directly modeling the interdependence that characterizes a multivariate dyadic data set. The interdependence, as captured by the intraclass correlation and related dyadic measures, plays an important role throughout these models.

3.4 CONCLUSIONS

This chapter had a modest goal. We wanted to illustrate a few analytic derivations of the parameter estimates and their standard errors for dyadic analysis. We demonstrated that increasingly easy-to-use computer programs that perform symbolic computations can make it efficient and educational to derive analytic results in a relatively painless way. The potential for this type of training and understanding is very high. In our simple examples we showed how the intraclass correlation, a standard measure of interdependence used in dyadic analysis, plays a major role not only in parameter estimates but also their standard errors. It is not clear that such an understanding would emerge from only studying simulations, even if they systematically varied the relevant parameters. There is value in analytic derivation in terms of the understanding that can emerge.

The contributions of Michael Browne to covariance analysis and SEM are much deeper than the simple examples that we demonstrated in this chapter. But if we have succeeded in exciting the reader about the role for analytic exercises in developing deeper understanding of models of covariance structures, and the use of computer software in achieving that goal, then we have played a small role in highlighting the deeper contribution that Michael Browne has made to quantitative psychology.

REFERENCES

Browne, M. W. (1974). Generalized least squares estimators in the analysis of covariance structures. *South African Statistics Journal, 8*, 1–24.

Browne, M. W. (1982). Covariance structures. In D. M. Hawkins (Ed.), *Topics in applied multivariate analysis* (pp. 72–141). Cambridge: Cambridge University Press.

Gonzalez, R., & Griffin, D. (1999). The correlational analysis of dyad-level data in the distinguishable case. *Personal Relationships, 6*, 449–469.

Gonzalez, R., & Griffin, D. (2001a). Testing parameters in structural equation modeling: Every "one" matters. *Psychological Methods, 6*, 258–269.

Gonzalez, R., & Griffin, D. (2001b). A statistical framework for modeling homogeneity and interdependence in groups. In M. Clark & G. Fletcher (Eds.), *Handbook of social psychology, Vol 2: Interpersonal processes* (pp. 505–534). Mass: Blackwell.

Griffin, D., & Gonzalez, R. (1995). The correlational analysis of dyad-level data: Models for the exchangeable case. *Psychological Bulletin, 118*, 430–439.

Kendall, M., & Stuart, A. (1966). *The advanced theory of statistics* (2nd ed.). London: Griffin.

Kenny, D. A., Kashy, D. A., & Cook, W. L. (2006). *Dyadic data analysis*. New York: Guilford.

Kenny, D. A., & La Voie, L. (1985). Separating individual and group effects. *Journal of Personality and Social Psychology, 48*, 339–348.

Rose, C., & Smith, M. (2002). *Mathematical statistics with Mathematica*. New York: Springer-Verlag.

Chapter 4

A Bayesian Hierarchical Model for Response Time Data Providing Evidence for Criteria Changes Over Time

Peter F. Craigmile, Mario Peruggia, and Trisha Van Zandt

4.1 INTRODUCTION

Empirical work in psychology and other social sciences frequently collects many measurements from individuals in an attempt to maximize statistical power. These sequential measurements, assumed to be influenced by one or more independent variables manipulated by the experimenter as well as a number of random and latent variables, are often treated as an independent and identically distributed (IID) sample. That is, considering each subject as a random-data-generating device like a coin, the measurements are collected over time just as one would write down the sequence of heads and tails obtained from independent flips of a coin.

The measurements obtained from a group of subjects must then be treated as nested within the random subject variable. A standard linear model with such repeated measurements accounts for correlations across subjects because of subject-specific factors, but not because the within-subject measurements themselves may be correlated over time. That is, the measurements are still considered IID within each subject. There is no good reason to assume that repeated measurements collected within a subject over time are IID. There are, in fact, many reasons to assume that they are not (for example: Browne & Nesselroade, 2005; Craigmile, Goodman, & Shiffrin, 2010). Indeed Browne and Nesselroade (2005) state that, *"a model for a dynamic, psychological process should include (a) variables representing the forces that cause change, (b) variables on which the changes (outcomes) are actually manifested, and (c) parameters that define, at least to some extent, a temporal flow in the relationship among and between the two kinds of variables."* This view inspires our modeling efforts in this chapter.

Response time (RT) data form the better part of empirical results in cognitive psychology. Models of human performance in tasks such as recognition memory, simple decision-making, perception, and many others are evaluated by observing changes in RT under different experimental conditions. People are asked to perform tasks repeatedly, and the RT, measured as the time between the onset of a stimulus and the subsequent response to that stimulus, is recorded. At the end of a number of trials, the RTs are typically analyzed in one of two ways. Either the mean RTs over trials and conditions are computed, reducing the data to a single measure for that subject in each within-subject condition, or the individual RTs are combined with those of other subjects for maximum likelihood or least-squares fits of a model, to obtain estimates of parameter values. These techniques almost exclusively assume that RTs within subjects are IID. Even Bayesian techniques, recently very popular in psychology, often assume conditionally IID observations within subjects.

Response time data have three features incompatible with the IID assumption. First, there are trial-by-trial correlations—slow responses tend to be followed by slow responses, and fast responses by fast responses—and the strength of these correlations varies over subjects. Second, over large numbers of trials, there tend to be many "outliers," very fast and very slow responses that do not seem to be tied to the cognitive process responsible for executing the task. Third, mean RT over trials tends to oscillate, sometimes in very complex ways. While long-range increases or decreases in mean RT over the course of the task can be tied to psychological and physiological processes, such as boredom, fatigue or learning, there are more rapid changes over shorter time scales that seem to be tied to more short-term cognitive mechanisms, such as attentional fluctuations, rhythmic response strategies, and perhaps other mysterious personal strategy shifts. In sum, a sample of RT data collected over time may be distributed as a complex mixture with strong autocorrelations across trials.

Our work has shown that such RT data can be well-modeled as time series (Craigmile et al., 2010; Perruggia, Van Zandt, & Chen, 2002). In Craigmile et al. (2010) we considered a three-component mixture model of log(RT). The first component is a Gaussian autoregressive process of order one (AR(1); to capture trial-by-trial dependence), with a trial-dependent mean modeled with a wavelet decomposition (to capture trend). The remaining two components of the mixture are heavy-tailed exponential distributions to explain extremely fast and slow RTs that cannot be accounted for by the Gaussian component—that is, RTs that would be considered outliers with respect to the Gaussian model. While the Gaussian component is always active, the two heavy-tailed components can turn on at random. More precisely, with some probability, on each trial, an exponential shock is subtracted (very fast RT) or added (very slow RT) to the Gaussian component.

Using a hierarchical Bayesian approach, we showed that there were effects of experimental factors on the different model components that were not at all obvious in the mean RTs or marginal RT distributions. Furthermore, the trends over trials were far more complex than is typically considered in RT data.

The assumption of a Gaussian AR(1) process for log(RT) data is not well-motivated theoretically. Most theoretically-based models of RT are based on the idea that information toward different possible responses is sampled over time from the environment, or from a mental representation of the environment. These sequential sampling models then take the form of a stochastic process in which one or more random variables increment (and/or decrement) over time until an absorbing state, representing "enough evidence," is reached. The state of the process after absorption determines the response, and the time to absorption determines the RT.

Sequential sampling models are difficult to model explicitly, especially in a Bayesian hierarchical time series. Following earlier work (Peruggia et al., 2002; Rouder, Lu, Speckman, Sun, & Jiang, 2005; Rouder, Sun, Speckman, Lu, & Zhou, 2003), we extend our modeling approach by using a Weibull likelihood as the key component of the data-generating process. While the Weibull itself is also a summary of the marginal RT distribution, it has an explicit link to sequential sampling models, specifically those models that assume that evidence accumulates on latent stochastically independent processes toward each response, and that the executed response is determined by the process (or "counter") that reaches its absorbing state first.

4.1.1 The Weibull Distribution

The Weibull distribution has been used to characterize distributions of chronometric measures since it was first introduced by Waloddi Weibull in 1939. It has been used primarily in studies of reliability, to model equipment failure times. A positive random variable Y follows the two-parameter Weibull(r, λ) distribution if its cumulative distribution function (cdf) is

$$F(y|r,\lambda) = 1 - \exp\{-\lambda y^r\}, \tag{4.1}$$

for $y > 0$, where $r > 0$ and $\lambda > 0$ are called the shape and rate parameters, respectively. The exponential distribution is a special case of the Weibull distribution with $r = 1$. One interpretation of the Weibull distribution is that it characterizes a power law transformation $Y = aX^r$ where X is exponentially distributed.

The expected value of Y is given by

$$\begin{aligned}\mu(r, \lambda) &= E(Y|r, \lambda) \\ &= \lambda^{-1/r}\,\Gamma(1 + 1/r).\end{aligned} \tag{4.2}$$

Therefore, for fixed r, $\mu(r, \lambda)$ decreases as λ increases. Denoting the digamma function by $\Psi(\cdot)$ (Abramowitz & Stegun, 1970), for fixed $\lambda \leq \exp(\Psi(1)) \approx 0.56$, $\mu(r, \lambda)$ is decreasing in r. For fixed $\lambda > \exp(\Psi(1))$, $\mu(r, \lambda)$ is decreasing in r up to $r(\lambda) = \left(\Psi^{-1}(\ln(\lambda)) - 1\right)^{-1}$ and increasing thereafter.

From a psychological modeling perspective, the Weibull is interpreted via its role as an extreme value (Type III) distribution. This means that, under certain constraints, the Weibull is the asymptotic distribution of a minimum statistic, the smallest of some number n of random variables, as n goes to infinity. Sequential sampling models of cognitive processing assume that response selection among n alternatives can be portrayed as a race among competing counters. In such a race, the first counter to finish determines the response to be made and the time at which it finishes determines the RT. Necessary and sufficient conditions for the RT to be asymptotically Weibull concern the distribution function of the finishing times for each counter.

The Weibull distribution was first seriously applied to RT data by Logan (1988, 1992). He demonstrated that RT data from an "alphabet arithmetic" task were well-fitted by a Weibull model, and that changes in the Weibull scale parameter over repeated stimulus exposures were well explained by his "Instance Theory," in which memory traces raced against each other to be retrieved. Although Colonius (1995) pointed out that Logan had not specified the distribution of the retrieval times, and hence not adequately defended his use of the Weibull as the appropriate extreme-value distribution for his model, the Weibull has continued to serve as an (at least asymptotically) appropriate RT model for certain tasks.

For convenience, Rouder et al. (2003, 2005) and Peruggia et al. (2002) used the Weibull as the likelihood function in a hierarchical Bayesian approach to modeling RT data. Perrugia et al. (2002) demonstrated that a hierarchical model incorporating fixed factors within a Weibull likelihood provided a useful framework for investigating effects not usually accounted for in traditional RT modeling, including learning, sequential dependencies, and individual differences. Furthermore, this framework provided a more sensitive analysis of experimental factors, which was able to detect reliable differences across experimental conditions when traditional ANOVA failed.

One simple model of response selection, the Poisson race model defines counters that have gamma-distributed processing times (Pike, 1973; Townsend & Ashby, 1983; Van Zandt, Colonius, & Proctor, 2000). The gamma distribution satisfies necessary and sufficient conditions for the minimum time of n IID racing counters to converge to a Weibull distribution (Galambos, 1987). Changes in the parameters of the Weibull over time can therefore be linked to changes in the parameters of the latent structure that underlies a potential response selection process. Cousineau, Goodman, and Shiffrin (2002) established a link between the limiting behavior of the race model processing times and the Weibull, pointing out that the parameters of the Weibull distribution that approximates the distribution of the minimum will be a function of n and the parameters of the racing gamma-distributed variables. Formally, for $T_n = \min\{X_1, X_2, \ldots, X_n\}$ where the X_i are IID gamma with shape α and scale β, letting $\lambda_n = n\beta^{-\alpha}/\alpha\Gamma(\alpha)$, we have that $\lambda_n^{1/\alpha} T_n$ converges in distribution to a Weibull(α,1) as n goes to infinity.

However, the rate of convergence to this limiting distribution can be very slow, depending on the value of α. In particular, if we want to use a Weibull($r_n = \alpha, \lambda_n$) to approximate the distribution of T_n for a finite value of n, this approximation will be unsatisfactory for large values of α. However, we have verified that, for α less than 20, we can satisfactorily approximate the distribution of T_n with a Weibull distribution in which *both* the shape parameter r_n^* and rate parameter λ_n^* depend on n. The values of r_n^* and λ_n^* are found by equating the cdfs of the minimum distribution and the Weibull(r_n^*, λ_n^*) distribution at two points p_1 and p_2.

For fixed values of α and β, we select r_n^* and λ_n^* to satisfy

$$p_1 = F(t_1, r_n^*, \lambda_n^*) \quad \text{and} \quad p_2 = F(t_2, r_n^*, \lambda_n^*), \tag{4.3}$$

where

$$t_1 = q\left(1 - (1 - p_1)^{1/n}; \alpha, \beta\right) \quad \text{and} \quad t_2 = q\left(1 - (1 - p_2)^{1/n}; \alpha, \beta\right), \tag{4.4}$$

and $q(\cdot, \alpha, \beta)$ is the quantile function of the gamma distribution with shape parameter α and scale parameter β. We found that $p_1 = .25$ and $p_2 = .75$, which establishes the interquartile range, yielded values of r_n^* and λ_n^* for which the Weibull approximation was very close to the distribution of T_n, even for large (greater than 2) values of α. The parameters are accurate even for small values of n. For small values of α, less than 5, the approximate Weibull distribution can be accurately estimated even for values of n less than 10. The values of λ_n^* that satisfy Equations (4.3) and (4.4) are very close to λ_n. The values of r_n^* are close to α for values of α less than 2 (as shown by the vertical contours in the left column of Figure 4.1), but they may differ considerably from α for larger values of α.

Above we discussed how the Weibull mean $\mu(r, \lambda)$ (Equation 4.2) behaves as a function of r and λ. This behavior can now be linked to changes in the shape and scale parameters of the underlying gamma(α, β) finishing times for each counter of the decision process, although not uniquely. Figure 4.1 shows a contour plot of the values of r_n^* and $\log(\lambda_n^*)$ as a function of α, β, and n. For a fixed value of β, as the shape parameter α increases, the approximating value λ_n^* will decrease; this can be seen by following the contours of λ_n^*. For a fixed value of α, changes in the scale parameter β will not affect r_n^*, but λ_n^* will decrease as β increases; this can also be seen by following

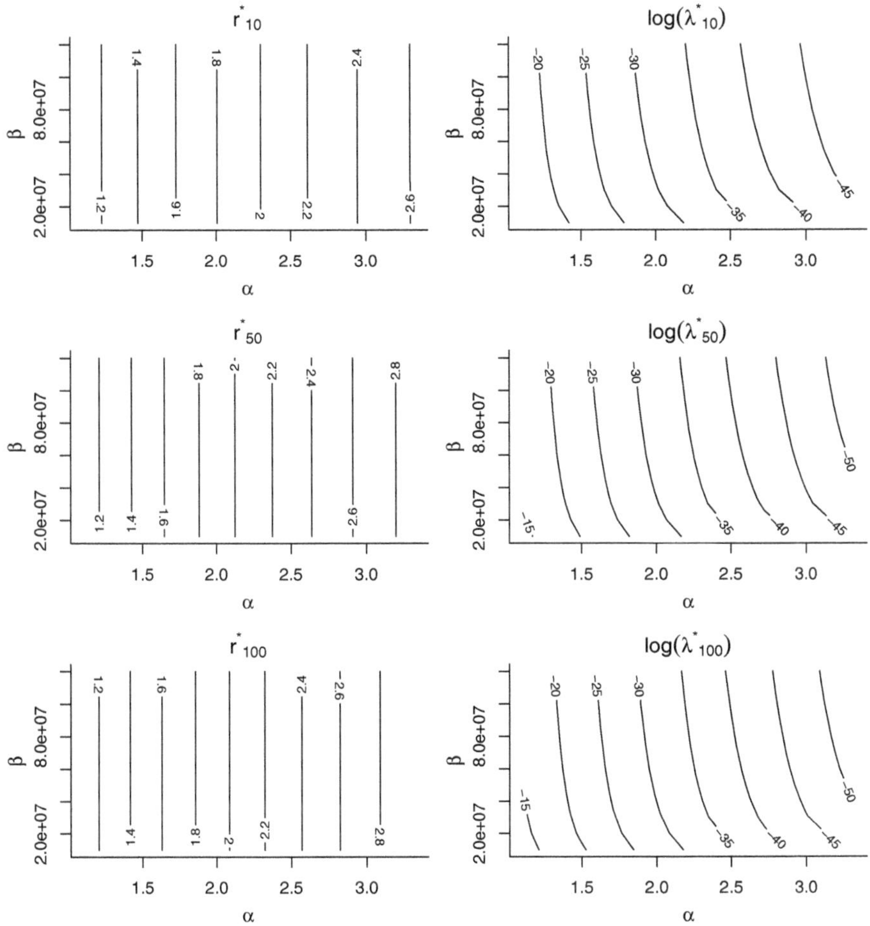

Figure 4.1 Values of r_n^* (left column) and λ_n^* (right column) as a function of the gamma shape parameter α, the gamma scale parameter β, and the number of racing gamma processes n (rows; $n = 10, 50, 100$ in the top, middle, and bottom rows, respectively). The values of α and β were selected to be close to the values that would be produced by the parameters we estimated for the Weibull model.

the contours of λ_n^*. For fixed β, and larger values of α (not shown), both r_n^* and λ_n^* will increase as n, the number of counters, increases.

Having now established a theoretical link between the Weibull likelihood and a theoretically motivated race model for RTs based on a latent structure, we can proceed by using the Weibull likelihood to construct a Bayesian hierarchical model.

Such a model allows us to consider important and usually overlooked factors of RT performance, such as sequential dependencies and individual differences. While we could incorporate such factors in a non-Bayesian model, it would not be trivial, and it would be computationally much more expensive. We could also use a theoretically motivated likelihood in the Bayesian hierarchy, perhaps the minimum of gammas directly, but again this would create computational difficulties. The Weibull likelihood is not only a good description of RT distributions, but reduces computational complexity considerably. This chapter, then, expands the work by Peruggia et al. (2002), by linking the Weibull distribution more closely to psychological theory, and hence justifying its use in Bayesian hierarchical models of RT.

We introduce the recognition memory experiment and data that we study in this chapter in Section 4.2. Our hierarchical Weibull model is outlined in Section 4.3, and in Section 4.4 we provide the results of fitting this Bayesian model using Markov chain Monte Carlo (MCMC). We refine the model for the shape parameter in Section 4.5, compare it to the first model, and close with a discussion and some conclusions in Section 4.6.

4.2 A RECOGNITION MEMORY EXPERIMENT

The data modeled were the practice trials from a previously published experiment exploring confidence in recognition memory (Van Zandt, 2000, Experiment 2). Recognition memory refers to those processes that allow people to discriminate between stimuli they have encountered before (usually in the context of a psychological experiment), and those that they have not. Recognition memory experiments are of broad interest in cognitive psychology, because they are very easy to conduct and they also provide a window on simple decision-making processes. In particular, the sequential sampling models discussed above have provided a powerful framework for explaining human performance in recognition memory tasks.

The original experiment requested people to study a list of 32 words randomly selected from a pool of common English words. Following the study phase, people were shown a test list of 40 words and asked to make keypress responses to each word, responses that identified it as an old (studied) or new (unstudied) word. The RTs for each response were recorded.

Four subjects performed a number of recognition memory tests on each of 10 days. Each test varied in the number of studied items that were tested, but the first two test lists presented on each day always contained an equal number (20) of studied and unstudied words. These tests were considered "practice" and were not analyzed in the original study. We chose to examine these practice data again for three reasons. First, the task was very simple, and there were few fixed experimental factors in these trials. Second, these practice tests, discarded because the subjects were still getting used to the task, gave us an opportunity to model fairly extensive changes in performance over time. Third, these data were used in Peruggia et al.'s original 2002 study establishing a Bayesian hierarchical framework for RT data, and so we can compare our new modeling approach to those published results.

4.3 THE HIERARCHICAL WEIBULL MODEL

We start by considering an enhanced version of the model presented by Peruggia et al. (2002). This expanded model was also used by Peruggia (2007) to investigate dependencies across sequences within word lists for each subject and day.

For subject i ($i = 1, 2, 3, 4$) on day d ($d = 1, \ldots, 10$) let $\mathrm{RT}_{i,d,l,w}$ denote the response time for the wth ($w = 1, \ldots, 40$) word on the lth list ($l = 1, 2$). Because we are not interested in location effects (Peruggia et al., 2002; Peruggia, 2007), we modeled the shifted RTs with the two-parameter Weibull discussed above by subtracting a proportion of the minimum RT from the RTs for each test list. Letting $M_{i,d,l}$ denote the minimum response time for subject i on day d and list l, we let

$$Y_{i,d,l,w} = \mathrm{RT}_{i,d,l,w} - 0.95M_{i,d,l},$$

be the shifted response time to word w. We could have modeled the unshifted RTs with a three-parameter Weibull but this introduces an unnecessary computational burden that we avoid here (Peruggia, 2007).

Assuming conditional independence across all the indices, we let

$$Y_{i,d,l,w} | r_{i,d,l}, \lambda_{i,d,l,w} \sim \mathrm{Weibull}(r_{i,d,l}, \lambda_{i,d,l,w}),$$

where the Weibull distribution has the cdf given by Equation (4.1). Thus, the shape $r_{i,d,l}$ is constant over the 40 words from a given subject, day, and list, but the rate parameter, $\lambda_{i,d,l,w}$, varies over subjects, days, lists, and responses. In Peruggia et al. (2002), the shape parameter was fixed over day, list, and word, and varied only with subject. The fits of the model with that restriction were not entirely satisfactory.

We now establish the relationships between the covariates and the Weibull parameters. To do this, we had to make a choice about the locus of the effects of the experimental covariates. These covariates are $\mathrm{Old}_{i,d,l,w}$, which has value one if the word was actually old and zero if the word was new, and $\mathrm{Right}_{i,d,l,w}$, which has value one if the response was correct and zero if the response was incorrect. We decided to isolate the effects of the covariates in the rate alone, using a mixed log-linear model. This model captures the fixed effects of the covariates, occurring at a population level, plus random effects that capture subject-specific changes. In addition, we consider correlation effects over words within lists for a given subject and day. Formally,

$$\log \lambda_{i,d,l,w} = \alpha_{i,d,l} + \beta_1 \mathrm{Old}_{i,d,l,w} + \beta_2 \mathrm{Right}_{i,d,l,w} + \beta_3(\mathrm{Old}_{i,d,l,w} \times \mathrm{Right}_{i,d,l,w}) + \eta_{i,d,l,w}. \tag{4.5}$$

Assuming independence over subjects, days, and lists, suppose that the $\alpha_{i,d,l}|\alpha_{0,d}, \tau_{\alpha,d} \sim N(\alpha_{0,d}, \tau_\alpha)$ where the $\alpha_{0,d}$, $d = 1, \ldots, 10$, are independent over days and $\alpha_{0,d}$ is $N(0, 1000)$. We let τ_α follow a diffuse inverse gamma prior with rate 0.1 and shape 0.1.

The fixed effect β_1 captures the effect of the type of word (old versus new) on the log rate parameter. The fixed effect β_2 denotes the effect of correctly identifying a word as old or new, relative to incorrect identifications. Finally, the fixed effect β_3 measures the strength of the interaction between the type of word and accuracy. We assume independence between the fixed effects β_j ($j = 1, 2, 3$), with $\beta_j \sim N(0,1000)$ for each j.

For each subject i, day d, and list l, we assume that the sequence of random effects $\{\eta_{i,d,l,w}\}_{w=1}^{40}$ is an AR(1) process. That is,

$$\eta_{i,d,l,1} \sim N(0, \sigma_1^2)$$

for the first word on each list, then

$$\eta_{i,d,l,w}|\eta_{i,d,l,w-1} \sim N(\phi_{i,d,l}\, \eta_{i,d,l,w-1}, \sigma^2),$$

for $w = 2, \ldots, 40$. We assume independence of the variance parameters σ_1^2 and σ^2 and the autocorrelation parameters $\{\phi_{i,d,l}\}$. Both σ_1^2 and σ^2 are given inverse gamma priors with rate 0.1 and shape 0.1. We assume that the autocorrelation parameters $\phi_{i,d,l}$ follow a standard normal distribution. The fact that the normal distribution is not constrained between -1 and 1 allows for possible non-stationarity of the AR(1) process (for example: West & Harrison, 1997; Du Toit & Browne, 2007). Allowing the autocorrelation parameter $\phi_{i,d,l}$ to vary over day and list as well as subject, as was done in Perrugia (2007), is a generalization of the model originally proposed by Peruggia et al. (2002).

Finally, we assume conditional independence of the shape parameters $r_{i,d,l}$ with

$$r_{i,d,l}|\tau_{r,d} \sim \exp(\tau_{r,d}),$$

where the $\tau_{r,d}|\tau_{r,0} \sim \exp(\tau_{r,0})$ are conditionally independent over days d and the rate parameter $\tau_{r,0}$ is exponentially distributed with rate of 0.001.

4.4 FITTING THE HIERARCHICAL WEIBULL MODEL

To fit the model described in Section 4.3, all computations were performed using the R software package (R Development Core Team, 2011) with the `rjags` library as an interface to the Markov chain Monte Carlo (MCMC) sampling capability of the JAGS software package (Plummer, 2010). For each fit we ran two parallel chains, assessing that convergence and mixing of the chains was adequate by examining trace plots and sample autocorrelation plots. From each chain, we discarded 10,000 draws as burn-in

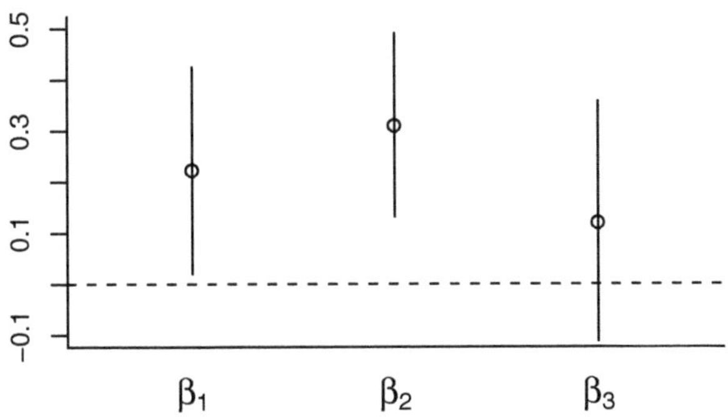

Figure 4.2 Marginal summaries of the posterior distributions of the fixed effects in the model for the log rate. For each parameter, the circle denotes the posterior mean, and the vertical line represents the equal-tail 95% credible interval.

and then obtained 50,000 subsequent draws. After subsampling every 20 draws in each chain and combining the samples over the two chains, our inference is based on 5,000 samples from the posterior distribution.

We start by summarizing the parameters characterizing the rate of the Weibull distribution (Equation 4.5). Figure 4.2 summarizes the posterior distributions of the fixed effect parameters $\beta_j, j = 1, 2, 3$. For each β_j, the circle denotes the posterior mean and the vertical line represents the 95% equal-tail credible interval. The intervals for β_1 and β_2 are positive and do not include zero. Because the bulk of the probability mass for β_3 is located above zero, there is some evidence that the interaction β_3 is also positive. This indicates that, on average, the log rate (and hence the rate) is increased when words are old and when responses are correct, with an additional increment for correct responses to old words; cf. Peruggia et al. (2002). Assuming a fixed shape and noting that larger rates imply faster mean RTs, the mean RT (see Equation 4.2) for both correct and incorrect responses is faster for old words than for new words, and for both old and new words is faster for correct than for incorrect responses. The fastest mean RT is for correct responses to old words. We discuss the combined effect of the rate parameter and shape on the mean RT later in this section.

Figure 4.3 displays a summary of the random intercepts $\alpha_{i,d,l}$ in the log-linear model of the rate (Equation 4.5). The first row shows in black the posterior mean of the intercept for the first list, $\alpha_{i,d,1}$, for each subject i as a function of day d. The gray regions denote pointwise equal-tail 95% credible intervals for this parameter. The second and third rows show the same summaries for the intercept for the second list, $\alpha_{i,d,2}$, and for the difference between the two, $\alpha_{i,d,1} - \alpha_{i,d,2}$. Compared to Subjects 1–3, the posterior distributions of $\alpha_{i,d,l}$ of Subject 4 tend to favor larger values. For a fixed shape, this would indicate that Subject 4 responds more slowly than the other subjects.

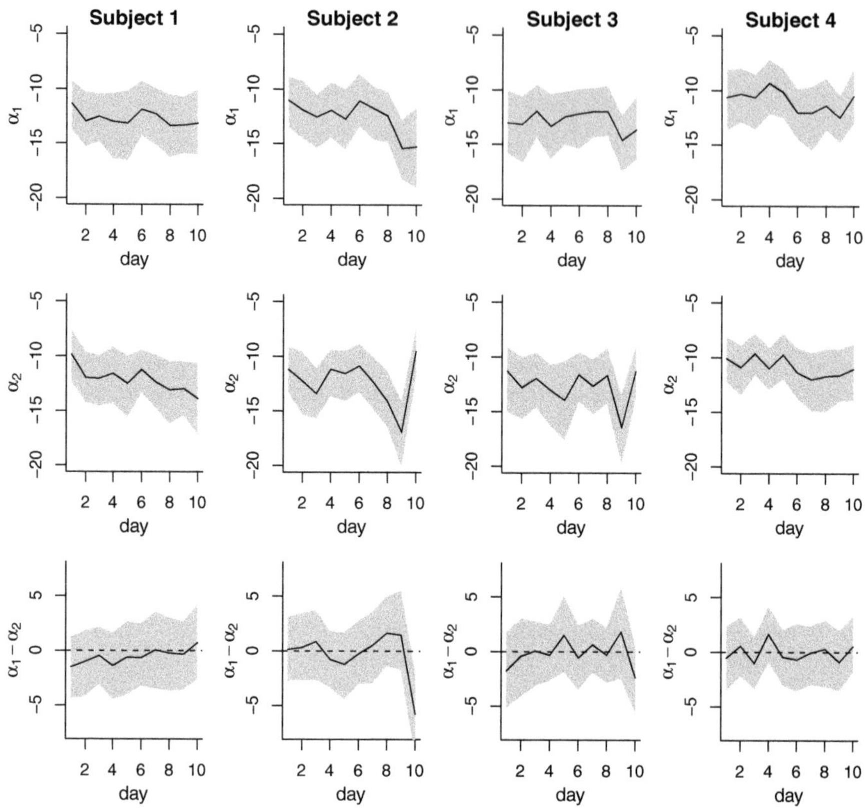

Figure 4.3 Marginal summaries of the posterior distributions of the intercepts for the log rates. The first and second rows show the intercepts for the first and second list, respectively, and the third row shows their differences. In each panel the posterior mean is shown as a black line and the pointwise equal-tail 95% posterior credible intervals are shaded in gray.

For each subject, the intercepts for both lists tend to decrease over day. A decreasing intercept, for a fixed shape, corresponds to faster mean RTs.

Next we summarize the autoregressive parameters $\phi_{i,d,l}$ of the AR(1) model for the sequence of random effects $\{\eta_{i,d,l,w}\}$. Figure 4.4 shows the posterior means and 95% equal-tail credible intervals of $\phi_{i,d,l}$ by subject and day for the first ($l = 1$; black) and second ($l = 2$; gray) lists. Most of the 95% credible intervals contain the value zero, and the posterior means of $\phi_{i,d,l}$ are close to zero, indicating little or no autocorrelation in the random effects. However, there are exceptions; for example, for Subject 1's first list on Day 2, there is a strong positive autocorrelation (close to 0.5). In the few cases where the autocorrelation is large, the autoregressive structure of the random

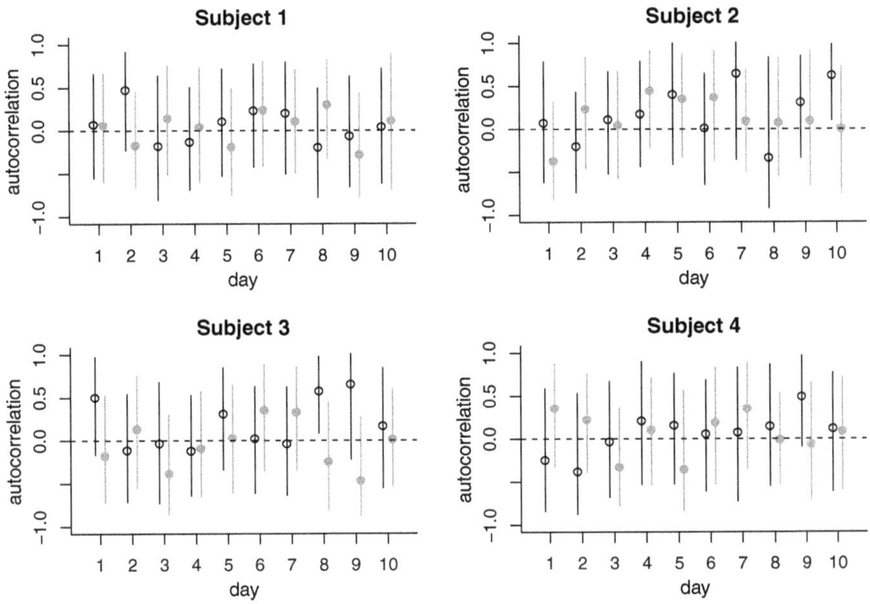

Figure 4.4 Marginal summaries of the posterior distributions of the residual autocorrelation in the log rate by day within subject. The circles denote the posterior means, and the vertical lines the equal-tail 95% credible intervals. The black and gray lines show the posteriors for the first and second list, respectively.

effects may compensate for possible inadequacies of the log-linear model for the rates (Peruggia, 2007).

One of our goals is to investigate more closely changes in the Weibull shape parameter over days and lists, changes suggested by deficiencies in the fits of the Weibull model with constant subject-specific shapes considered by Peruggia et al. (2002). Figure 4.5 shows posterior summaries of the log shapes, $\log r_{i,d,l}$, by day for different subjects (columns) and lists. The first and second rows show the posterior summaries for the first and second list, respectively, and the third row shows their difference (the log ratio). For all subjects the shapes tend to increase, possibly linearly, by day. The credible intervals for the differences, including zero and the posterior mean differences are, for the most part, close to zero, indicating relatively little difference between the shape parameters for the two lists.

Comparing Figures 4.3 and 4.5, changes in the posterior mean shape parameter $r_{i,d,l}$ are mirrored by changes in the posterior mean $\alpha_{i,d,l}$, such that increases in $r_{i,d,l}$ are paired with decreases in $\alpha_{i,d,l}$. We used the sampled parameter values to compute posterior summaries for the mean RTs under the Weibull distribution (Equation 4.2). In our calculations we ignored the autocorrelation effects between the words in a list

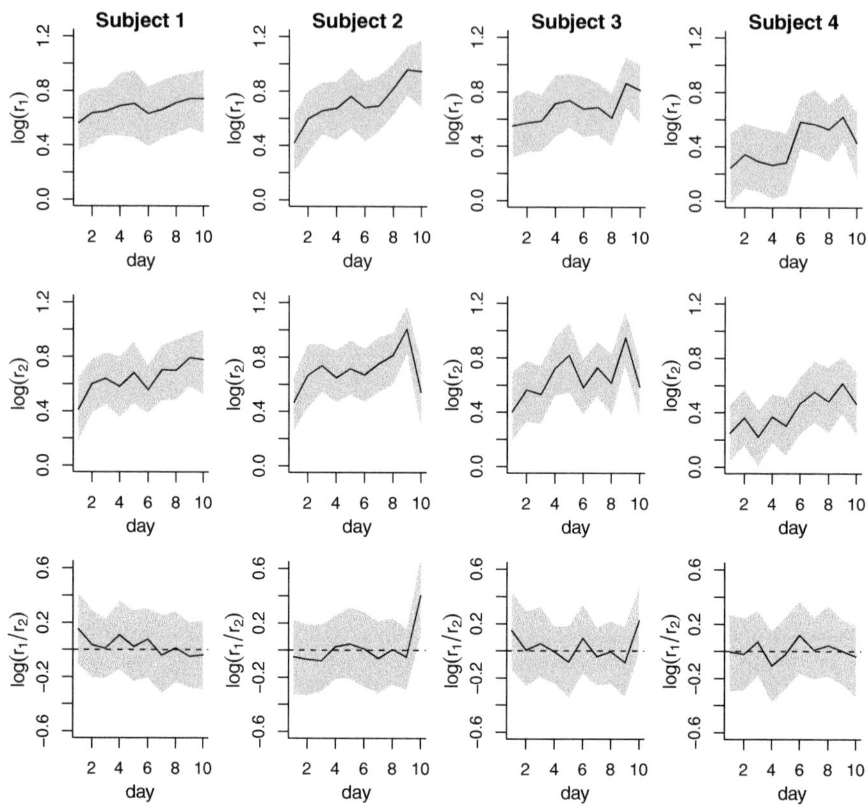

Figure 4.5 Marginal summaries of the posterior distributions of the log shapes by day, subject (columns), and list (rows 1 and 2). The third row shows the posterior log shape differences. The black lines are the posterior means and the pointwise equal-tail 95% posterior credible regions are shaded in gray.

(determined by $\eta_{i,d,l,w}$ in the equation for the rate parameter (4.5)), and conditioned on the four Old by Right covariate combinations. Figure 4.6 summarizes the posterior distributions of the Weibull means for the first (black) and second (gray) lists over days. The means of the posterior distributions are plotted as circles and the equal-tail 95% posterior credible intervals as lines.

We observe a generally decreasing trend over day in the means of the posterior Weibull mean RTs $\mu(r_{i,d,l}, \lambda_{i,d,l,\cdot})$. At the same time, the shape parameters $r_{i,d,l}$ tend to increase and the intercepts of the log rates $\alpha_{i,d,l}$ tend to decrease over day. (The decreases in $\alpha_{i,d,l}$ correspond to decreases in log rate.) Subject 4, who has the highest rates and the lowest shapes, has the slowest mean posterior mean RT and the largest change over day. There are small differences between posterior mean RTs for lists 1

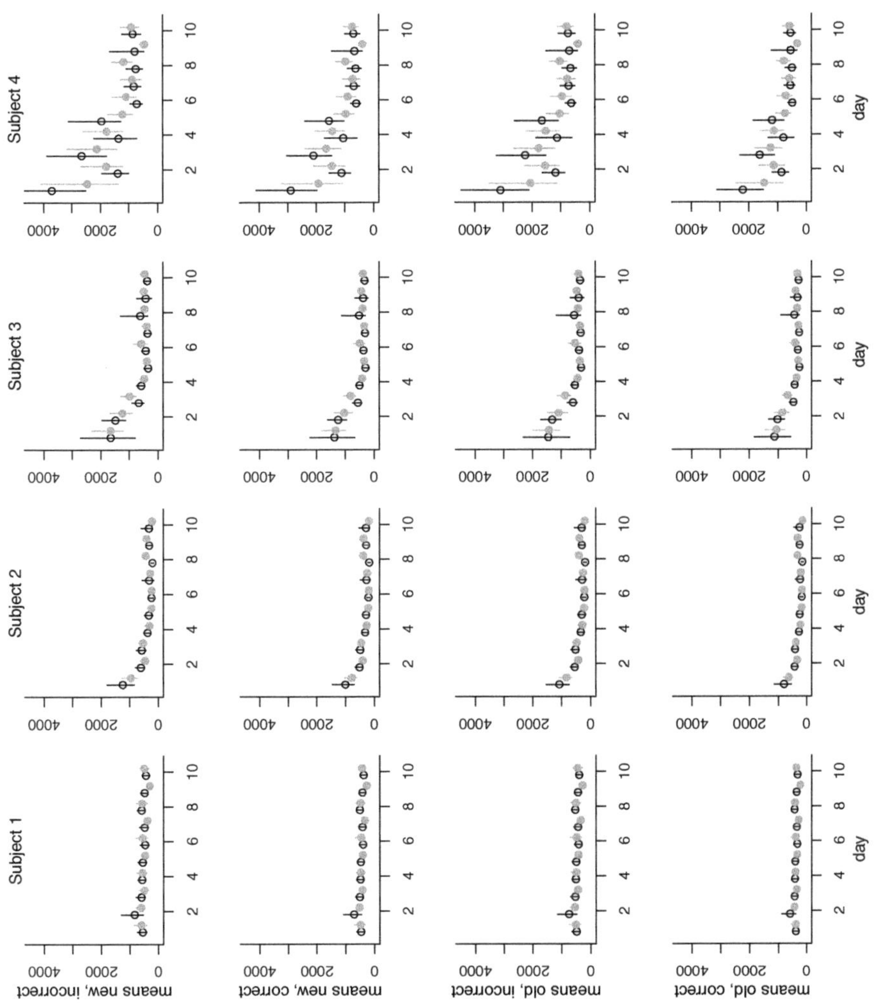

Figure 4.6 Posterior summaries of the Weibull means by day for the four covariate combinations (rows) and subjects (columns). The black and gray circles denote respectively the posterior mean of the Weibull mean RT $\mu(r, \lambda)$ for word lists 1 and 2, and the vertical lines show the equal-tail pointwise 95% credible intervals.

and 2. The posteriors are more variable for list 1, and the mean posterior mean RT fluctuates more over day for list 1.

Referring back to Figure 4.1, we can now suggest a theoretical explanation for the decreases in mean RT over day. For all subjects, there is a negative correlation between the day-specific Weibull posterior mean shapes and the Weibull posterior mean rates. This correlation is consistent with an explanation of the changes in mean RTs that follows from changes in α, the shape parameter of the gamma-distributed racing counters underlying the latent decision process while the number of counters n remains constant (left panel of Figure 4.1).

Increases in $r_{i,d,l}$ could result from either an increase in α or an increase in n. (This dependence on n becomes more evident as the values of α become larger—see Figure 4.1.) Decreases in $\lambda_{i,d,l}$ could also result from an increase in α or a increase in β. While we can't state with any certainty yet exactly what is changing in the latent structures over day, we can indulge in some informed speculation. In particular, we described above Logan's (1988, 1992) Instance Theory, which suggests that the more a task is performed (or, in our case, the more words are studied), the more memory traces race to be recalled. All other parameters remaining fixed, if the number of counters increases with increasing exposure to the words used in the recognition memory task, then the increase in n should result in increases in $\lambda_{i,d,l,w}$. Instead, $\lambda_{i,d,l,w}$ (as seen in $\alpha_{i,d,l}$), decreases over day. Therefore, α or β (or both) must be increasing for us to observe this result.

We note that Logan (1992) reported changes in the best-fitting Weibull shape parameter with task experience, changes that cannot be explained in the context of his Instance Theory. We argue that these changes might arise because of increases in α, the counter thresholds, over day. To explain why α might be increasing we can look to work by Usher, Olami, and McClelland (2002), who used a sequential sampling model to explain performance in an RT task where the number of response options increased, requiring an increase in the number of theoretical counters n entering the decision process. They found that as the number of response options increased, the thresholds of the counters increased exponentially to keep response accuracy at a reasonable level.

Therefore, if the number of racing counters n increases with experience, then it is reasonable to assume that subjects would also increase the response thresholds α for each to prevent fast errors. This behavior would result in the changes in $r_{i,d,l}$ and $\alpha_{i,d,l}$ that we observe over days.

4.5 REFINING THE WEIBULL MODEL TO INCORPORATE SERIAL SHAPE EFFECTS OVER DAYS

If the shape parameter α of the latent decision process increases exponentially with n or day (Usher et al., 2002), then $\log(r_n)$ (for small α) should be approximately linear over day. Because of this, and because of the evidence for linearity displayed in Figure 4.5, we now consider an alternative parametric model for the log shape that incorporates a subject-specific linear trend. For subject i, day d, and list l, we suppose that

$$\log r_{i,d,l} = \gamma_{1,i} + \gamma_{2,i}(d - 5.5) + \epsilon_{i,d,l}.$$

The quantity $(d - 5.5)$ recenters the day variable around the midpoint of the day range 1–10. We assume that the intercept $\gamma_{1,i}$ is independent over subjects i with $\gamma_{1,i}|\gamma_{1,0}, \kappa_1 \sim N(\gamma_{1,0}, \kappa_1)$. We assume that the prior mean, $\gamma_{1,0}|\gamma_1^*, \kappa_1^* \sim N(\gamma_1^*, \kappa_1^*)$, with $\gamma_1^* \sim N(0, 1000)$; and the variance terms κ_1 and κ_1^* follow a diffuse inverse gamma distribution with rate 0.01 and shape 0.01. We assume a similar model for the slope parameters $\gamma_{2,i}$.

To complete the model we assume that, over subjects, days, and lists, the errors $\{\epsilon_{i,d,l}\}$ are drawn from an independent sequence of $N(0, \sigma_\epsilon^2)$ random variables. We assume an inverse gamma prior with rate 0.01 and shape 0.01 for σ_ϵ^2.

But for a slight reduction in variability, the posterior distributions for the fixed effects $\beta_j, j = 1, 2, 3$ and the autoregressive parameters $\phi_{i,d,l}$ were little changed from the previous posteriors, so we do not represent them here. The larger change is in the posterior distributions of the log Weibull shapes $r_{i,d,l}$ (Figure 4.7). Whereas previously the posteriors for these parameters were quite variable, with considerable fluctuation in the posterior mean over day, the linear structure induces less variability and much smoother changes in the posterior mean over day.

Because of the strong correlation between the Weibull log shapes and log rate intercepts, these changes in the posterior of the log shape are mirrored in the posterior distributions of the intercept of the log rate (see Figure 4.8).

4.6 DISCUSSION AND CONCLUSIONS

Anyone who has interacted regularly with Michael Browne can attest to his fascination with models of statistical processes rooted in psychological mechanisms, especially when these mechanisms are used to motivate experiments, the data from which can then verify predictions afforded by psychological theories. The quote from Browne and Nesselroade (2005) with which we began this chapter embodies much of his concern that statistical models do more than describe the data, but also tell us something about how the data came to be.

Our goal in this chapter was to expand a descriptive model, the Weibull model for RTs, which has rather poorly motivated interpretations of its parameters. That is, when the Weibull shape or rate parameters change across conditions, we have no good ideas about what might be changing in the data-generating process. The lack of such interpretations for these Weibull parameters means that when, during the estimation process, we observe systematic changes in their values (over day or condition), we are often unable to provide an explanation for those changes.

In this chapter we fit an expanded Weibull model to RTs from a recognition memory task, a model suggested by deficiencies in a similar model explored by Peruggia et al. (2002), and proposed by Peruggia (2007). Whereas Peruggia et al. (2002) held the Weibull shape parameter for each subject constant over day and

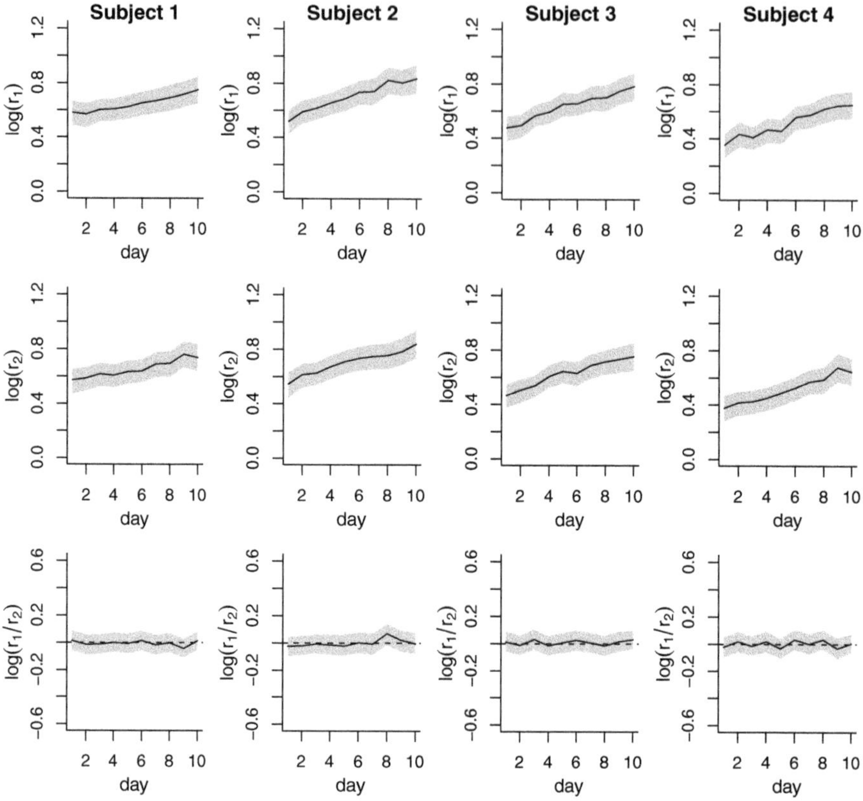

Figure 4.7 Marginal summaries of the posterior distributions of the log shapes by day, subject (columns), and list (rows 1 and 2) for the parameterized shape model. The third row shows the posterior log shape differences. The black lines are the posterior means and the pointwise equal-tail 95% posterior credible regions are shaded in gray.

list, we and Peruggia (2007) permitted changes in this parameter over day and list. These changes allowed us to exploit the statistical connection between the Weibull distribution and the distribution of a minimum statistic.

The most important psychological models of simple decision-making, such as the "old"/"new" decision in recognition memory, are based on the idea that information toward response alternatives is stored and accumulated over time. A class of these models stores information about different responses (and potentially stimuli) on separate counters that "race" toward a threshold. The first counter to reach its threshold, the minimum finishing time of n counters, wins the race and determines the response. Making the reasonable assumption that finishing times are distributed as gamma(α, β) variables, the RT (the minimum of the finishing times) tends to a Weibull distribution as n increases.

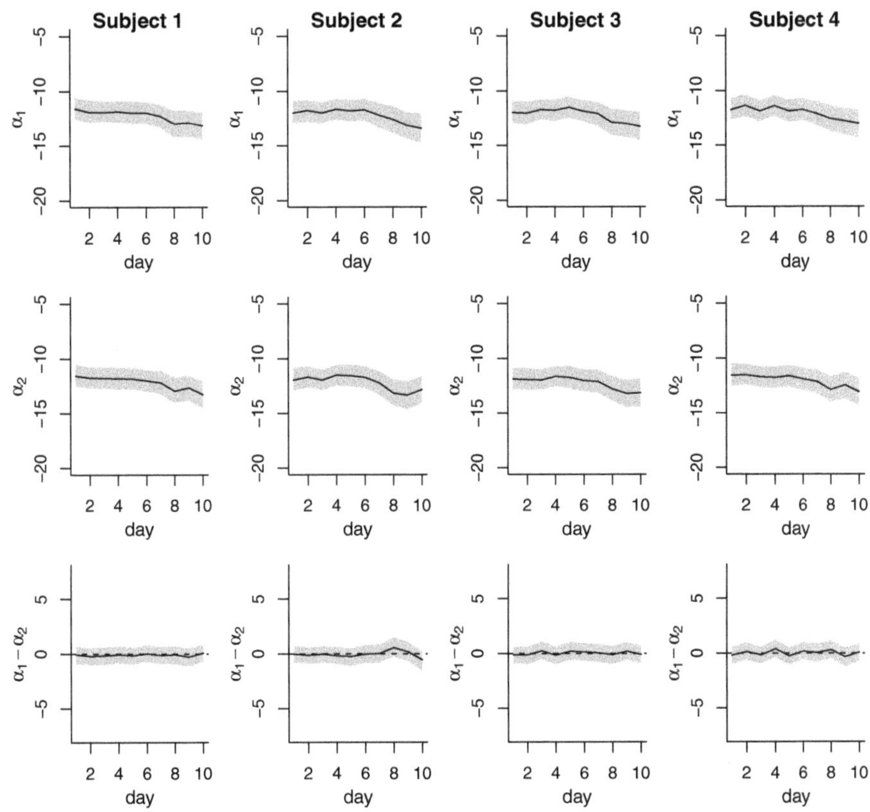

Figure 4.8 Marginal summaries of the posterior distributions of the intercepts for the log rates for the parameterized shape model. The first and second rows show the intercepts for the first and second list, respectively, and the third row shows their differences. In each panel the posterior mean is shown as a black line and the pointwise equal-tail 95% posterior credible intervals are shaded in gray.

This structure permits us to link the values of the Weibull distribution to the values of the latent decision-making process. Increases in the Weibull shape parameter, together with decreases in the Weibull rate, result in decreasing mean RTs over time. But why would the shape parameter increase? Logan (1992), arguing that the shape parameter should remain constant with increases in n, found that decreases in mean RT were well-captured by increases in the Weibull rate, explained only by increases in n. However, his fits of the Weibull to RT data were improved by allowing the shape parameter to vary over conditions. (He did not permit the shape parameter to vary over time.) He suggested that these unexplained variations in shape may have been due to insufficient observations.

Our results, together with additional findings by Usher et al. (2002), permit an alternative interpretation. As n increases with experience, subjects respond by increasing α, the response threshold. Increases in the response threshold prevent responses from being made too quickly (by preventing racing counters from winning too quickly) as additional counters are added to the decision process. We cannot fully investigate this claim here; we note only that we could not have come to this hypothesis without the link between the descriptive Weibull model and the unobservable latent structure generating the RTs.

An additional benefit of this approach was realized in the parameterization of the Weibull shape over days, a parameterization consistent with theoretical work and empirical findings in which sequential sampling race models are applied to tasks where n is explicitly manipulated by the experimenter (Usher et al., 2002). The log-linear model for the Weibull shape results in smoother estimates that make it easier to recognize the possible connection between the Weibull model and the components of the latent decision-making process.

Past research has demonstrated that the Weibull distribution fits the marginal RT distributions well (Logan, 1992), and similarly that the minimum of gamma variables fits both the marginal RT distributions and the response frequency data (proportion of "old"/"new" responses over condition) well (Van Zandt et al., 2000). However, we are exploring changes in the parameters of these models over time, which is a novel approach. A strong connection between the Weibull model and the latent decision process requires that we demonstrate that the dynamic Weibull model and the race process fit the data well. Our preliminary explorations of model fit give us no reason to suspect that there were serious deficiencies in the Weibull model, but a thorough treatment of this issue will require that we fit both the Weibull and the gamma race model to the data, and compare the estimates of the parameters obtained in both cases.

ACKNOWLEDGMENTS

This work is supported by the National Science Foundation under award number SES-1024709.

REFERENCES

Abramowitz, M., & Stegun, I. A. (1970). *Handbook of mathematical functions with formulas, graphs and mathematical tables*. Washington: National Bureau of Standards.

Browne, M. W., & Nesselroade, J. R. (2005). Representing psychological processes with dynamic factor models: Some promising uses and extensions of ARMA time series models. In A. Maydeu-Olivares and J. J. McArdle (Eds.) *Advances in psychometrics: A festschrift to Roderick P. McDonald* (pp. 415–451). Mahwah, NJ: Lawrence Erlbaum Associates.

Colonius, H. (1995). The instance theory of automaticity: Why the Weibull? *Psychological Review, 102*, 744–750.

Cousineau, D., Goodman, V. W., & Shiffrin, R. M. (2002). Extending statistics of extremes to distributions varying in position and scale and the implications for race models. *Journal of Mathematical Psychology, 46*, 431–454.

Craigmile, P. F., Peruggia, M., & Van Zandt, T. (2010). An autocorrelated mixture model for sequences of response time data. *Psychometrika, 75*, 613–632.

Du Toit, S. H. C., & Browne, M. W. (2007). Structural equation modeling of multivariate time series. *Multivariate Behavioral Research, 42*, 67–101.

Galambos, J. (1987). *The asymptotic of extreme order statistics*. Malabar, Florida: R. E. Krieger.

Logan, G. D. (1988). Toward an instance theory of automatization. *Psychological Review, 95*, 492–527.

Logan, G. D. (1992). Shapes of reaction-time distributions and shapes of learning curves: A test of the instance theory of automaticity. *Journal of Experimental Psychology: Learning, Memory and Cognition, 18*, 883–914.

Peruggia, M. (2007). Bayesian model diagnostics based on artificial autoregressive errors. *Bayesian Analysis, 2*, 817–842.

Peruggia, M., Van Zandt, T., & Chen, M. (2002). Was it a car or a cat I saw? An analysis of response times for word recognition. In *Case studies in Bayesian statistics*, volume 6, pp. 319–334. New York: Springer.

Pike, R. (1973). Response latency models for signal detection. *Psychological Review, 80*, 53–68.

Plummer, M. (2010). *JAGS Version 2.1.0 manual.* Available at http://mcmc-jags.sourceforge.net/

R Development Core Team (2011). *R: A Language and environment for statistical computing.* Vienna: R Foundation for Statistical Computing.

Rouder, J. N., Lu, J., Speckman, P., Sun, D., & Jiang, Y. (2005). A hierarchical model for estimating response time distributions. *Psychonomic Bulletin and Review, 12*, 195–223

Rouder, J. N., Sun, D., Speckman, P. L., Lu, J., & Zhou, D. (2003). A hierarchical Bayesian statistical framework for response time distributions. *Psychometrika, 68*, 589–606.

Townsendm, J. T., & Ashby, F. G. (1983). *Stochastic modeling of elementary psychological processes*. New York: Cambridge University Press.

Usher, M., Olami, Z., & McClelland, J. L. (2002). Hick's law in a stochastic race model with speed-accuracy tradeoff. *Journal of Mathematical Psychology, 46*, 704–715.

Van Zandt, T. (2000). ROC curves and confidence judgments in recognition memory. *Journal of Experimental Psychology: Learning, Memory, and Cognition, 26*, 582–600.

Van Zandt, T., Colonius, H., & Proctor, R. W. (2000). A comparison of two response time models applied to perceptual matching. *Psychonomic Bulletin and Review, 7*, 208–256.

West, M., & Harrison, J. (1997). *Bayesian forecasting and dynamic models*. New York: Springer.

Chapter 5
A Review of Estimation Methods for Latent Variable Models

Irini Moustaki

5.1 INTRODUCTION

This chapter reviews estimation methods for the generalized linear latent variable models (GLLVM) for cross-sectional and longitudinal data. The theoretical formulation of the GLLVM for cross-sectional data is discussed in Moustaki and Knott (2000) and Bartholomew et al. (2011), and for longitudinal data in Dunson (2003) and Cagnone et al. (2009).

The generalized linear framework allows the analysis of metric and categorical, and a mixture of those variables. The models developed within the psychometric literature known as item response theory are special cases of the GLLVM framework. For example, there are the two-parameter logistic model (Birnbaum, 1968; Bock & Aitkin, 1981), the three-parameter model (Birnbaum, 1968), the model for nominal responses (Bock, 1972), the partial credit model (Masters, 1982), the graded response model (Samejima, 1969), and the generalized partial credit model (Muraki, 1992). A review of those models and many more can be found in van der Linden and Hambleton (1997).

The recent advances in estimation methods such as the E–M algorithm (Dempster et al., 1977), Bayesian estimation methods, and composite likelihood methods, as well as advances in computers, have made latent variable models with many factors and complex structures easier to estimate than 40 years ago. In this chapter, I discuss the E-M algorithm and Markov Chain Monte Carlo (MCMC) methods for estimating the GLLVM for cross-sectional data, and discuss under what conditions pairwise likelihood estimation can be a useful estimation tool. As it turns out, pairwise likelihood estimation does not reduce the computational complexity of the GLLVM for cross-sectional data, but it does offer an advantage when the GLLVM framework is extended to longitudinal data (see Vasdekis et al., 2012). In general, composite likelihood estimation methods can be extremely useful when an underlying variable (UV) approach is followed for the factor analysis for categorical data. Specifically, Jöreskog and Moustaki (2001) implemented the pairwise likelihood estimation method for fitting a factor analysis model to ordinal data, using the UV approach that assumes that each observed ordinal variable is a manifestation of an underlying continuous variable. Under the UV approach, full information maximum likelihood (FIML) estimation requires a p-dimensional integration and becomes intractable for a relatively small number of items. However, pairwise likelihood or limited information maximum likelihood (LIFM) reduces the integral dimension to two. Recently, Katsikatsou et al. (2012) extended the pairwise estimation method for fitting a structural equation model to any type of observed variables. However, within the FIML estimation, Cai (2010b) extended Gibbons and Hedeker's idea of item bifactor analysis (Gibbons & Hedeker, 1992) to fit latent variable models to longitudinal data. His method provides a significant dimension reduction. Finally in a similar spirit to mine, Wirth and Edwards (2007) provide an overview of recently developed estimation methods for factor analysis type models for ordinal items.

This chapter is organized as follows. I begin with the theoretical framework that is being used to model latent variables and I also discuss the models within a generalized linear latent variable framework. I then review various estimation methods and discuss goodness-of-fit matters. Finally, we present an example.

5.2 THEORETICAL FRAMEWORK FOR CROSS-SECTIONAL DATA

Let us denote with $\mathbf{y}$ the $p \times 1$ vector of manifest variables and with $\mathbf{z}$ the $q \times 1$ vector of latent variables. As only $\mathbf{y}$ can be observed any inference must be based on the joint distribution of $\mathbf{y}$:

$$f(\mathbf{y}) = \int_{R_{\mathbf{z}}} g(\mathbf{y} \mid \mathbf{z})\varphi(\mathbf{z})d\mathbf{z}, \tag{5.1}$$

where $\varphi(\mathbf{z})$ is the prior distribution of the latent variables $\mathbf{z}$ and $g(\mathbf{y} \mid \mathbf{z})$ is the conditional distribution of $\mathbf{y}$ given $\mathbf{z}$. Note that the density functions $\varphi(\mathbf{z})$ and $g(\mathbf{y} \mid \mathbf{z})$ are not uniquely determined.

Modelling manifest variables as functions of latent variables implies that if correlations among the ys can be explained by a set of latent variables, then when all zs are accounted for the ys will be independent (local independence). Therefore, q must be chosen so that:

$$g(\mathbf{y} \mid \mathbf{z}) = \prod_{i=1}^{p} g(y_i \mid \mathbf{z}).$$

If this condition holds then (5.1) is written as:

$$f(\mathbf{y}) = \int_{R_{\mathbf{z}}} \prod_{i=1}^{p} g_i(y_i \mid \mathbf{z}) \varphi(\mathbf{z}) d\mathbf{z}. \tag{5.2}$$

For a random sample of size n the marginal log-likelihood is written as:

$$l = \sum_{h=1}^{n} \log f(\mathbf{y}_h). \tag{5.3}$$

5.3 GENERALIZED LINEAR LATENT VARIABLE MODELS

One can adopt a generalized linear model framework to specify latent variable models for categorical and metric responses (see Moustaki & Knott, 2000; Moustaki, 2003; Bartholomew et al., 2011). A generalized linear model consists of three components:

1. The random component in which each of the p random response variables, $(y_1, \ldots, y_p)$ has a distribution from the exponential family (such as Bernoulli, Poisson, Multinomial, Normal, Gamma).
2. The systematic component in which latent variables $\mathbf{z}' = (z_1, \ldots, z_q)$ produce a linear predictor η_i corresponding to each y_i:

$$\eta_i = \alpha_{i0} + \sum_{j=1}^{q} \alpha_{ij} z_j, \qquad i = 1, \ldots, p. \tag{5.4}$$

3. The links between the systematic component and the conditional means of the random component distributions:

$$\eta_i = v_i(\mu_i(\mathbf{z})),$$

where

$$\mu_i(\mathbf{z}) = E(y_i \mid \mathbf{z}),$$

and $v_i(.)$ is called the link function which can be any monotonic differentiable function, and may be different for different manifest variables $y_i, i = 1, \cdots, p$.

I shall, in fact, assume that $(y_1, y_2, \ldots, y_p)$ denotes a vector of p manifest variables where each variable has a distribution in the exponential family taking the form:

$$g_i(y_i; \theta_i, \phi_i) = \exp\left\{\frac{y_i\theta_i - b_i(\theta_i)}{\phi_i} + d_i(y_i, \phi_i)\right\}, \qquad i = 1, \ldots, p, \tag{5.5}$$

where $b_i(\theta_i)$ and $d_i(y_i, \phi_i)$ are specific functions taking a different form depending on the distribution of the response variable y_i. All the distributions discussed in this paper have canonical link functions with $\theta_i = \eta_i$ except in the case of ordinal manifest variables; ϕ_i is a scale parameter.

For example, in the case of binary responses the manifest binary variable has a Bernoulli distribution with expected value $\pi_i(\mathbf{z})$. The link function is defined to be either the logit:

$$v(\pi_i(\mathbf{z})) = \theta_i(\mathbf{z}) = \text{logit}\pi_i(\mathbf{z}) = \ln\left(\frac{\pi_i(\mathbf{z})}{1 - \pi_i(\mathbf{z})}\right) = \alpha_{i0} + \sum_{j=1}^{q} \alpha_{ij} z_j, \tag{5.6}$$

or the probit

$$v(\pi_i(\mathbf{z})) = \text{probit}\pi_i(\mathbf{z}) = \Phi^{-1}(\pi_i(\mathbf{z})) = \alpha_{i0} + \sum_{j=1}^{q} \alpha_{ij} z_j, \tag{5.7}$$

where Φ^{-1} is the inverse of the normal distribution function.

Similarly one can define a model for normal, nominal, ordinal, count, or skewed data, and details for those models can be found in Bartholomew et al. (2011).

I give here the E–M algorithm as it is described in Bartholomew et al. (2011) for the one-factor model to keep the notation simple. The E-M starts from the log-likelihood l_c for the complete data $(y_{ih}, z_h), i = 1, \ldots, p, h = 1, \ldots, n$, where z_h is the value of the latent variable for individual h.

$$l_c = \sum_{h=1}^{n} \ln f(\mathbf{y}_h, z_h) = \sum_{h=1}^{n} \left[\sum_{i=1}^{p} \ln g(y_{ih} \mid z_h) + \ln \phi(z_h)\right]. \tag{5.8}$$

The E–M algorithm alternates E-steps, which evaluate $E[l_c \mid \mathbf{y}_1, \ldots, \mathbf{y}_n]$, and M-steps which maximize $E[l_c \mid \mathbf{y}_1, \ldots, \mathbf{y}_n]$ over the parameters. The expectation of the E-step uses current estimates for the parameters, and the M-step finds new estimates for the parameters that achieve the maximization. Since $\phi(z)$ does not depend on the parameters, there is no need to use the complete data likelihood; we can replace l_c by

$$L = \sum_{h=1}^{n} \ln g(\mathbf{y}_h \mid z_h) = \sum_{h=1}^{n} \left[\sum_{i=1}^{p} \ln g(y_{ih} \mid z_h)\right], \tag{5.9}$$

and at each E-step find $E[L \mid \mathbf{y}]$.

The conditional density can be written as:

$$g(y_{ih} \mid z_h) = [\pi_i(z_h)]^{y_{ih}}[1 - \pi_i(z_h)]^{(1-y_{ih})} = (1 - \pi_i(z_h)) \exp\{y_{ih}(\alpha_{i0} + \alpha_{i1} z_h)\}, \quad (5.10)$$

and so

$$L = \sum_{h=1}^{n} \ln g(\mathbf{y}_h \mid z_h) = \sum_{h=1}^{n} \left[\sum_{i=1}^{p} (\ln(1 - \pi_i(z_h)) + y_{ih}(\alpha_{i0} + \alpha_{i1} z_h)) \right]$$

$$= \sum_{i=1}^{p} \left[\sum_{h=1}^{n} \ln(1 - \pi_i(z_h)) + \alpha_{i0} \sum_{h=1}^{n} y_{ih} + \alpha_{i1} \sum_{h=1}^{n} y_{ih} z_h \right]. \quad (5.11)$$

The E-step is therefore

$$E[L \mid \mathbf{y}] = \sum_{i=1}^{p} \left[\sum_{h=1}^{n} E(\ln(1 - \pi_i(z_h)) \mid \mathbf{y}_1, \ldots, \mathbf{y}_n) + \alpha_{i0} \sum_{h=1}^{n} y_{ih} + \alpha_{i1} \sum_{h=1}^{n} y_{ih} E(z_h \mid \mathbf{y}_1, \ldots, \mathbf{y}_n) \right]$$

$$= \sum_{i=1}^{p} \left[\sum_{h=1}^{n} E(\ln(1 - \pi_i(z_h)) \mid \mathbf{y}_h) + \alpha_{i0} \sum_{h=1}^{n} y_{ih} + \alpha_{i1} \sum_{h=1}^{n} y_{ih} E(z_h \mid \mathbf{y}_h) \right]. \quad (5.12)$$

For the M-step one can maximize separately for each item i

$$\sum_{h=1}^{n} E(\ln(1 - \pi_i(z)) \mid \mathbf{y}_h) + \alpha_{i0} \sum_{h=1}^{n} y_{ih} + \alpha_{i1} \sum_{h=1}^{n} x_{ih} E(z \mid \mathbf{y}_h). \quad (5.13)$$

One can write (5.13) as

$$\left[\ln(1 - \pi_i(z)) N(z) + \alpha_{i0} \sum_{h=1}^{n} y_{ih} + \alpha_{i1} R_i(z) \right] dz, \quad (5.14)$$

where

$$R_i(z) = \sum_{h=1}^{n} y_{ih} h(z \mid \mathbf{x}_y),$$

$$N(z) = \sum_{h=1}^{n} h(z \mid \mathbf{y}_h). \quad (5.15)$$

Bock and Aitkin (1981) used Gauss–Hermite quadrature to approximate $h(z \mid \mathbf{y}_h)$, which would work best if that distribution were standard normal. Suppose that the Gauss–Hermite quadrature uses the discrete distribution at values $z^1, \ldots, z^k$ with probabilities $h(z^1), \ldots, h(z^k)$. Then approximating the integrals in (5.15) gives

$$R_{it} = \sum_{h=1}^{n} y_{ih} h(z^t \mid \mathbf{x}_y)$$

$$N_t = \sum_{h=1}^{n} h(z^t \mid \mathbf{y}_h). \quad (5.16)$$

This leads to the M-step used by Bock and Aitkin.

It may be helpful to give an interpretation of R_{it} and N_t. N_t is the expected number of individuals at z_t and R_{it} is the expected number of those predicted to be at z_t who will respond positively.

Note that $E[L \mid \mathbf{y}]$ is the log-likelihood of the observed data, and can be directly maximized, without fixing the parameters in the conditional expectation. That can allow an interplay between the E–M algorithm and a maximization of the observed data log-likelihood. Newton–Raphson algorithms may be used for direct maximization. They are known for being fast but sensitive to initial values. On the contrary, the E–M is not sensitive to the initial values, but convergence can be slow in high dimensional problems. Combination of the two methods can give an improvement. For example, the software ltm (Rizopoulos, 2006) uses a combination of Newton–Raphson and E–M algorithms, while the software GLAMM (Rabe-Hesketh et al., 2001) uses a Newton–Raphson maximization algorithm.

5.3.1 Other Methods for Approximating the Integral

Gauss–Hermite quadrature has been found to lose accuracy as the number of factors increases. Other methods, which are particularly valuable for large q, have been found more reliable, such as the adaptive quadrature (Schilling & Bock, 2005), Monte Carlo methods (Sammel et al., 1997), and Laplace approximations (Huber et al., 2004).

Adaptive quadrature is an improved version of the Gauss–Hermite quadrature described above. The values $z^1, \dots, z^k$ and probabilities $h(z^1), \dots, h(z^k)$ are chosen so that the normal distributions used match the location and dispersion of the posterior distribution $h(z \mid \mathbf{y}_h)$ for each response pattern $\mathbf{y}_h$ during every E-step. Schilling and Bock (2005) show that when there are several factors, this method of numerical integration can produce fast and accurate solutions using k as low as 2 for each factor.

The Monte Carlo method known in the literature as a Monte Carlo E–M (MCEM) algorithm, works by generating a large sample of $z^{(1)}, \dots, z^{(R)}$ values from $h(z \mid \mathbf{y}_h)$, which distribution is not usually available in a closed form. Expectation over the latter is replaced by averaging over the large sample. A large number of draws R is required to obtain precise estimates, which slows down convergence of the E–M algorithm. Meng and Schilling (1996) proposed drawing from the posterior distribution $h(z \mid \mathbf{y}_h)$ by using a Gibbs sampler.

5.4 BAYESIAN ESTIMATION: MARKOV CHAIN MONTE CARLO METHODS

The GLLVM can be estimated with Bayesian estimation and in particular with the MCMC methodology. Bayesian inference is based on the posterior distribution of the vector of all random parameters conditionally on the data, and MCMC is primarily used to help to make draws from that posterior distribution. In most cases the posterior distribution is not in a closed form. MCMC methods, such as the Metropolis–Hastings algorithm and its special case Gibbs sampling (Gelfand & Smith, 1990) can be used to estimate the parameters of latent variable models. In particular, Metropolis–Hastings within Gibbs sampling is used for estimating the logit model in (5.6), whereas Gibbs sampling can be used for the probit model in (5.7). Computational details can be found in relevant papers cited in the sections below and in Bartholomew et al. (2011). Recently Cai (2010a) adopted the Metropolis–Hastings Robbins–Monro algorithm for estimating a confirmatory factor analysis model. I outline here the Gibbs sampling and the Metropolis–Hastings for special cases of the GLLVM.

5.4.1 Gibbs Sampling

Gibbs sampling (Geman & Geman, 1984; Gelfand & Smith, 1990), is a way of sampling from complex joint posterior distributions. Consider the posterior distribution $h(\mathbf{v} \mid \mathbf{y})$, where it will be convenient here to consider partitioning the vector of unknown parameters $\mathbf{v}$ into two components namely, $\mathbf{v}_1 = \mathbf{z}$ and $\mathbf{v}_2 = \boldsymbol{\alpha}$. Other partitions might be more convenient for other problems. When it is computationally difficult to take draws from the joint posterior of all parameters it may be easier to simulate from the full conditional distributions, here $h(\mathbf{z}^1 \mid \boldsymbol{\alpha}^0, \mathbf{y})$ and $h(\boldsymbol{\alpha}^1 \mid \mathbf{z}^1, \mathbf{y})$. Gibbs sampling produces a sequence of iterations $\mathbf{v}^0, \mathbf{v}^1, \ldots, \mathbf{v}^k$ that form a Markov Chain that eventually converges to its stationary distribution, which is the posterior distribution. It can be summarized in the following two steps:

1. Start with initial guesses for $\mathbf{z}^0, \boldsymbol{\alpha}^0$.
2. Then simulate in the following order:

   - draw $\mathbf{z}^1$ from $h(\mathbf{z} \mid \boldsymbol{\alpha}^0, \mathbf{y})$,
   - draw $\boldsymbol{\alpha}^1$ from $h(\boldsymbol{\alpha} \mid \mathbf{z}^1, \mathbf{y})$

where the conditional distributions

$$h(\mathbf{z} \mid \boldsymbol{\alpha}, \mathbf{y}) = \frac{g(\mathbf{y} \mid \mathbf{z}, \boldsymbol{\alpha})h(\mathbf{z}, \boldsymbol{\alpha})}{\int g(\mathbf{y} \mid \mathbf{z}, \boldsymbol{\alpha})h(\mathbf{z}, \boldsymbol{\alpha})d\mathbf{z}}, \tag{5.17}$$

$$h(\boldsymbol{\alpha} \mid \mathbf{z}, \mathbf{y}) = \frac{g(\mathbf{y} \mid \mathbf{z}, \boldsymbol{\alpha})h(\mathbf{z}, \boldsymbol{\alpha})}{\int g(\mathbf{y} \mid \mathbf{z}, \boldsymbol{\alpha})h(\mathbf{z}, \boldsymbol{\alpha})d\boldsymbol{\alpha}}. \tag{5.18}$$

Step 2 is iterated t times generating the sample ($\mathbf{v}^t = \mathbf{v}_1^t, \mathbf{v}_2^t$). For suitably large t, the joint distribution of $\mathbf{v}^t$ will be close to the joint distribution of $\mathbf{v}$. So that will be considered as one simulated value from the posterior distribution of $\mathbf{v}$. Note that all the draws up to t are discarded (burn-in period) and t should be large enough to guarantee convergence. Repeating the above whole process s times gives a sample of size s from the posterior distribution. That sample can be used to approximate the mean and variance of the posterior distribution of each model parameter.

Note that each distribution used at step 2 is called a full conditional distribution, it is multivariate and, depending on the number of items and number of sample members, can be heavily multi-dimensional. That problem can be easily solved by arranging the blocks of parameters to have smaller dimensions. In latent trait models, one might block the parameters ($\boldsymbol{\alpha}_i$, $i = 1, \ldots, p$) separately for each item (p blocks), and further form n blocks, one for each subject's latent vector ($\mathbf{z}_h$, $h = 1, \ldots, n$). In that case, the vector of parameters $\mathbf{v}$ consists of $(n + 2 \times p)$ components, and the number of transitions required under step 2 will be equal to $(n + p)$. The elements of each of those components will not then usually be larger than three to four dimensions. Gibbs sampling is particularly useful when the full conditional distributions given in (5.17) and (5.18) exist in closed form. However, the normalising constants that appear in the denominator of the posterior distributions are not often easy to compute. Methods based on data augmentation, rejection sampling, and Metropolis–Hastings have been used with Gibbs sampling to simplify or avoid the computation of the normalizing constants. Albert (1992) proposed a data augmentation scheme within Gibbs for estimating the parameters of the probit of (5.7). The data augmentation scheme borrows the idea of the underlying variable approach, which indicates that the observed binary variable, y_i, is assumed to be a manifestation of an underlying continuous variable y_i^*. Details can be found in Bartholomew et al. (2011).

5.4.2 Metropolis–Hastings

A method that altogether avoids the computation of the normalising constant is the Metropolis–Hastings algorithm (M–H) (Metropolis et al., 1953; Hastings, 1970). For the M–H algorithm at each step, the next state is chosen by sampling a candidate from a proposal distribution $q(.)$ that can be of any form. Independently of the form of the proposal distribution, the chain will converge to the desired distribution $\pi(\mathbf{z}, \boldsymbol{\alpha})$. For a given value $(\mathbf{z}^0, \boldsymbol{\alpha}^0)$ the proposal distribution $q[(\mathbf{z}^0, \boldsymbol{\alpha}^0), (\mathbf{z}^*, \boldsymbol{\alpha}^*)]$ is a density function of $(\mathbf{z}^1, \boldsymbol{\alpha}^1)$ for the given $(\mathbf{z}^0, \boldsymbol{\alpha}^0)$. If the candidate is accepted, then $(\mathbf{z}^1, \boldsymbol{\alpha}^1) = (\mathbf{z}^*, \boldsymbol{\alpha}^*)$, otherwise we generate another candidate. One can show that under general conditions the sequence $(\mathbf{z}^0, \boldsymbol{\alpha}^0), (\mathbf{z}^1, \boldsymbol{\alpha}^1), (\mathbf{z}^2, \boldsymbol{\alpha}^2), \ldots$ is a Markov chain with stationary distribution $\pi(\mathbf{z}, \boldsymbol{\alpha})$.

In theory one can choose any proposal distribution. Depending on the choice of the proposal distribution q, the algorithm might converge to its stationary distribution after very few or very many steps. Usually, known distributions such as the uniform, Beta, Gamma, Normal, or t-distribution are chosen. The number of iterations until convergence or, in other words, the number of candidate states being rejected will increase with the number of parameters.

To avoid the slow convergence of the M–H algorithm and the problem of computing the normalising constant required in the Gibbs sampling, Patz and Junker (1999) discuss rejection sampling techniques such as the M–H, within the Gibbs sampling. In essence, an M–H sampling scheme is used within the Gibbs for drawing samples from the complete conditionals, which need only be defined up to a normalising constant. M–H avoids calculation of the normalising constants by performing draws using rejection sampling. The steps are provided in the above mentioned papers for the logit model. Furthermore, Edwards (2010) develops an efficient MCMC estimation method for confirmatory factor analysis models.

In Bayesian inference, one needs to define prior distributions for all model parameters. Diffuse or vague priors are needed in most applications, so that the likelihood is emphasized rather than the prior, unless one would like to impose some restrictions on the parameters, such as positive loadings or values within a specific range.

MCMC methods are very computationally intensive especially for models with many parameters, such as the factor analysis type models. Convergence diagnostics already developed within the MCMC framework need to be used here too. In addition, label switching remain an unsolved problem.

5.5 THEORETICAL FRAMEWORK FOR LONGITUDINAL DATA

The linear predictor given in (5.4) can be extended to allow associations among items over time as follows

$$\eta_i = \alpha_{i0} + \sum_{j=1}^{q} \alpha_{ij} z_j + u_i, \qquad i = 1, \ldots, p. \tag{5.19}$$

where the u_i terms are item-individual random effects that account for the dependencies of the same item across time. An AR(1) autoregressive model is used to model the relationships of the latent variable across time.

The joint density of the full response pattern $\mathbf{y}$ becomes

$$f(\mathbf{y}) = \int_{R_{\mathbf{z}}} \int_{R_{\mathbf{u}}} g(\mathbf{y} \mid \mathbf{z}, \mathbf{u}) \varphi_1(\mathbf{z}) \varphi_2(\mathbf{u}) d\mathbf{z} d\mathbf{u} \tag{5.20}$$

where $\varphi_1(\mathbf{z})$ is the prior distribution of the latent variables $\mathbf{z}$, as before, and $\varphi_2(\mathbf{u})$ is the distribution of the random effects assumed to have a multivariate normal with mean 0 and a diagonal covariance matrix, and $g(\mathbf{y} \mid \mathbf{z}, \mathbf{u})$ is the conditional distribution of $\mathbf{y}$ given $\mathbf{z}$, $\mathbf{u}$.

This model has been estimated with MCMC methodology in Dunson (2003) and with the E–M in Cagnone et al. (2009). The full maximum likelihood very quickly becomes intractable as the number of items increases. In the longitudinal case the number of integrals required is equal to the number of latent variables and the number of items, since both the latent variables and the random effects need to be integrated out in (5.20). MCMC is also extremely computationally demanding. Alternative methods such as pairwise likelihoods have been found to work much more satisfactorily (see Vasdekis et al., 2012).

The composite likelihood is obtained by replacing the joint likelihood of the data $\mathbf{y}$ with a suitable product, of conditional, or marginal densities (Besag, 1974; Lindsay, 1988; Jöreskog & Moustaki, 2001). In factor analysis type models, bivariate marginal distributions used by Feddag and Bacci (2009) and Varin and Vidoni (2005) have been found to work satisfactorily.

Let us denote with $L_i(\boldsymbol{\alpha})$ and $L_{ij}(\boldsymbol{\alpha})$ the univariate and bivariate log-likelihoods. According to Cox and Reid (2004) we should maximize

$$pl(\boldsymbol{\alpha}) = \sum_{i>j} L_{ij}(\boldsymbol{\alpha}) - c\sum_{i} L_i(\boldsymbol{\alpha}) , \tag{5.21}$$

where c is a suitable constant. pl should be maximized with respect to the parameter vector $\boldsymbol{\alpha}$. In the binary case, let

$$L_i = (p_s^{(i)} \ln \pi_s^{(i)} - p_s^{(i)} \ln p_s^{(i)}) , \tag{5.22}$$

$$L_{ij} = \sum_{s=1}^{2}\sum_{m=1}^{2}(p_{sm}^{(ij)} \ln \pi_{sm}^{(ij)} - p_{sm}^{(ij)} \ln p_{sm}^{(ij)}), \tag{5.23}$$

where $p_s^{(i)}$ and $\pi_s^{(i)}$ are the observed and theoretical proportions of responding into category s to item i, and $p_{sm}^{(ij)}$ and $\pi_{sm}^{(ij)}$ are the pairwise observed and theoretical proportions of responding into category s to item i and category m to item j, respectively, where $s, m = 1, 2$. Since pl depends on all the parameters in $\boldsymbol{\alpha}$, it has to be maximized with respect to all parameters simultaneously. The theoretical probabilities can be obtained from (5.6) for the logit model and from (5.7) for the probit model.

Since each component of Equation (5.21) is a likelihood object, it can be claimed that $\nabla pl(\boldsymbol{\alpha}) = 0$ is unbiased under regularity conditions (Varin & Vidoni, 2005). For the pairwise likelihood defined in (5.21), the maximum pairwise likelihood estimator $\hat{\boldsymbol{\alpha}}_{PL}$ converges in probability to the true parameter value $\boldsymbol{\alpha}_0$, and

$$\hat{\boldsymbol{\alpha}}_{PL} \rightarrow N_q\left(\boldsymbol{\alpha}_0, J^{-1}(\boldsymbol{\alpha}_0)H(\boldsymbol{\alpha}_0)J^{-1}(\boldsymbol{\alpha}_0)\right) , \tag{5.24}$$

where $J(\boldsymbol{\alpha}) = E_y\left\{\nabla^2 pl(\boldsymbol{\alpha}; \mathbf{y})\right\}$ and $H(\boldsymbol{\alpha}) = var(\nabla pl(\boldsymbol{\alpha}; \mathbf{y}))$. The maximization of the likelihood in (5.21) can be performed using the E–M algorithm or any other maximization, and details can be found in Vasdekis et al. (2012).

5.6 GOODNESS-OF-FIT AND MODEL SELECTION CRITERIA

Testing how well a latent variable model fits the data is one of the most difficult problems in evaluating a model. In principle, one could use overall goodness-of-fit statistics such as the Pearson X^2 (chi-square) or the likelihood ratio statistic G^2. When the sample size n is large and p small the above statistics (under the hypothesis that the model fits) follow a chi-square distribution, with degrees of freedom the number of response patterns minus the number of independent parameters minus one. As the number of items increases the chi-square approximation to the distribution of either goodness-of-fit statistic ceases to be valid. Parameter estimates are still valid but it is difficult to assess the model. Problems arising from the sparseness of the multi-way contingency tables in the binary case are discussed in Reiser and VandenBerg (1994). Bartholomew and Tzamourani (1999) proposed alternative ways for assessing the goodness-of-fit of the model for binary variables based on Monte Carlo methods and parametric bootstrapping.

Bartholomew and Tzamourani (1999), Jöreskog and Moustaki (2001), Moustaki et al. (2004), and Bartholomew et al. (2008) used as a measure of fit the differences in the one-, two-, and three-way margins between the observed and expected frequencies under the model. The differences were computed using a X^2 or a G^2 statistic. The above discrepancy measures can be used to measure discrepancies in the margins. Those statistics, also known as residuals, are not independent and so no formal testing can be done. Residuals analysis of that type can be used as a diagnostics procedure for identifying pairs of variables that the model does not fit well, and to suggest ways in which the scales constructed through the model may be improved.

In addition, one can compute for a given response pattern, standardized residuals (Cochran, 1954) or adjusted residuals (Haberman, 1973). The latter under certain regularity conditions follows the standard normal distribution.

Formal limited information tests that are based on the univariate and bivariate margins have been developed by Christoffersson (1975), Muthén (1978), Reiser and VandenBerg (1994), Bartholomew and Leung (2002), Cai et al. (2006), Maydeu-Olivares and Joe (2005, 2006), and Joe and Maydeu-Olivares (2010). In those tests the null hypothesis is stated as $H_0 : \epsilon = 0$ where ϵ can be a vector of first- and second-order margins with elements:

$$\epsilon_i = \sqrt{n}(p_i - \pi_i) \text{ and } \epsilon_{ij} = \sqrt{n}(p_{ij} - \pi_{ij}),$$

where p_i and p_{ij} are the true first- and second-order proportions, and π_i and π_{ij} are the ones under the model.

Most tests are based on a Wald type statistic defined as:

$$W = \mathbf{e}'\hat{\Sigma}^{-1}\mathbf{e},$$

where $\hat{\Sigma}$ is the estimated covariance matrix of $\boldsymbol{\epsilon}$ and

$$e_i = \sqrt{n}(\hat{p}_i - \hat{\pi}_i) \text{ and } e_{ij} = \sqrt{n}(\hat{p}_{ij} - \hat{\pi}_{ij}),$$

where $\hat{p}_i$ and $\hat{p}_{ij}$ are the observed first- and second-order proportions, and $\hat{\pi}_i$ and $\hat{\pi}_{ij}$ are the estimated ones under the model.

Differences among the tests are with respect to the way the covariance matrix is estimated, whether they account for first- and second-order margins or even higher-order margins, whether they use standardized or unstandardized residuals, and finally whether the parameters are taken to be known or not.

Finally, instead of testing the goodness-of-fit of a specified model we could alternatively use a criterion for selecting among a set of different models. This procedure gives information about the goodness-of-fit for each model in comparison with other models. This is particularly useful for determination of the number of factors required. Sclove (1987) gives a review of some of the model selection criteria used in multivariate analysis, such as those due to Akaike, Schwarz, and Kashap. These criteria take into account the value of the likelihood at the maximum likelihood solution and the number of parameters estimated. Furthermore, Varin and Vidoni (2005) introduce model selection criteria that are based on composite likelihoods. Finally, within the Bayesian framework Sinharay et al. (2006) have used posterior predictive model checking for assessing goodness-of-fit.

5.7 APPLICATION

I present here an example where we compare the E–M and MCMC estimation methods. Comparisons between full maximum likelihood and composite likelihood methods can be found in Jöreskog and Moustaki (2001) and Vasdekis et al. (2012).

This example is taken from a major study of "social life feelings" reported in Schuessler (1982) and Krebs and Schuessler (1987). The aim was to establish scales of measurement comparable to those used in ability testing. Further analyses of these data will be found in Bartholomew et al. (1997). Here I take scale 10 called "Economic self-determination." We use the "intercultural version" derived from a sample of 1490 Germans. Yes or no responses were obtained to the following five statements.

1. Anyone can raise his standard of living if he is willing to work at it.
2. Our country has too many poor people who can do little to raise their standard of living.
3. Individuals are poor because of the lack of effort on their part.
4. Poor people could improve their lot if they tried.
5. Most people have a good deal of freedom in deciding how to live.

Table 5.1 Parameter estimates and standard errors for the logit model fitted to the Social Life Feelings scale 10 with the E–M and MCMC

	E–M			MCMC	
Item	$\hat{\alpha}_{i0}$	$\hat{\pi}_{i0}$	$\hat{\alpha}_{i1}$	$\hat{\alpha}_{i0}$	$\hat{\alpha}_{i1}$
1	−2.35(0.13)	0.09	1.20(0.14)	−2.37(0.14)	1.21(0.15)
2	0.80(0.06)	0.69	0.71(0.09)	0.79(0.06)	0.70(0.09)
3	0.99(0.09)	0.73	1.53(0.17)	0.99(0.09)	1.52(0.17)
4	−0.67(0.13)	0.34	2.55(0.41)	−0.74(0.20)	2.86(0.81)
5	−1.10(0.07)	0.25	0.92(0.10)	−1.10(0.07)	0.92(0.11)

Table 5.1 gives the maximum likelihood with the E–M parameter estimates and their asymptotic standard errors as well as the MCMC estimates and standard deviations obtained from BUGS after 10,000 iterations. BUGS uses Gibbs sampling within M–H. There is little to choose between the two sorts of estimates. For the E–M solution, we have given both the $\hat{\alpha}_{i0}$s and the $\hat{\pi}_{i0}$s, but with standard errors for the $\hat{\alpha}_{i0}$s only. The $\hat{\pi}_{i0}$s are the estimated probabilities of a positive response for the median individual ($z = 0$). The $\hat{\pi}$s vary much more in this example as one might expect. The $\hat{\alpha}$s also show considerable variation, with statements 2 and 3 having the greatest discriminating power.

5.8 CONCLUSION

This chapter reviews some recent estimation methods for fitting latent variable models under the generalized linear model framework. It makes a distinction between full information and limited information maximum likelihood methods as well as Bayesian estimation methods. Maximization is achieved by the use of the E–M or the Newthon–Raphson algorithm. All methods produce similar results both in terms of parameter estimates and standard errors. Bayesian inference has the advantage of providing information for the whole posterior distribution of each parameter. At the same time, Bayesian inference can be quite complex due to the large number of parameters involved in factor analysis type models. Further research needs to be done in the area of rotation and label switching.

Full maximum likelihood methods depend on high-dimensional integrations that can be approximated with a high accuracy using adaptive quadrature points. Research has shown that a small number of points will be sufficient. However, when the number of factors increases, estimation becomes very slow.

Finally, composite likelihood estimation methods can be a flexible alternative to full maximum likelihood methods for factor analysis models for longitudinal responses, and for factor analysis models that adopt the underlying variable approach.

REFERENCES

Albert, J. H. (1992). Bayesian estimation of normal ogive item response curves using Gibbs sampling. *Journal of Educational Statistics, 17*, 251–269.

Bartholomew, D. J., Knott, M., & Moustaki, I. (2011). *Latent variable models and factor analysis: a unified approach* (3rd ed.). London: John Wiley.

Bartholomew, D. J., & Leung, S. O. (2002). A goodness of fit test for sparse 2^p contingency tables. *British Journal of Mathematical and Statistical Psychology, 55*, 1–15.

Bartholomew, D. J., Menezes, L. de, & Tzamourani, P. (1997). Latent trait and latent class models applied to survey data. In J. Rost & R. Langeheine (Eds.), *Application of latent trait and latent class models in the social sciences.* Münster/New York: Waxmann.

Bartholomew, D. J., Steele, F., Moustaki, I., & Galbraith, J. (2008). *Analysis of multivariate social science data* (2nd ed.). London/Florida: Chapman & Hall/CRC.

Bartholomew, D. J., & Tzamourani, P. (1999). The goodness-of-fit of latent trait models in attitude measurement. *Sociological Methods and Research, 27*, 525–546.

Besag, J. E. (1974). Spatial interaction and the statistical analysis of lattice systems (with discussion). *Journal of the Royal Statistical Society, Series B, 36*, 192–236.

Birnbaum, A. (1968). Some latent trait models and their use in inferring an examinee's ability. In F. M. Lord & M. R. Novick (Eds.), *Statistical theories of mental test scores* (pp. 425–435). Reading, MA: Addison-Wesley.

Bock, R. D. (1972). Estimating item parameters and latent ability when responses are scored in two or more nominal categories. *Psychometrika, 37*(1), 29–51.

Bock, R. D., & Aitkin, M. (1981). Marginal maximum likelihood estimation of item parameters: application of an EM algorithm. *Psychometrika, 46*(4), 443–459.

Cagnone, S., Moustaki, I., & Vasdekis, V. (2009). Latent variable models for multivariate longitudinal ordinal responses. *The British Journal of Mathematical and Statistical Psychology, 62*, 401–415.

Cai, L. (2010a). Metropolis–Hastings and Robbins–Monro algorithm for confirmatory factor analysis. *Journal of Educational and Behavioral Statistics, 35*, 307–335.

Cai, L. (2010b). A two-tier full-information item factor analysis model with applications. *Psychometrika, 75*, 581–612.

Cai, L., Maydeu-Olivares, A., Coffman, L., & Thissen, D. (2006). Limited information goodness of fit testing of item response theory models for sparse 2^p tables. *British Journal of Mathematical and Statistical Psychology, 59*, 173–194.

Christoffersson, A. (1975). Factor analysis of dichotomized variables. *Psychometrika, 40*, 5–32.

Cochran, G. (1954). Some methods for strengthening the common χ^2 tests. *Biometrics, 10*, 417–451.

Cox, D. R., & Reid, N. (2004). A note on pseudolikelihood constructed from marginal densities. *Biometrika, 91*, 729–737.

Dempster, A. P., Laird, N. M., & Rubin, D. B. (1977). Maximum likelihood from incomplete data via the EM algorithm. *Journal of The Royal Statistical Society, Series B, 39*, 1–38.

Dunson, D. B. (2003). Dynamic latent trait models for multidimensional longitudinal data. *Journal of the American Statistical Association, 98*, 555–563.

Edwards, M. C. (2010). A Markov Chain Monte Carlo approach to confirmatory item factor analysis. *Psychometrika, 75*, 474–497.

Feddag, M., & Bacci, S. (2009). Pairwise likelihood for the longitudinal mixed Rasch model. *Computational Statistics and Data Analysis, 53*, 1027–1037.

Gelfand, A., & Smith, A. (1990). Sampling-based approaches to calculating marginal densities. *Journal of the American Statistical Association, 85*, 398–409.

Geman, S., & Geman, D. (1984). Stochastic relaxation, Gibbs distributions, and the Bayesian restoration of images. *IEEE Transactions Pattern Analysis and Machine Intelligence, 6*, 721–741.

Gibbons, R. D., & Hedeker, D. (1992). Full-information item bifactor analysis. *Psychometrika, 57*, 423–436.

Haberman, S. (1973). The analysis of residuals in cross-classified tables. *Biometrics, 29*, 205–220.

Hastings, W. (1970). Monte Carlo sampling methods using Markov chains and their applications. *Biometrika, 57*, 97–109.

Huber, P., Ronchetti, E., & Victoria-Feser, M.-P. (2004). Estimation of generalized linear latent variable models. *Journal of the Royal Statistical Society, Series B, 66*, 893–908.

Joe, H., & Maydeu-Olivares, A. (2010). A general family of limited information goodness-of-fit statistics for multinomial data. *Psychometrika, 75*, 393–419.

Jöreskog, K. G., & Moustaki, I. (2001). Factor analysis of ordinal variables: A comparison of three approaches. *Multivariate Behavioral Research, 36*, 347–387.

Katsikatsou, M., Moustaki, I., Yang-Wallentin, F., & Jöreskog, K. G. (2012). Pairwise likelihood estimation for factor analysis models with ordinal data. *Computational Statistics and Data Analysis, 56*, 4243–4258.

Krebs, D., & Schuessler, K. F. (1987). *Soziale Empfindungen: ein interkultureller Skalenvergleich bei Deutschen und Amerikanern (Social Attitudes: an intercultural Scale Comparison for Germans and Americans)*. Frankfurt/Main: Campus Verlag.

Lindsay, B. (1988). Composite likelihood methods. In N. U. Prabhu (Ed.), *Statistical inference from stochastic processes* (pp. 221–239). Providence, RI: Americal Mathematical Society.

Masters, G. N. (1982). A Rasch model for partial credit scoring. *Psychometrika, 47*, 149–174.

Maydeu-Olivares, A., & Joe, H. (2005). Limited and full-information estimation and goodness-of-fit testing in 2^n contingency tables: A unified framework. *Journal of the American Statistical Association, 6*, 1009–1020.

Maydeu-Olivares, A., & Joe, H. (2006). Limited information goodness-of-fit testing in multidimensional contingency tables. *Psychometrika, 71*, 713–732.

Meng, X.-L., & Schilling, S. (1996). Fitting full-information item factor models and an empirical investigation of Bridge sampling. *Journal of the American Statistical Association, 91*, 1254–1267.

Metropolis, N., Rosenbluth, A., Rosenbluth, M., Teller, A., & Teller, E. (1953). Equations of state calculations by fast computing machine. *Journal of Chemical Physics, 21*, 1087–1091.

Moustaki, I. (2003). A general class of latent variable models for ordinal manifest variables with covariate effects on the manifest and latent variables. *British Journal of Mathematical and Statistical Psychology, 56*, 337–357.

Moustaki, I., Jöreskog, K. G., & Mavridis, D. (2004). Factor models for ordinal variables with covariate effects on the manifest and latent variables: a comparison of Lisrel and IRT approaches. *Structural equation modelling, 11*, 487–513.

Moustaki, I., & Knott, M. (2000). Generalized latent trait models. *Psychometrika, 65*, 391–411.

Muraki, E. (1992). A generalized partial credit model: application of an EM algorithm. *Applied Psychological Measurement, 16*, 159–176.

Muthén, B. (1978). Contributions to factor analysis of dichotomous variables. *Psychometrika, 43*, 551–560.

Patz, R. J., & Junker, B. W. (1999). A straightforward aproach to Markov Chain Monte Carlo methods for item response models. *Journal of Educational and Behavioral Statistics, 24*, 146–178.

Rabe-Hesketh, S., Pickles, A., & Skrondal, A. (2001). *GLLAMM: Stata program to fit generalised linear latent and mixed models* (Technical Report). Institute of Psychiatry, King's College, London.

Reiser, M., & VandenBerg, M. (1994). Validity of the chi-square test in dichotomous variable factor analysis when expected frequencies are small. *British Journal of Mathematical and Statistical Psychology, 47*, 85–107.

Rizopoulos, D. (2006). ltm: An R package for latent variable modelling and item response theory analyses. *Journal of Statistical Software, 17*, 1–25.

Samejima, F. (1969). Estimation of latent ability using a response pattern of graded scores. *Psychometrika, Monograph Supplement No. 17, 34*.

Sammel, R. D., Ryan, L. M., & Legler, J. M. (1997). Latent variable models for mixed discrete and continuous outcomes. *Journal of the Royal Statistical Society, B, 59*, 667–678.

Schilling, S., & Bock, R. (2005). High-dimensional maximum marginal likelihood item factor analysis by adaptive quadrature. *Psychometrika, 70*, 533–555.

Schuessler, K. F. (1982). *Measuring social life feelings*. San Francisco: Jossey-Bass.

Sclove, S. (1987). Application of model-selection criteria to some problems of multivariate analysis. *Psychometrika, 52*, 333–343.

Sinharay, S., Johnson, M. S., & Stern, H. S. (2006). Posterior predictive assessment of item response theory models. *Applied Psychological Measurement, 30*, 298–321.

van der Linden, W., & Hambleton, R. E. (1997). *Handbook of item response theory*. Berlin: Springer.

Varin, C., & Vidoni, P. (2005). A note on composite likelihood inference and model selection. *Biometrika, 92*, 519–528.

Vasdekis, V., Cagnone, S., & Moustaki, I. (2012). A pairwise likelihood inference in latent variable models for ordinal longitudinal responses. *Psychometrika, 77*, 425–441.

Wirth, R. J., & Edwards, M. C. (2007). Item factor analysis: Current approaches and future directions. *Psychological Methods, 12*, 58–79.

Chapter 6

Standard Errors for Ordinary Least Squares Estimates of Parameters in Structural Equation Modeling

Guangjian Zhang[1] and Chun-Ting Lee

> A further approach has been investigated theoretically but has not yet been tried out empirically. This is to make use of convenient estimates, such as normal theory maximum likelihood or ordinary least squares, and use Propositions 4 and 2 to obtain an asymptotically distribution-free test statistic and standard errors.
>
> Browne (1984)

One of the common features of Michael Browne's works is pragmatism. He regards statistics as a discipline neighboring engineering. Needless to say, he is talented at mathematics, but he never lets mathematics take over. He always uses his mathematical talents to solve practical problems.

[0] We are grateful for many helpful conversations with Michael W. Browne. We thank Michael C. Edwards and Kristopher J. Preacher for their comments on this chapter.

6.1 INTRODUCTION

Latent variable models are suitable for use in the social and behavioral sciences where many constructs are defined conceptually and their error-free measurement does not exist. Estimation of latent variable models is a process determining how well manifest variables represent latent factors and how latent factors affect each other. Although estimation of latent variable models appears an easy task using computer software, the process is indeed nontrivial. Lack of convergence and boundary estimates occur frequently in practice. These estimation problems are more likely to occur for small samples, non-Gaussian data, and complex models. Some estimation methods are more susceptible to these estimation problems than other estimation methods, however.

In this chapter, we consider ordinary least squares (OLS) estimation of parameters in structural equation modeling (SEM). Advantages of OLS estimation include computational robustness and easy implementation. In simulation studies on exploratory factor analysis with Gaussian data, Browne (1968) found that Heywood cases are less likely to occur with OLS estimation than with maximum Wishart likelihood estimation. OLS estimation is also preferred in confirmatory factor analysis (CFA) with ordinal data (Forero, Maydeu-Olivares, & Gallardo-Pujol, 2009; Yang-Wallentin, Jöreskog, & Luo, 2010). OLS estimation has been suggested (Browne & Zhang, 2007) for process factor analysis with lagged correlation matrices of time series data (Browne & Nesselroade, 2005; Nesselroade, McArdle, Aggen, & Meyers, 2002), which explicitly models dependence of observations at adjacent time points.

It is important to supplement OLS estimates with standard error estimates and confidence intervals. Proposition 2 of Browne (1984) is a generic formula computing standard errors of generalized least squares (GLS) estimates of parameters in SEM. This proposition can be readily adapted for OLS estimates in a variety of modeling conditions. Our goal in this chapter is to demonstrate the versatility of this standard error procedure. We do not propose new methods. We present two empirical examples where this proposition is adapted to estimate standard errors of OLS estimates. The first empirical example involves time series data and the second empirical example involves ordinal manifest variables. The adaptation of this proposition for time series data is presented in Zhang, Chow, and Ong (2011); adaptations of this proposition for ordinal manifest variables are presented in Muthén (1993) and Yang-Wallentin et al. (2010). Additionally, Browne and Tateneni (2008) adapted this proposition to compute standard errors for OLS estimates of parameters of EFA with Gaussian variables.

The rest of the chapter is organized as follows. We first describe Proposition 2 of Browne (1984), which computes standard errors for GLS estimates with sample covariances, and its adaptations for OLS estimation with sample correlations. Two empirical examples of such adaptations are then presented. We conclude the chapter with a few remarks.

6.2 ADAPTATIONS OF PROPOSITION 2 IN BROWNE (1984) FOR OLS ESTIMATION

6.2.1 Standard Errors of GLS Estimates with Sample Covariances

Let $\boldsymbol{S}$ be a $p \times p$ sample manifest variable covariance matrix. Let $\boldsymbol{\Sigma}(\boldsymbol{\gamma})$ be a manifest variable covariance structure implied by a SEM model. The $q \times 1$ vector $\boldsymbol{\gamma}$ contains model parameters. GLS estimates, $\widehat{\boldsymbol{\gamma}}_{\text{gls}}$, are obtained by minimizing the discrepancy function

$$f_{\text{gls}}(\boldsymbol{S}, \boldsymbol{\Sigma}(\boldsymbol{\gamma})|\boldsymbol{U}) = (\boldsymbol{s} - \boldsymbol{\sigma}(\boldsymbol{\gamma}))'\boldsymbol{U}^{-1}(\boldsymbol{s} - \boldsymbol{\sigma}(\boldsymbol{\gamma})). \tag{6.1}$$

Here $\boldsymbol{s}$ is a $p(p+1)/2$ component vector containing distinct elements of $\boldsymbol{S}$, and $\boldsymbol{\sigma}(\boldsymbol{\gamma})$ is a $p(p+1)/2$ component vector containing distinct elements of $\boldsymbol{\Sigma}(\boldsymbol{\gamma})$. The weight matrix $\boldsymbol{U}$ can be a stochastic matrix or a constant matrix. The asymptotic covariance matrix of $\widehat{\boldsymbol{\gamma}}_{\text{gls}}$ was derived in Browne (1984, Proposition 2),

$$\text{Cov}\left(\widehat{\boldsymbol{\gamma}}, \widehat{\boldsymbol{\gamma}}'\right) \approx N^{-1}(\boldsymbol{\Delta}_0'\overline{\boldsymbol{U}}^{-1}\boldsymbol{\Delta}_0)^{-1}(\boldsymbol{\Delta}_0'\overline{\boldsymbol{U}}^{-1}\boldsymbol{\Upsilon}\overline{\boldsymbol{U}}^{-1}\boldsymbol{\Delta}_0)(\boldsymbol{\Delta}_0'\overline{\boldsymbol{U}}^{-1}\boldsymbol{\Delta}_0)^{-1}. \tag{6.2}$$

Here $\boldsymbol{\Delta}_0$ is the model Jacobian matrix $\frac{\partial}{\partial \boldsymbol{\gamma}'}\boldsymbol{\sigma}(\boldsymbol{\gamma})$; $\boldsymbol{\Upsilon}$ is the asymptotic covariance matrix of sample covariances $\boldsymbol{s}$; $\overline{\boldsymbol{U}}$ is the limit matrix of a stochastic matrix $\boldsymbol{U}$.

6.2.2 Standard Errors of OLS Estimates with Sample Correlations

OLS estimates of parameters in SEM with sample correlations are obtained by minimizing the discrepancy function,

$$f_{\text{ols}}(\boldsymbol{R}, \boldsymbol{P}(\boldsymbol{\gamma})) = (\boldsymbol{r} - \boldsymbol{\rho}(\boldsymbol{\gamma}))'(\boldsymbol{r} - \boldsymbol{\rho}(\boldsymbol{\gamma})). \tag{6.3}$$

Here $\boldsymbol{r}$ is a vector collecting distinct elements of the sample correlation matrix $\boldsymbol{R}$, and $\boldsymbol{\rho}(\boldsymbol{\gamma})$ is a vector collecting distinct elements of the model-implied correlation matrix $\boldsymbol{P}(\boldsymbol{\gamma})$. GLS estimation with sample covariances of Equation (6.1) becomes OLS estimation with sample correlations of Equation (6.3) if the weight matrix $\boldsymbol{U}$ is replaced by an identity matrix and sample covariances $\boldsymbol{s}$ by sample correlations $\boldsymbol{r}$. We deliberately omit details of $\boldsymbol{r}$ and $\boldsymbol{\rho}(\boldsymbol{\gamma})$ in Equation (6.3), because OLS estimation admits many modeling possibilities. For example, factor analysis models can be estimated with product moment correlations of Gaussian variables, or polychoric correlations of ordinal variables, or lagged correlations of time series data.

The asymptotic covariance matrix of OLS estimates with sample correlations is a reduced form of Equation (6.2),

$$\mathrm{Cov}\,(\widehat{\boldsymbol{\gamma}}, \widehat{\boldsymbol{\gamma}}') \approx N^{-1}(\boldsymbol{\Delta}_0'\boldsymbol{\Delta}_0)^{-1}(\boldsymbol{\Delta}_0'\boldsymbol{\Upsilon}\boldsymbol{\Delta}_0)(\boldsymbol{\Delta}_0'\boldsymbol{\Delta}_0)^{-1}. \tag{6.4}$$

Both the model Jacobian matrix $\boldsymbol{\Delta}_0 = \frac{\partial}{\partial \boldsymbol{\gamma}'}\boldsymbol{\rho}(\boldsymbol{\gamma})$ and the asymptotic covariance matrix of sample correlations $\boldsymbol{\Upsilon}$ need to be modified to accommodate different modeling contexts.

Standard errors of OLS estimates with sample correlations are the square roots of diagonal elements of Equation (6.4). A $(1-\alpha) \times 100\%$ confidence interval on a parameter γ_i can be constructed using the OLS estimate and its standard error estimate

$$\left(\gamma_i^{\text{lower}}, \gamma_i^{\text{upper}}\right) = \left(\widehat{\gamma}_i - z_{\alpha/2} \times \mathrm{SE}_{\widehat{\gamma}_i}, \widehat{\gamma}_i + z_{\alpha/2} \times \mathrm{SE}_{\widehat{\gamma}_i}\right). \tag{6.5}$$

Here $z_{\alpha/2}$ is the $\alpha/2$th percentile of the standard Gaussian distribution. Some parameters may have natural bounds. For example, communalities in a factor analysis model should be between 0 and 1. Browne (1982, pp. 95–97) described a procedure for constructing confidence intervals on parameters such that the confidence intervals are inside admissible regions.

Before presenting the two examples demonstrating OLS estimates and the associated standard error estimates and confidence intervals, we digress to compare and contrast OLS estimation and GLS estimation.

6.2.3 OLS Estimation and GLS Estimation

OLS estimation of Equation (6.3) is easier to implement than GLS estimation of Equation (6.1). GLS estimation requires inverting a matrix $\boldsymbol{U}$, but the matrix inversion is unnecessary in OLS estimation. The matrix $\boldsymbol{U}$ is a big matrix. For example, $\boldsymbol{U}$ is 300×300 when an SEM model is estimated with a 25×25 correlation matrix. Inversion of a large matrix is computationally intensive and numerically unstable. Furthermore, the matrix $\boldsymbol{U}$ often needs to be estimated from sample data. Estimates of $\boldsymbol{U}$ are less accurate than estimates of correlations. A popular choice of $\boldsymbol{U}$ is an estimate of the asymptotic covariance matrix of the sample correlations. Let $\boldsymbol{\Upsilon}$ denote this asymptotic covariance matrix. Elements of $\boldsymbol{\Upsilon}$ are the fourth order moments of data: they are variances and covariances of sample correlations. In contrast, sample correlations correspond to the second order moments. Inaccuracy in estimating the big matrix $\boldsymbol{U}$ and its inversion make OLS estimation an attractive alternative to GLS estimation for small to moderately sized samples.

When the matrix $\boldsymbol{U}$ is a consistent estimate of the asymptotic covariance matrix $\boldsymbol{\Upsilon}$ in the GLS estimation of Equation (6.1), GLS estimates are asymptotically efficient. Properties of GLS estimation with Gaussian data were described in Browne (1974); properties of GLS estimation with data of arbitrary distributions (asymptotically distribution free methods) were described in Browne (1984). In addition, maximum Wishart likelihood estimation was equivalent to GLS estimation where the matrix $\boldsymbol{U}$ is a function of model-implied covariance matrix $\boldsymbol{\Sigma}(\boldsymbol{\gamma})$ (Browne, 1974, Proposition 6).

It is informative to quantify the amount of loss in asymptotic efficiency of OLS estimation against asymptotically efficient methods. Browne (1984, Proposition 4) provided a test statistic which is appropriate for OLS estimation. The test statistic consists of two parts. The first part is a weighted sum of residuals when OLS estimation is considered; the second part is a measure of loss of efficiency of OLS estimation. When both OLS estimation and asymptotically efficient estimation methods are feasible, their respective standard errors directly reflect efficiency. For example, Browne and Tateneni (2008) found that asymptotic standard errors of OLS estimates were larger than asymptotic standard errors of maximum Wishart likelihood estimates of exploratory factor analysis with Gaussian data, but the differences were negligible.

The notion that "all models are wrong but some are useful" (Box, 1976; MacCallum, 2003) has implications for comparisons between OLS and asymptotically efficient estimation methods like GLS estimation and maximum Wishart likelihood estimation. It is unrealistic to assume that our model can fully capture the complex world. Consequently, any useful models contain model error. When an imperfect model is fitted to a population covariance matrix, different estimation methods may produce different sets of parameter values (Browne, 1982; MacCallum, 2003). Additionally, OLS estimation was found to recover weak factors more accurately than maximum Wishart likelihood estimation under different conditions (Briggs & MacCallum, 2003; MacCallum, Browne, & Cai, 2007; MacCallum, Tucker, & Briggs, 2001).

We next present two examples in which OLS estimation is used for SEM with sample correlations. In the first example, a confirmatory dynamic factor analysis model is considered for multiple time series; in the second example, a CFA model is considered for ordinal manifest variables.

6.3 STANDARD ERRORS OF DYNAMIC FACTOR ANALYSIS

6.3.1 Process Factor Analysis with Daily Ratings

This first application involves multiple time series. It is part of a study on widowhood adjustment.[1] Participants recorded self-ratings on a number of items daily over a period of 98 days. Daily ratings on the "positive affect" scale and the "anxiety" scale from one individual are used in this illustration. The "positive affect" scale consists of

[1] We are grateful to Cindy Bergeman for making the data available to us.

11 four-point ordinal items. An example is "Today I felt the future was hopeful and promising." The "anxiety" scale consists of nine four-point ordinal items. An example is "Today I was anxious and worried." Three parcels are formed from the 11 items of the "positive affect"; three parcels are formed from the nine items of the "anxiety" scale.

A process factor analysis model (Browne & Nesselroade, 2005; Browne & Zhang, 2007), displayed in Figure 6.1, is considered for the six-variate time series. Specification of this process factor analysis model consists of two parts: the factor analysis part and the time series part. The factor analysis part specifies that manifest variables at time t are linear combinations of the latent factors and measurement error variables,

$$\boldsymbol{y}_t = \boldsymbol{\Lambda f}_t + \boldsymbol{u}_t. \tag{6.6}$$

Here $\boldsymbol{\Lambda}$ is the factor loading matrix indicating the influence of $\boldsymbol{f}_t$ on $\boldsymbol{y}_t$. The time series part specifies a vector autoregressive process,

$$\boldsymbol{f}_t = \boldsymbol{A}_i \boldsymbol{f}_{t-1} + \boldsymbol{z}_t. \tag{6.7}$$

The current "positive affect" and "anxiety" scores are affected by these two factors on the previous day and the concurrent random shocks. This process is a vector AR1 process, which is a special case of a vector autoregressive process with moving average random shocks (VARMA).

Dynamic factor analysis is often estimated with manifest variable lagged correlation matrices due to the flexibility of SEM (Browne & Zhang, 2007; Molenaar, 1985; Nesselroade et al., 2002). In particular, Browne and Zhang (2007) described OLS estimation specially developed for process factor analysis. OLS estimation is easier to implement than maximum Wishart likelihood estimation and GLS estimation in process factor analysis. Maximum Wishart likelihood estimation of process factor analysis requires arranging manifest variable lagged correlation matrices into a large block Toeplitz matrix, and addresses element duplications by imposing a large number of equality constraints. OLS estimation avoids making this large Toeplitz matrix. Additionally, usual desirable properties of maximum likelihood estimates may not be applicable to maximum Wishart likelihood estimates of a process factor analysis model because a Wishart distribution may be inappropriate for elements of the large Toeplitz matrix. GLS estimation requires inversion of a weight matrix, which is unnecessary in OLS estimation. The weight matrix is of $(p(p-1))/2 + p^2L$ by $(p(p-1))/2 + p^2L$, where p is the number of manifest variables and L is the maximum time lag of interest. The illustrative example involves 6 variables and the time lag $L = 1$: the weight matrix is 51 by 51. The size of the weight matrix grows rapidly when p and L increase. For a model of 30 variables and time lag $L = 2$ (Browne & Zhang, 2007, Table 13-1), the weight matrix will be 2235 by 2235. Inversion of the weight matrix becomes impractical for a moderately large model.

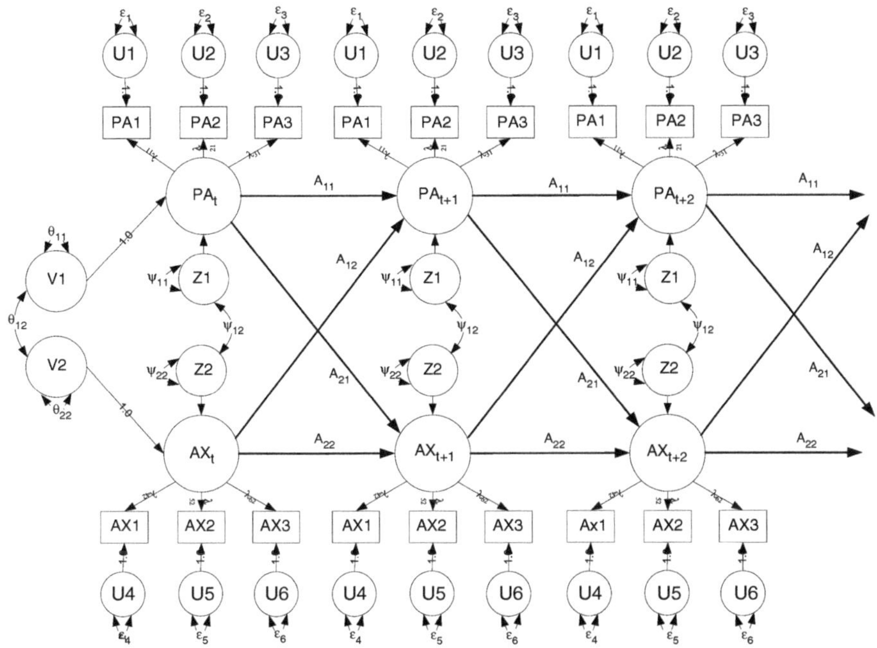

Figure 6.1 Process factor analysis with daily ratings of positive affect and anxiety. PA stands for positive affect and AX for anxiety. PA1, PA2, and PA3 are three parcels formed from indicators of PA; AX1, AX2, and AX3 are three parcels formed from indicators of AX.

6.3.2 Adaptation of Proposition 2 of Browne (1984) for Dynamic Factor Analysis

Adaption of Proposition 2 of Browne (1984) for OLS estimates of confirmatory process factor analysis is reported by Zhang et al. (2011). The adaptation involves derivation of the asymptotic covariance matrix of lagged manifest variable correlations. Let $\rho_{m,ij}$ denote the population lagged correlation of the jth component of the time series at time t, with the ith component of the time series at time $t+m$. Let $r_{m,ij}$ denote the corresponding sample lagged correlation. A typical element of the asymptotic covariance matrix is the covariance between two lagged correlations $r_{m,ij}$ and $r_{n,kl}$, which can be estimated using

$$\mathrm{Lcov}(r_{m,ij}, r_{n,kl}) = \sum_{u=-\infty}^{\infty} \left\{ \frac{1}{2}\, \rho_{m,ij}\rho_{n,kl}\left[\rho_{u,ik}^2 + \rho_{u,jk}^2 + \rho_{u,il}^2 + \rho_{u,jl}^2\right] \right.$$
$$- \rho_{n,kl}\left[\rho_{u,jk}\rho_{u+m,ik} + \rho_{u,jl}\rho_{u+m,il}\right] \qquad (6.8)$$

$$- \rho_{m,ij} \left[\rho_{u,il}\rho_{u-n,ik} + \rho_{u,jl}\rho_{u-n,jk}\right]$$

$$+ \left[\rho_{u,jl}\rho_{u-n+m,ik} + \rho_{u-n,jk}\rho_{u+m,il}\right] \Big\} .$$

Equation (6.8) involves population lagged correlations $\rho_{m,ij}$. When it is used to estimate the asymptotic covariance matrix of $\boldsymbol{r}$ in a sample, these population correlations need to be replaced by their estimates. One option is to use sample lagged correlations $r_{m,ij}$. A disadvantage of this option is that sample lagged correlations may contain sampling irregularities for short time series. Another option is to fit the process factor analysis model first, and to use model-implied manifest variable correlations. This option was suggested by Browne and Strydom (1997) in a different modeling context.

In applications of Equation (6.8), model-implied lagged correlations $\widehat{\boldsymbol{\rho}}$ may be closer to the population lagged correlations $\boldsymbol{\rho}$ than are the sample lagged correlations $\boldsymbol{r}$. Model-implied lagged correlations $\widehat{\boldsymbol{\rho}}$ are functions of parameter estimates $\widehat{\boldsymbol{\gamma}}$, which themselves are functions of sample lagged correlations $\boldsymbol{r}$. Each model-implied lagged correlation can thus be regarded as an estimate, pooling together information from all sample lagged correlations. When this pooling process is undertaken with an appropriate model, sampling irregularities in sample lagged correlations are filtered out. This is likely to occur when model error is smaller than sampling error. For example, we expect that model-implied lagged correlations $\widehat{\boldsymbol{\rho}}$ are better estimates of the population lagged correlations $\boldsymbol{\rho}$ than the sample lagged correlations $\boldsymbol{r}$, when a process factor analysis model is fitted to a short time series, and the test of close fit (Browne & Cudeck, 1993) is not rejected.

Standard errors of OLS estimates of process factor analysis with Equation (6.4) require the model Jacobian matrix $\boldsymbol{\Delta}_0$. It contains partial derivatives of manifest variable lagged correlations with respect to time series parameters and factor analysis parameters. This model Jacobian matrix has been involved in point estimation of OLS estimation of process factor analysis. Thus use of this matrix for standard error estimation involves no extra computation. OLS estimates and associated standard errors of process factor analysis with lagged correlation matrices can be obtained using the computer program DyFA 3 (Browne & Zhang, 2010).

6.3.3 Results of the Empirical Illustration

Table 6.1 reports OLS estimates of a process factor analysis model with the daily ratings of positive affect and anxiety. Also reported are 90% confidence intervals constructed using the point estimates and their standard error estimates. Confidence intervals of the six factor loadings indicate that these manifest variables well represent the latent factors "positive affect" and "anxiety." The smallest lower bound of these

Table 6.1 Process factor analysis with daily positive affect and anxiety

	Factor loading matrix			*AR matrix*	
	PA	*AX*		*PA*	*AX*
PA1	.96(.92, 1.00)		PA	.15(−.11, .40)	−.24(−.50, .01)
PA2	.91(.86, .96)		AX	−.11(−.36, .14)	.34(.09, .59)
PA3	.87(.82, .93)				
AX1		.95(.91, .98)			
AX2		.99(.97, 1.00)			
AX3		.84(.78, .91)			
	Ψ			Θ	
	PA	*AX*		*PA*	*AX*
PA	.87(.70, .95)	−.58(−.69, −.46)	PA	.13(.05, .30)	−.15(−.27, −.04)
AX	−.58(−.69, −.46)	.82(.65, .92)	AX	−.15(−.27, −.04)	.18(.08, .35)
	Φ_0			Φ_1	
	PA	*AX*		*PA*	*AX*
PA	1.00	−.73(−.81, −.63)	PA	.32(.14, .48)	−.35(−.51, −.17)
AX	−.73(−.81, −.63)	1.00	AX	−.36(−.52, −.18)	.42(.25, .57)

Note: Confidence intervals are presented in parentheses. PA stands for positive affect and AX for anxiety. PA1, PA2, and PA3 are three parcels formed from indicators of PA; AX1, AX2, and AX3 are three parcels formed from indicators of AX.

six confidence intervals is 0.87. Among confidence intervals of the four AR weights, only one does not overlap with zero. The more anxious the participant was on a certain day, the more anxious the participant tended to be on the following day. Elements of the shock covariance matrix are substantially larger than elements of the initial state vector covariance matrix. Daily random shocks contribute to concurrent "positive affect" and "anxiety" more than process factors and random shocks of previous days. The two factors "positive affect" and "anxiety" have a correlation of −0.73 with a confidence interval of (0.63, 0.81). This high correlation is surprising. Note that the process factor analysis model is used to understand unique patterns of positive affect and anxiety of this participant. Other participants can have different patterns.

Table 6.2 Ordinal variables of avoidance and anxiety

1	I don't feel entirely comfortable opening up to my partner.
2	I am nervous when my partner gets too close to me.
3	I tell my partner just about everything.
4	I usually discuss my problems and concerns with my partner.
5	I don't mind asking my partner for comfort, advice, or help.
6	At times, I worry about being abandoned.
7	I am concerned that my partner may not care about me as much as I care about him/her.
8	I need a lot of reassurance that I am loved by my partner.
9	When my partner disapproves of me, I tend to feel bad about myself.

Note: Variables 1 to 5 are indicators of the factor avoidance and variables 6 to 9 are indicators of the factor anxiety. Variables 3, 4, and 5 are reverse scored.

6.4 STANDARD ERRORS OF FACTOR ANALYSIS WITH ORDINAL VARIABLES

6.4.1 CFA with Ordinal Variables

In this illustration, a CFA model is considered for nine seven-point ordinal variables. The illustration data set is part of a personality study in which 370 undergraduate students in a medium size public university participated to fulfill a course requirement.[2] These nine variables are presented in Table 6.2. They are indicator variables of the two factors "avoidance" and "anxiety," which are the two dimensions of adult attachment theory.

Figure 6.2 displays a CFA model with ordinal variables. Note that manifest variables X_1 to X_9 are ordinal variables. A continuous variable X_i^*, represented by a circle, underlies the ordinal manifest variable X_i. The factor analysis model is appropriate for these continuous underlying variables, but is inappropriate for ordinal manifest variables.

Although these underlying continuous variables are not directly observed, their correlations can be estimated using contingency tables formed from pairs of ordinal manifest variables (Olsson, 1979). Parameter estimates of the CFA model with a polychoric correlation matrix are obtained using OLS estimation for its simple implementation and computational robustness. Maximum Wishart likelihood estimation is often infeasible, because the polychoric correlation matrix may not be

[2] We are grateful to Shanhong Luo for making the data available to us.

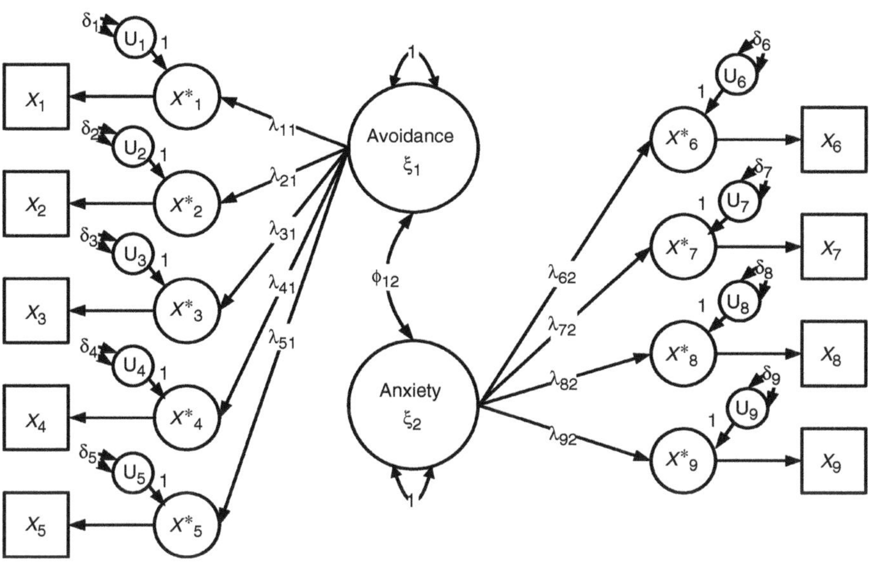

Figure 6.2 Confirmatory factor analysis with ordinal data.

positive definite. GLS estimation (Jöreskog & Sörbom, 1996) and a closely related procedure, weighted least squares estimation (Muthén & Muthén, 2007), often suffer low convergence rates at sample sizes encountered by applied researchers (Dolan, 1994). A $p(p-1)/2$ by $p(p-1)/2$ stochastic matrix, the asymptotic covariance matrix of polychoric correlations, is inverted in GLS estimation. Inversion of this large weight matrix contributes to the implementation difficulties of GLS estimation in this modeling context. Inversion of the large matrix is unnecessary in OLS estimation, however. Recent simulation studies (Forero et al., 2009; Yang-Wallentin et al., 2010) suggest that OLS estimates are more satisfactory than GLS estimates at moderate sample sizes.

6.4.2 Adaption of Proposition 2 of Browne (1984) for Factor Analysis with Ordinal Variables

Adaptation of Proposition 2 of Browne (1984) for OLS estimates of CFA with ordinal variables involves replacing $\boldsymbol{\Upsilon}$ by the asymptotic covariance matrix of polychoric correlations and replacing $\boldsymbol{\Delta}_0$ by the model Jacobian matrix of CFA. Although the asymptotic covariance matrix of polychoric correlations is needed in estimation of standard errors of OLS estimates, its inversion is unnecessary. Estimation of the asymptotic covariance matrix of polychoric correlations was described in Jöreskog (1994) and Muthén (1984). Jöreskog (1994) made use of the property that polychoric correlations can be expressed as linear functions of two-way contingency tables asymptotically, and showed that variances and covariances of polychoric correlations can be expressed as functions of four-way contingency tables. Muthén (1984) obtained estimates of polychoric correlations by minimizing a composite likelihood function that is a sum of likelihood functions involving pairs of ordinal manifest variables. The asymptotic covariance matrix of polychoric correlations is then obtained from the inverse of the information matrix. Muthén and Satorra (1995) provided additional details. Yang-Wallentin et al. (2010) observed that two methods produce very similar estimates for the asymptotic covariance matrix.

The model Jacobian matrix $\mathbf{\Delta}_0$ of CFA is evaluated at OLS estimates. OLS estimates and associated standard error estimates of CFA with polychoric correlations can be obtained using LISREL 8 (Jöreskog & Sörbom, 1996) or Mplus 5 (Muthén & Muthén, 2007).

6.4.3 Results of the Empirical Illustration

Table 6.3 reports the point estimates and associated 90% confidence intervals of the two factor model with nine seven-point ordinal variables. Some categories were combined to avoid sparseness of contingency tables. The two factors are well represented by the manifest variables. Among the five confidence intervals on the factor loadings of avoidance, the smallest lower bound is 0.65. Note that variables 3, 4, and 5 are reverse scored. Among the four confidence intervals on the factor loadings of anxiety, the smallest lower bound is 0.44. The correlation between avoidance and anxiety is 0.25 with a confidence interval (0.15, 0.35).

6.5 CONCLUDING REMARKS

Estimation of SEM parameters is a nontrivial task for non-Gaussian data, dependent data, and small samples. OLS estimation is worth considering in these less -than-ideal situations. Proposition 2 of Browne (1984) is a generic formula which can be readily adapted to provide standard errors of OLS estimation in a variety of conditions. In this chapter, we demonstrated the flexibility of this formula using two examples. In the first example, standard errors for OLS estimates of parameters in process factor analysis with time series data are computed; in the second example, standard errors for OLS estimates of parameters in CFA with ordinal variables are computed.

It is unrealistic to assume that a model of interest contains no model error in the population. On the other hand, a model is of little use if it contains too much

Muthén, B., & Satorra, A. (1995). Technical aspects of Muthén's LISCOMP approach to estimation of latent variable relations with a comprehensive measurement model. *Psychometrika, 60*, 489–503.

Muthén, L. K., & Muthén, B. O. (2007). *Mplus user's guide* (5 ed.). Los Angeles, CA: Muthén & Muthén.

Nesselroade, J. R., McArdle, J. J., Aggen, S. H., & Meyers, J. M. (2002). Dynamic factor analysis models for representing process in multivariate time-series. In D. Moskowitz & S. L. Hershberger (Eds.), *Modeling intraindividual variability with repeated measures data: Methods and applications* (pp. 235–265). Mahwah, NJ: Erlbaum.

Olsson, U. (1979). Maximum likelihood estimation of the polychoric correlation coefficient. *Psychometrika, 44*, 443–460.

White, H. (1980). Using least squares to approximate unknown regression functions. *International Economic Review, 21*, 149–170.

Yang-Wallentin, F., Jöreskog, K. G., & Luo, H. (2010). Confirmatory factor analysis of ordinal variables with misspecified models. *Stuctural Equation Modeling, 17*, 392–423.

Yuan, K., & Hayashi, K. (2006). Standard errors in covariance structure models: asymptotics versus bootstrap. *British Journal of Mathematical and Statistical Psychology, 59*, 397–417.

Zhang, G., & Browne, M. W. (2006). Bootstrap fit testing, confidence intervals, and standard error estimation in the factor analysis of polychoric correlation matrices. *Behaviormetrika, 33*, 61–74.

Zhang, G., & Browne, M. W. (2010). Bootstrap standard error estimates in dynamic factor analysis. *Multivariate Behavioral Research, 45*, 453–482.

Zhang, G., Chow, S., & Ong, A. (2011). A sandwich-type standard error estimator of SEM models with single time series. *Psychometrika, 76*, 77–96.

Table 6.3 Confirmatory factor analysis with ordinal variables

	Factor loading matrix	
	Avoidance	*Anxiety*
Avoidance1	.75(.68, .82)	
Avoidance2	.72(.65, .79)	
Avoidance3	.77(.70, .84)	
Avoidance4	.79(.72, .86)	
Avoidance5	.79(.74, .84)	
Anxiety1		.77(.69, .85)
Anxiety2		.80(.72, .88)
Anxiety3		.68(.60, .76)
Anxiety4		.52(.44, .60)
	Factor correlations	
	Avoidance	*Anxiety*
Avoidance	1.00	.25(.15, .35)
Anxiety	.25(.15, .35)	1.00

Note: Confidence intervals are presented in parentheses.

model error. Standard errors of Proposition 2 (Browne, 1984) are appropriate if model error is no larger than random sampling error. In contrast, standard errors based on the information matrix require that the model be correctly specified. Thus the assumption for Proposition 2 is less restrictive than the one made by information-based standard error estimators.

Although we focus on standard error estimates for OLS estimation of SEM parameters, it was not our intention to recommend OLS estimation under all circumstances. OLS estimation uses less information than GLS, ADF, and ML estimation. When their respective assumptions are appropriate, GLS, ADF, and ML estimation are expected to be more efficient than OLS estimation. On the other hand, when these assumptions appear inappropriate, or more sophisticated methods like GLS, ADF, and ML are difficult to implement, OLS estimation provides an easily implemented and computationally robust method. Standard errors for OLS estimates of SEM parameters should be useful in these situations.

Standard errors of OLS estimates of parameters in SEM can be estimated using the bootstrap (Efron & Tibshirani, 1993). Specifics of different modeling contexts need to be addressed in the bootstrap. Zhang and Browne (2010) employed a parametric bootstrap and a moving block bootstrap to deal with time series data in OLS estimation of process factor analysis. Zhang and Browne (2006) employed a parametric bootstrap to deal with ordinal variables in OLS estimation of parameters in CFA. Asymptotic standard errors considered in the present chapter require much less computation than bootstrap methods. In addition, boundary estimates and nonconvergence of bootstrap samples need to be properly addressed. Bootstrap methods often require fewer assumptions than asymptotic methods, however.

Proposition 2 in Browne (1984) appears different from other methods (White, 1980; Yuan & Hayashi, 2006) for computing standard errors. Proposition 2 involves the first order derivatives of the model with respect to parameters; the other methods involve both the first order and second order derivatives of the model with respect to parameters. Thus Proposition 2 is easier to implement. Easy implementation is an important consideration for complex models like dynamic factor analysis in which second order derivatives are intractable. The assumption that model error is no larger than random sampling error is unnecessary for standard error methods involving second order derivatives, however. These standard error methods may outperform standard error estimates with first order derivatives alone when the sample size is huge and model mis-specification is severe. Future studies need to address the relative performance of these two types of standard errors in more realistic conditions (small to moderately large samples and mild model mis-specification).

REFERENCES

Box, G. E. P. (1976). Science and statistics. *Journal of the American Statistical Association, 71*, 791–799.

Briggs, N. E., & MacCallum, R. C. (2003). Recovery of weak common factors by maximum likelihood and ordinary least squares estimation. *Multivariate Behavioral Research, 38*, 25–56.

Browne, M. W. (1968). A comparison of factor analytic techniques. *Psychometrika, 33*, 267–334.

Browne, M. W. (1974). Generalized least squares estimators in the analysis of covariance structure. *South African Statistical Journal, 8*, 1–24.

Browne, M. W. (1982). Covariance structures. In D. M. Hawkins (Ed.), *Topics in applied multivariate analysis* (pp. 72–141). Cambridge, England: Cambridge University Press.

Browne, M. W. (1984). Asymptotically distribution-free methods for the analysis of covariance structures. *British Journal of Mathematical and Statistical Psychology, 37*, 62–83.

Browne, M. W., & Cudeck, R. (1993). Alternative ways of assessing model fit. In K. A. Bollen & J. S. Long (Eds.), *Testing structural equation models* (pp. 136–162). Newbury Park, CA: Sage.

Browne, M. W., & Nesselroade, J. R. (2005). Representing psychological processes with dynamic factor models: Some promising uses and extensions of ARMA time series models.

In A. Maydeu-Olivares & J. J. McArdle (Eds.), *Advances in psychometrics: A festschrift for Roderick P. McDonald* (pp. 415–452). Mahwah, NJ: Erlbaum.

Browne, M. W., & Strydom, H. F. (1997). Non-iterative fitting of the direct product model for multitrait–multimethod matrices. In M. Berkane (Ed.), *Latent variable modeling with application to causality* (pp. 229–245). New York: Springer-Verlag.

Browne, M. W., & Tateneni, K. (2008). *Standard errors for OLS estimates in exploratory factor analysis.* Annual Meeting of the Society of Multivariate Experimental Psychology, Montreal, Canada.

Browne, M. W., & Zhang, G. (2007). Developments in the factor analysis of individual time series. In R. Cudeck & R. C. MacCallum (Eds.), *Factor analysis at 100: Historical developments and future directions* (pp. 265–291). Mahwah, NJ: Erlbaum.

Browne, M. W., & Zhang, G. (2010). *DyFA 3.00 user guide.* Retrieved from http://quantrm2.psy.ohiostate.edu/browne/

Dolan, C. V. (1994). Factor analysis of variables with 2, 3, 5, and 7 response categories: A comparison of categorical variable estimators using simulated data. *British Journal of Mathematical and Statistical Psychology, 47*, 309–326.

Efron, B., & Tibshirani, R. J. (1993). *An introduction to the bootstrap.* New York: Chapman and Hall.

Forero, C. G., Maydeu-Olivares, A., & Gallardo-Pujol, D. (2009). Factor analysis with ordinal indicators: a Monte Carlo study comparing DWLS and ULS estimation. *Structural Equaiton Modeling, 16*, 625–641.

Jöreskog, K. G. (1994). On the estimation of polychoric correlations and their asymptotic covariance matrix. *Psychometrika, 59*, 381–389.

Jöreskog, K. G., & Sörbom, D. (1996). *Lisrel 8: User's reference guide.* Chicago: Scientific Software.

MacCallum, R. C. (2003). Working with imperfect models. *Multivariate Behavioral Research, 38*, 113–139.

MacCallum, R. C., Browne, M. W., & Cai, L. (2007). Factor analysis models as approximatior In R. Cudeck & R. C. MacCallum (Eds.), *Factor analysis at 100: Historical developme and future directions* (pp. 153–175). Mahwah, NJ: Erlbaum.

MacCallum, R. C., Tucker, L. R., & Briggs, N. E. (2001). An alternative perspectiv parameter estimation in factor analysis and related mothods. In R. Cudeck & S. D (Eds.), *Structural equation modeling: Present and future* (pp. 39–57). Chicago: Sci Software.

Molenaar, P. C. M. (1985). A dynamic factor analysis model for the analysis of mu time series. *Psychometrika, 50*, 181–202.

Muthén, B. (1984). A general structural equation model with dichotomous, ordered c and continuous latent variable indicators. *Psychometrika, 49*, 115–132.

Muthén, B. (1993). Goodness of fit with categorical and other nonnormal variab Bollen & J. S. Long (Eds.), *Testing structural equation models* (pp. 205–2 Park, CA: Sage.

Chapter 7

Three Cheers for the Asymptotically Distribution Free Theory of Estimation and Inference: Some Recent Applications in Linear and Nonlinear Latent Variable Modeling

Li Cai[1]

7.1 INTRODUCTION

According to the Google Scholar, Browne's (1984) paper titled "Asymptotically Distribution Free Methods for the Analysis of Covariance Structures" is the most highly cited piece that the *British Journal of Mathematical and Statistical Psychology*

[0] Part of this research is made possible by grants from the Institute of Education Sciences (R305B080016 and R305D100039) and grants from the National Institute on Drug Abuse (R01DA026943 and R01DA030466).

has ever published. The present chapter pays homage to Browne's landmark paper by offering an incomplete examination of its impact through a tiny sampling of the recent applications of its Proposition 4.

The original context of Browne's (1984) paper is the modeling of covariance structures, which can be extended to the case of general moment structure modeling. Consider the linear case first. The basic setup involves a sample of size N for n observed variables. Let $\mathbf{Y}$ be an $N \times n$ matrix of observations, whose rows are independent with mean vector $\boldsymbol{\mu}$ and covariance matrix $\boldsymbol{\Sigma}$. Let $\hat{\boldsymbol{\mu}}$ be the sample mean vector and $\hat{\boldsymbol{\Sigma}}$ be the sample covariance matrix. Let the mean structure model be $\boldsymbol{\mu}(\boldsymbol{\theta})$, and let $\boldsymbol{\Sigma}(\boldsymbol{\theta})$ be the covariance structure model, where $\boldsymbol{\theta} \in \boldsymbol{\Theta}$ is a $d \times 1$ vector of free structural parameters in a subset of $\mathbb{R}^d$. The combined mean and covariance structure model can be written as

$$\boldsymbol{\sigma}(\boldsymbol{\theta}) = \begin{pmatrix} \boldsymbol{\mu}(\boldsymbol{\theta}) \\ \text{vech}[\boldsymbol{\Sigma}(\boldsymbol{\theta})] \end{pmatrix}, \tag{7.1}$$

where vech denotes the half-vectorization operator that stacks the $n(n+1)/2$ unique elements of a symmetric matrix into a vector. Let the sample counterpart of $\boldsymbol{\sigma}$ be

$$\mathbf{s} = \begin{pmatrix} \hat{\boldsymbol{\mu}} \\ \text{vech}(\hat{\boldsymbol{\Sigma}}) \end{pmatrix}. \tag{7.2}$$

Under appropriate regularity conditions, Browne (1984) showed that the estimation of the structural parameters $\boldsymbol{\theta}$ can be achieved by minimizing a quadratic form discrepancy function

$$F(\boldsymbol{\theta}; \mathbf{s}) = [\mathbf{s} - \boldsymbol{\sigma}(\boldsymbol{\theta})]'\mathbf{V}^{-1}[\mathbf{s} - \boldsymbol{\sigma}(\boldsymbol{\theta})], \tag{7.3}$$

where $\mathbf{V}$ is the asymptotic covariance matrix of $\mathbf{s}$. Let the resulting estimate be $\hat{\boldsymbol{\theta}}$. By Browne's (1984) Propositions 1 and 2, it is consistent and asymptotically normal.

Furthermore, Browne (1984) showed that for a consistent and asymptotically normal estimator $\hat{\boldsymbol{\theta}}$, the following test statistic

$$T_B = (N-1)[\mathbf{s} - \boldsymbol{\sigma}(\hat{\boldsymbol{\theta}})]'\boldsymbol{\Gamma}(\mathbf{V})[\mathbf{s} - \boldsymbol{\sigma}(\hat{\boldsymbol{\theta}})]$$

is asymptotically chi-square distributed with $n + n(n+1)/2 - d$ degrees of freedom under the null hypothesis that the model fits exactly (Proposition 4). The weight matrix

$$\boldsymbol{\Gamma}(\mathbf{V}) = \mathbf{V}^{-1} - \mathbf{V}^{-1}\boldsymbol{\Delta}(\boldsymbol{\Delta}'\mathbf{V}^{-1}\boldsymbol{\Delta})^{-1}\boldsymbol{\Delta}'\mathbf{V}^{-1} = \boldsymbol{\Delta}_c(\boldsymbol{\Delta}_c'\mathbf{V}\boldsymbol{\Delta}_c)^{-1}\boldsymbol{\Delta}_c', \tag{7.4}$$

is itself a weighted projection operator, with $\mathbf{V}$ serving as the weight, where $\boldsymbol{\Delta} = \partial\boldsymbol{\sigma}(\hat{\boldsymbol{\theta}})/\partial\boldsymbol{\theta}$ is the Jacobian of the model, a matrix of all first-order partial derivatives of the means and covariances with respect to the structural parameters. The second expression of $\boldsymbol{\Gamma}$ utilizes an orthogonal complement $\boldsymbol{\Delta}_c$ of $\boldsymbol{\Delta}$ such that $\boldsymbol{\Delta}_c'\boldsymbol{\Delta} = \mathbf{0}$.

In this most abstract form, $\mathbf{V}$ can be estimated from a sample of data, in a completely non-parametric manner, as is done in Browne (1984), to provide a consistent means of estimation and model fit testing even under arbitrary forms of nonnormality of the observed variables. However, the dominating insight here is that Browne's (1984) Proposition 4 is far more general than its original context would imply. Even as **s** and **V** may be replaced by completely different identities, a common pattern remains unchanged, leading to a number of different applications in linear and nonlinear latent variable modeling. This is the topic of the remainder of this chapter.

7.2 AN EM-BASED TWO-STAGE ESTIMATOR

In the applications of mean and covariance structure modeling, missing data is an ubiquitous phenomenon. Excellent books have been written on this subject (e.g., Allison, 2001; Little & Rubin, 1987; Schafer, 1997) and introductory papers on missing data techniques for structural equation modeling abound (e.g., Allison, 2003). Cai (2008) and Cai and Lee (2009) considered a popular and computationally simple approach for fitting mean and covariance structure models under missing data. They refer to this approach as the EM2S (EM-based Two-Stage) estimator. In stage one, the EM algorithm (Dempster, Laird, & Rubin, 1977) is used to produce a maximum likelihood estimate of the observed mean vector and covariance matrix under missing data, usually assuming a multivariate normal model for the observed variables. In stage two, the EM-estimated means and covariance matrix are used as input in ordinary mean and covariance structure modeling software programs, whereupon the structural parameters are estimated by routine maximum likelihood estimation for complete data analysis.

More formally, one may regard the EM algorithm as a pre-processor of an incomplete data matrix $\mathbf{Y}$. What it produces is the incomplete data maximum likelihood estimate of the mean vector $\tilde{\boldsymbol{\mu}}$ and the covariance matrix $\tilde{\boldsymbol{\Sigma}}$. Consider the analogue of the vector $\mathbf{s}$ as in Equation (7.2),

$$\tilde{\boldsymbol{\sigma}} = \begin{pmatrix} \tilde{\boldsymbol{\mu}} \\ \text{vech}(\tilde{\boldsymbol{\Sigma}}) \end{pmatrix}. \tag{7.5}$$

In the EM2S estimator, the second stage typically estimates the structural parameters in $\boldsymbol{\theta}$ by minimizing the following maximum Wishart likelihood discrepancy function

$$F(\boldsymbol{\theta}; \tilde{\boldsymbol{\sigma}}) = \text{tr}\{\tilde{\boldsymbol{\Sigma}}[\boldsymbol{\Sigma}(\boldsymbol{\theta})]^{-1}\} - \log|\tilde{\boldsymbol{\Sigma}}[\boldsymbol{\Sigma}(\boldsymbol{\theta})]^{-1}| + [\tilde{\boldsymbol{\mu}} - \boldsymbol{\mu}(\boldsymbol{\theta})]'[\boldsymbol{\Sigma}(\boldsymbol{\theta})]^{-1}[\tilde{\boldsymbol{\mu}} - \boldsymbol{\mu}(\boldsymbol{\theta})] - n. \tag{7.6}$$

Let the resulting estimator be $\tilde{\boldsymbol{\theta}}$. While this estimator is consistent and asymptotically normal (Yuan & Bentler, 2000), the naive goodness-of-fit test statistic obtained by multiplying the minimized value of F by $(N - 1)$, i.e.,

$$T_F = (N - 1)F(\hat{\boldsymbol{\theta}}; \tilde{\boldsymbol{\sigma}}),$$

is not asymptotically chi-square distributed. Its distribution is a mixture of single degree-of-freedom chi-square variates.

Despite the convenience of EM2S, numerous authors view the lack of a chi-square distributed test statistic as a fatal flaw of the EM2S estimator. Enders and Peugh (2004) have proposed *ad hoc* remedies by correcting the scalar multiplier based on the number of missing cases. Their correction hinges on the notion that $\tilde{\boldsymbol{\sigma}}$ is derived from a data set containing less information than a full complete $N \times n$ data matrix would imply, yet the maximum Wishart likelihood discrepancy function in Equation (7.6) assumes that $\tilde{\boldsymbol{\sigma}}$ is obtained from the complete data means and the usual sample covariance matrix. This explanation, however intuitive, fails to take the unevenly distributed fraction of missing information (cf. missing cases) across the $n + n(n+1)/2$ components of $\tilde{\boldsymbol{\sigma}}$ into account. Cai (2008) reasoned that a matrix-valued correction factor would be required.

When viewed from the angle of Browne's (1984) results, the lack of chi-squaredness of T_F has a simple explanation: The weight matrix that is implied by the maximum Wishart likelihood discrepancy function is not the same as the inverse of the asymptotic covariance matrix of the EM-estimated means and covariances $\tilde{\boldsymbol{\sigma}}$. Because the EM2S estimator is consistent and normal, to construct an asymptotically chi-square distributed goodness-of-fit test statistic, one would only need to find the covariance matrix of $\tilde{\boldsymbol{\sigma}}$ and apply Proposition 4. Cai (2008) and Cai and Lee (2009) showed that it is also simple to compute the covariance matrix of $\tilde{\boldsymbol{\sigma}}$ when the EM algorithm is used and when the iteration history of EM is saved. The basic concept is to apply the Supplemented EM algorithm (Meng & Rubin, 1991) that seeks to find a matrix-valued correction factor that is essentially an expression of the "fraction" of missing information. More formally, let the complete data covariance matrix of the means and covariances be $\mathbf{V}$. The Supplemented EM corrected asymptotic covariance matrix of $\tilde{\boldsymbol{\sigma}}$ is

$$\mathbf{W} = \mathbf{V}(\mathbf{I} - DM)^{-1},$$

where $\mathbf{I}$ is an identity matrix of order $n + n(n+1)/2$, and $\mathbf{I} - DM$ is the matrix fraction of missing information, supplied by the Supplemented EM algorithm, serving to inflate the complete data covariance matrix $\mathbf{V}$ to appropriately acknowledge the effect of missing data. Consequently, the test statistic

$$T_{BM} = (N-1)[\tilde{\boldsymbol{\sigma}} - \boldsymbol{\sigma}(\tilde{\boldsymbol{\theta}})]'\boldsymbol{\Gamma}(\mathbf{W})[\tilde{\boldsymbol{\sigma}} - \boldsymbol{\sigma}(\tilde{\boldsymbol{\theta}})]$$

is chi-square distributed with $n + n(n+1)/2 - d$ degrees-of-freedom under the null hypothesis that the model fits exactly. The weight matrix $\boldsymbol{\Gamma}(\mathbf{W})$ is defined analogously as in Equation (7.4). Browne's (1984) Proposition 4 shows that instead of correcting the multiplier, it is the weight matrix that requires correction.

Cai and Lee (2009) use both simulated and empirical data to show that T_{BM} has the desired distribution under the null hypothesis, and EM2S is at least as good as the full-information maximum likelihood estimator. Furthermore, because EM2S uses means and covariances as the input, it facilitates research reporting and simplifies calculations-of-fit indices such as the Root Mean Square Error of Approximation (RMSEA).

7.3 LIMITED-INFORMATION GOODNESS-OF-FIT TESTING FOR ITEM RESPONSE THEORY

Taking a sweeping turn to nonlinear latent variable modeling, Browne's (1984) Proposition 4 has yielded unexpected discoveries in Item Response Theory (IRT) modeling as well. IRT is widely used in educational and psychological testing (Thissen & Wainer, 2001). It can be characterized as a test theory with accompanying statistical models for handling categorical item-level response data. In the simplest case, consider a sample of N respondents' responses to n dichotomously scored items, e.g., right–wrong scores on a math achievement test. Let the ith observed response pattern be $\mathbf{y}_i = (y_{i1}, \ldots, y_{ij}, \ldots, y_{in})$, where y_{ij} can be either 0 (incorrect/not endorsing) or 1 (correct/endorsing). A popular IRT model is the unidimensional two-parameter logistic model,[1]

$$\pi(y_{ij} = 1|\eta, \boldsymbol{\theta}) = \frac{1}{1 + \exp(-\alpha_j - \beta_j \eta)},$$

which expresses the probability of a correct/endorsement response to item j as a function of a latent variable η, and item parameters α_j and β_j (both α_j and β_j are components of $\boldsymbol{\theta}$). Consequently, $\pi(y_{ij} = 0|\eta, \boldsymbol{\theta}) = 1 - \pi(y_{ij} = 1|\eta, \boldsymbol{\theta})$. The IRT model expresses the response pattern probability as the integral

$$\pi_i(\boldsymbol{\theta}) = \int \prod_{j=1}^{n} \pi(y_{ij}|\eta, \boldsymbol{\theta}) h(\eta) d\eta, \tag{7.7}$$

where $h(\eta)$ is the (assumed) distribution of the latent variable.

Statistically, the IRT model is defined on an underlying contingency table made up of the full 2^n cross-classifications of the response patterns. Let the 2^n (unstructured) multinomial cell probabilities be $\boldsymbol{\pi} = (\pi_1, \ldots, \pi_i, \ldots, \pi_{2^n})'$ and observed cell proportions be $\mathbf{p} = (p_1, \ldots, p_i, \ldots, p_{2^n})'$. Of course, $\sum_{i=1}^{2^n} \pi_i = 1$ and $\sum_{i=1}^{2^n} p_i = 1$. It is a standard discrete multivariate analysis result (see, e.g., Bishop,

[1] Notation here follows the statistical convention that the parameters are represented by $\boldsymbol{\theta}$. This conflicts with notational conventions in IRT. However, because IRT is only a subset of the range of models discussed in this chapter, the more general notation is used here.

Fienberg, & Holland, 1975) that the asymptotic distribution of $\mathbf{p} - \boldsymbol{\pi}$ is normal with zero means and covariance matrix

$$\Xi = \text{diag}(\boldsymbol{\pi}) - \boldsymbol{\pi}\boldsymbol{\pi}'.$$

Upon introducing the IRT model, the multinomial cell probabilities become structured as $\boldsymbol{\pi}(\boldsymbol{\theta}) = [\pi_1(\boldsymbol{\theta}), \ldots, \pi_i(\boldsymbol{\theta}), \ldots, \pi_{2^n}(\boldsymbol{\theta})]'$. Fitting this model to the observed proportions by maximum likelihood yields the solution $\hat{\boldsymbol{\theta}}$. Bock and Lieberman (1970) refer to $\hat{\boldsymbol{\theta}}$ as the maximum marginal likelihood estimator (because of the integral in Equation 7.7). Under standard conditions, the cell residual vector $\mathbf{e} = \mathbf{p} - \boldsymbol{\pi}(\hat{\boldsymbol{\theta}})$ is also asymptotically normally distributed with zero means and covariance matrix

$$\boldsymbol{\Omega} = \Xi - \boldsymbol{\Delta}\mathcal{F}^{-1}\boldsymbol{\Delta}'.$$

where $\boldsymbol{\Delta} = \partial\boldsymbol{\pi}(\boldsymbol{\theta})/\partial\boldsymbol{\theta}$ is (once again) the Jacobian matrix, and $\mathcal{F}$ denotes the Fisher information matrix (see, e.g., Cai, Maydeu-Olivares, Coffman, & Thissen, 2006 for details).

While algorithms developed by Bock and Aitkin (1981) and Thissen (1982) have made IRT parameter estimation practical, rigorous tests of the goodness-of-fit of the IRT model have eluded psychometricians for decades. This is due to a well-known problem in categorical data analysis, namely, sparseness of cells. The number of cells in the contingency table is exponential in the number of items, which quickly exceeds any sample size that is conceivable in the social and behaviorial sciences. Thus, for more than a dozen or so items, the underlying contingency table is sparse almost by definition, with many cells having near zero expected probabilities. The usual Pearson chi-square test statistic

$$X^2 = \sum_{i=1}^{2^n} \frac{[p_i - \pi_i(\hat{\boldsymbol{\theta}})]^2}{\pi_i},$$

or its large-sample equivalent, the likelihood ratio test statistic

$$G^2 = 2N \sum_{i}^{2^n} p_i \log\left(\frac{p_i}{\pi_i(\hat{\boldsymbol{\theta}})}\right),$$

can no long achieve the limiting chi-square distribution under sparseness, resulting in severely miscalibrated Type I error rates and less than ideal power for detection of model misspecification.

Recently, researchers have turned increasingly toward limited-information tests as a potential solution to the sparseness issue (e.g., Cai et al., 2006; Maydeu-Olivares & Joe, 2005). In these limited-information tests, marginal subtables are used instead of the full contingency table of response pattern cross-classifications. Typically, marginal residuals up to order 2 are used, i.e., first and second order margins. The marginal residuals are obtained via operator matrices, e.g.,

$$\mathbf{e}_2 = \mathbf{L}_2\mathbf{e} = \mathbf{L}_2\mathbf{p} - \mathbf{L}_2\boldsymbol{\pi}(\hat{\boldsymbol{\theta}}) = \mathbf{p}_2 - \boldsymbol{\pi}_2(\hat{\boldsymbol{\theta}}),$$

where $\mathbf{L}_2$ is a particular $n(n+1)/2 \times 2^n$ fixed matrix of 0s and 1s that can collapse the cell residuals into marginal residuals (see e.g., Cai et al., 2006). The linearity of $\mathbf{e}_2$ implies that they are asymptotically normally distributed with zero means and covariance matrix

$$\boldsymbol{\Omega}_2 = \boldsymbol{\Xi}_2 - \boldsymbol{\Delta}_2\mathcal{F}^{-1}\boldsymbol{\Delta}_2',$$

where $\boldsymbol{\Xi}_2 = \mathbf{L}_2\boldsymbol{\Xi}\mathbf{L}_2'$, $\boldsymbol{\Delta}_2 = \mathbf{L}_2\boldsymbol{\Delta} = \partial\boldsymbol{\pi}_2(\boldsymbol{\theta})/\partial\boldsymbol{\theta}$.

A subtle dilemma arises. Fit statistics based on $\mathbf{e}_2$ must directly involve $\mathcal{F}$. For the multinomial sampling model, computation of $\mathcal{F}$ requires matrix products running over all 2^n (see, e.g., Rao, 1973), which is impractical if n is large. The direct remedy was provided by Cai (2008), who proposes computation of $\mathcal{F}$ via an approximation based on the Supplemented EM algorithm. An earlier, much more elegant, solution was proposed by Maydeu-Olivares and Joe (2005), which turns out to be an implicit application of Browne's (1984) Proposition 4.

Maydeu-Olivares and Joe (2005) noticed that if the model is locally identified from the marginal moments, i.e., $\boldsymbol{\Delta}_2$ has full column rank, then there exists an orthogonal complement such that $\boldsymbol{\Delta}_{2c}'\boldsymbol{\Delta}_2$ is zero. Furthermore, $\boldsymbol{\Delta}_{2c}\mathbf{e}$ is normally distributed with zero means and covariance matrix $\boldsymbol{\Delta}_{2c}\boldsymbol{\Xi}_2\boldsymbol{\Delta}_{2c}'$, which eliminates the need to compute $\mathcal{F}$ altogether! Consequently, the statistic

$$M_2 = \mathbf{e}'\boldsymbol{\Delta}_{2c}'(\boldsymbol{\Delta}_{2c}\boldsymbol{\Xi}_2\boldsymbol{\Delta}_{2c}')^{-1}\boldsymbol{\Delta}_{2c}\mathbf{e}$$

is asymptotically chi-square distributed with $n(n+1)/2 - d$ degrees of freedom for any consistent estimator of $\boldsymbol{\theta}$, the maximum likelihood estimator included. Despite sweeping differences in the modeling framework, the similarity of M_2 to T_B is evident. Maydeu-Olivares and Joe (2005) showed that M_2 not only maintains adequate Type I error rate calibration, but (counter-intuitively) can be more powerful than the full-information tests.

7.4 ALTERNATIVE MULTIPLE-IMPUTATION INFERENCE

This section returns to the linear case. Model-based multiple imputation is recognized as one of the most flexible methods for handling missing responses in practice (Schafer, 1997). Multiple imputation is built around the concept of replacing the unobserved values by $M > 1$ random draws from a posterior predictive distribution, proceeding with standard tools for complete data (e.g., regression), and then combining the M sets of complete data inferences by averaging the point estimates and properly adjusting the standard errors for the added uncertainty due to missing data. In some contexts, multiple imputation is simply required. For instance, the *plausible value* methodology (Mislevy, Beaton, Kaplan, & Sheehan, 1992) is central to large-scale educational assessments such as the National Assessment of Educational Progress (NAEP). The plausible values are multiple imputations that reflect the uncertainty in individual students' scores so that consistent group-level inferences can be drawn (e.g., for schools and States).

However, when mean and covariance structure models must be fitted to multiply imputed data sets, a number of difficulties arise. Chief among these is again the lack of a convenient procedure for computing an overall goodness-of-fit test statistic. In response, a new alternative approach to multiple-imputation inference is proposed here. The new estimator may be called the Multiple Imputation Two-Stage (MI2S) estimator.

It is shown here that for linear mean and covariance structure modeling, the combination of multiple imputations can occur in the beginning, before any structural models are even fitted. The underlying statistical theory is a direct extension of the EM2S results obtained by Cai (2008) and Cai and Lee (2009). An asymptotically chi-square distributed overall model fit statistic is derived along the same lines as Browne's (1984) residual-based test (Proposition 4 at work again).

7.4.1 Conventional Multiple Imputation Estimation and Inference

Making a return to the notation of Section 7.1, if missing values are present, or if some or all variables in $\mathbf{Y}$ have more than one copy of plausible values, estimation of $\boldsymbol{\theta}$ can be carried out in the following manner. Consider the mth imputation (or plausible value) $\mathbf{Y}_m$, $m = 1, \ldots, M$. Based on this complete data set, the sample mean vector $\hat{\boldsymbol{\mu}}_m$ and sample covariance matrix $\hat{\boldsymbol{\Sigma}}_m$ are computed. For each imputation, the maximum Wishart likelihood discrepancy function for the mean and covariance structure model is $F(\boldsymbol{\theta}; \hat{\boldsymbol{\sigma}}_m)$, where $\hat{\boldsymbol{\sigma}}_m = [\hat{\boldsymbol{\mu}}'_m, \text{vech}(\hat{\boldsymbol{\Sigma}}_m)']'$. Minimization of $F(\boldsymbol{\theta}; \hat{\boldsymbol{\sigma}}_m)$ leads to the parameter estimate $\hat{\boldsymbol{\theta}}_m$ based on the mth imputation. The covariance matrix of $\hat{\boldsymbol{\theta}}_m$ is $Q_m = \text{var}(\hat{\boldsymbol{\theta}}_m)$.

Standard multiple imputation formulas can be used to combine the point estimates and variability information (Schafer, 1997). The final point estimate is the average of the multiple imputation estimates: $\check{\boldsymbol{\theta}} = M^{-1} \sum_{m=1}^{M} \hat{\boldsymbol{\theta}}_m$, and the multiple imputation variance approximation is $\text{var}(\check{\boldsymbol{\theta}}) = \bar{Q} + (1 + M^{-1})B$, where $\bar{Q} = M^{-1} \sum_{m=1}^{M} Q_m$ is the average of the within-imputation covariance matrices and $B = (M-1)^{-1} \sum_{m=1}^{M} \times (\hat{\boldsymbol{\theta}}_m - \check{\boldsymbol{\theta}})(\hat{\boldsymbol{\theta}}_m - \check{\boldsymbol{\theta}})'$ is the between-imputation component.

7.4.2 The New Two-Stage Estimator

7.4.2.1 Stage One: Combining Multiple Imputations Recall that the estimate of $\boldsymbol{\sigma}$ from imputation m is $\hat{\boldsymbol{\sigma}}_m$. The combined estimate of $\boldsymbol{\sigma}$ across the M imputations is the average

$$\bar{\boldsymbol{\sigma}} = \frac{1}{M}\sum_{m=1}^{M}\hat{\boldsymbol{\sigma}}_m. \tag{7.8}$$

The above equation is a direct application of the standard multiple imputation combination rule to the unstructured first and second order moments of the complete data. In essence, Equation (7.8) is a stochastic counterpart of the first stage of the EM2S estimator discussed earlier in Section 7.2.

Let the covariance matrix of $\hat{\boldsymbol{\sigma}}_m$ be $N^{-1}[\mathfrak{F}(\hat{\boldsymbol{\sigma}}_m)]^{-1}$, where $\mathfrak{F}(\hat{\boldsymbol{\sigma}}_m)$ is the Fisher information matrix for the unstructured multivariate normal model (see e.g., Mardia, Kent, & Bibby, 1979). An estimate of the covariance matrix of $\bar{\boldsymbol{\sigma}}$ is

$$\text{var}(\bar{\boldsymbol{\sigma}}) = \frac{1}{N}\mathbf{U} = \left[\frac{1}{M}\sum_{m=1}^{M}\frac{1}{N}[\mathfrak{F}(\hat{\boldsymbol{\sigma}}_m)]^{-1}\right] + \left[\frac{1+M^{-1}}{M-1}\sum_{m=1}^{M}(\hat{\boldsymbol{\sigma}}_m - \bar{\boldsymbol{\sigma}})(\hat{\boldsymbol{\sigma}}_m - \bar{\boldsymbol{\sigma}})'\right]. \tag{7.9}$$

A basic tenet of the theory behind multiple imputation is that if the imputations are proper, and if M is large, $\bar{\boldsymbol{\sigma}}$ provides a consistent estimate of $\boldsymbol{\sigma}$ (Rubin, 1996). Furthermore, when M is large, the limiting distribution of $\sqrt{N}(\bar{\boldsymbol{\sigma}} - \boldsymbol{\sigma})$ is normal, with zero means, and an asymptotic covariance matrix that can be consistently estimated by $\mathbf{U}$.

7.4.2.2 Stage Two: Estimation and Inference The second stage of MI2S is entirely analogous to the EM2S estimator. The structural parameters $\boldsymbol{\theta}$ are found by minimizing the maximum likelihood discrepancy function $F(\boldsymbol{\theta}; \bar{\boldsymbol{\sigma}})$ as defined in Equation (7.6). Let the MI2S estimator be denoted as $\bar{\boldsymbol{\theta}}$. Using Yuan and Jennrich's (1998) asymptotic distribution theory of generalized estimating equations, it can be shown that $\bar{\boldsymbol{\theta}}$ is consistent and asymptotically normal.

A familiar pattern emerges here. The naive statistic

$$T_M = (N-1)F(\bar{\boldsymbol{\theta}}; \bar{\boldsymbol{\sigma}})$$

is not distributed as a chi-square variable. Due to the presence of multiple imputations, the asymptotic covariance matrix of $\bar{\boldsymbol{\sigma}}$ is not of a standard form that is assumed by the maximum likelihood discrepancy function. Equation (7.9) makes this point amply clear.

To remedy the situation, Browne's (1984) residual-based statistic can be used. Let $\boldsymbol{\sigma}(\bar{\boldsymbol{\theta}})$ be the model-implied moments. The following test statistic

$$T_{BM2} = N[\bar{\boldsymbol{\sigma}} - \boldsymbol{\sigma}(\bar{\boldsymbol{\theta}})]'\boldsymbol{\Gamma}(\mathbf{U})[\bar{\boldsymbol{\sigma}} - \boldsymbol{\sigma}(\bar{\boldsymbol{\theta}})]$$

is asymptotically distributed as a central chi-square variable with $n + n(n+1)/2 - d$ degrees-of-freedom under the null hypothesis of exact fit, where $\mathbf{\Gamma}(\mathbf{U})$ is as defined in Equation (7.4).

7.4.3 A Simulation Study

This small simulation study is conducted to show that the proposed MI2S fit test statistic T_{BM2} is indeed chi-square distributed under the null hypothesis. The data generating process is a confirmatory factor analysis model with six manifest variables and two correlated factors. The generating factor loading matrix is

$$\mathbf{\Lambda}' = \begin{pmatrix} 0.8 & 0.8 & 0.8 & 0 & 0 & 0 \\ 0 & 0 & 0 & 0.7 & 0.7 & 0.7 \end{pmatrix},$$

and the factor covariance matrix is

$$\mathbf{\Psi} = \begin{pmatrix} 1.0 & \\ .4 & 1.0 \end{pmatrix}.$$

The unique factor variances are given by $\mathbf{\Phi} = \text{diag}(\phi_{11}, \ldots, \phi_{66})$, where all the diagonal elements are equal to 0.5. For this model, $q = 13$ free parameters make up $\boldsymbol{\theta}$. They are the six non-zero loadings, the factor correlation ψ_{21}, and the six unique variances. The degrees-of-freedom is equal to 8.

Complete multivariate normal data having the generating factor analysis covariance structure are first simulated. To simulate missingness, the complete data sets are then subject to two kinds of missing data mechanisms: missing completely at random (MCAR) and missing at random (MAR). For the MCAR condition, the same procedure as in Cai and Lee (2009) is used. First, for each row in the simulated data matrix, a fair coin is flipped to decide whether there should be any missing values. Next, for a case that is chosen to contain missing observations, two uniform integers between 1 and 6, inclusive, are generated and the values of the manifest variables indexed by the two integers are set to missing. This leaves about 16% of all observations missing. For the MAR condition, the first three manifest variables are set to missing if the means of the last three manifest variables are less than -0.67. Thus about 25% of all cases contain missing observations under MAR. As a benchmark, a condition with no missing data (NOMIS) is also included.

The three missing data conditions (NOMIS, MCAR, and MAR) are crossed with three sample sizes: $N = 100, 300, 500$, resulting in nine simulation conditions. In each condition, 500 replications are attempted. For MI2S, the $M = 20$ multiple imputations are generated under the multivariate normal model using a data augmented Gibbs sampler described by Schafer (1997). The code for data generation, imputation, and model fitting are programmed in GAUSS (Aptech Systems, Inc., 2003).

Because the fitted model is correctly specified, T_{BM2} should be distributed as a central chi-square variable with eight degrees of freedom regardless of the amount or kind of missing data. On the other hand, the naive fit statistic T_M should not be chi-square distributed under MAR or MCAR. Under NOMIS, both statistics are distributed as central chi-square variable with eight degrees-of-freedom.

Table 7.1 presents a summary of the results of this simulation study. First, under NOMIS, both T_M and T_{BM2} are clearly behaving as they should. The means of the empirical distribution of these two test statistics are very close to 8, which is the theoretical expected value of a central chi-square variable with eight degrees

Table 7.1 Type I error rates

							Significance level		
N	*Converged*	*Statistic*	*Mean*	*Var*	*Min*	*Max*	*0.01*	*0.05*	*0.10*
				NOMIS					
100	500	T_M	8.13	17.38	.99	28.78	.011	.045	.113
		T_{BM2}	8.07	14.72	1.05	23.77	.006	.044	.100
300	500	T_M	8.16	17.20	.69	31.80	.013	.049	.102
		T_{BM2}	8.14	16.70	.71	33.73	.012	.049	.097
500	500	T_M	8.13	16.35	.81	29.59	.009	.057	.108
		T_{BM2}	8.12	16.03	.83	32.56	.009	.054	.107
				MCAR					
100	500	T_M	12.55	42.49	1.30	38.52	.116	.298	.390
		T_{BM2}	8.02	14.84	.81	23.31	.006	.044	.098
300	500	T_M	11.69	34.85	.78	35.91	.086	.222	.318
		T_{BM2}	8.14	17.47	.47	25.87	.020	.054	.110
500	500	T_M	11.55	31.45	1.82	32.74	.094	.212	.312
		T_{BM2}	8.25	18.05	1.40	29.50	.020	.054	.114
				MAR					
100	499	T_M	11.82	36.72	1.39	42.77	.096	.220	.321
		T_{BM2}	7.78	13.46	1.00	20.26	.002	.048	.090
300	499	T_M	11.64	34.42	1.28	41.29	.088	.240	.331
		T_{BM2}	7.99	14.30	1.10	21.03	.006	.040	.090
500	500	T_M	12.00	32.43	.89	31.56	.100	.230	.340
		T_{BM2}	8.46	16.47	.62	25.53	.014	.056	.126

Note: The entries in the "Converged" column refer to the number of converged replications in each condition.

of freedom. The variances are also close to the theoretical value of 16. The empirical Type I error rates at the .01, .05, and .10 levels closely follow the nominal alpha level. Next, for the MCAR condition, the naive test statistic T_M is obviously not chi-square distributed. Its means and variances are too large, and the empirical rejection rates are far above the nominal level. On the other hand, T_{BM2} maintains adequate control over Type I error rates, and the mean-variance relation much better approximates that of a central chi-square variable with eight degrees of freedom. The MAR condition shows a similar pattern as MCAR.

7.5 DISCUSSION

Browne's (1984) paper was written more than a quarter of a century ago. Yet, as this chapter shows, it had and continues to have tremendous impact on methodologists working on statistical issues surrounding latent variable modeling. From nonlinear item response modeling to treatment of missing data, a number of recent results can be thought of as natural continuation or applications of Browne's (1984) Proposition 4. Describing Browne (1984) as influential would be an understatement. It deserves a standing ovation with three cheers.
Hip Hip Hooray! Hip Hip Hooray! Hip Hip Hooray!

REFERENCES

Allison, P. D. (2001). *Missing data.* Sage University Papers Series on Quantitative Applications in the Social Sciences, 07–136. Thousand Oaks, CA: Sage.

Allison, P. D. (2003). Missing data techniques for structural equation modeling. *Journal of Abnormal Psychology, 112*, 545–557.

Aptech Systems, Inc. (2003). GAUSS (Version 6.08) [Computer software]. Maple Valley, WA.

Bishop, Y. M. M., Fienberg, S. E., & Holland, P. W. (1975). *Discrete multivariate analysis: Theory and practice.* Cambridge, MA: MIT Press.

Bock, R. D., & Aitkin, M. (1981). Marginal maximum likelihood estimation of item parameters: Application of an EM algorithm. *Psychometrika, 46*, 443–459.

Bock, R. D., & Lieberman, M. (1970). Fitting a response model for *n* dichotomously scored items. *Psychometrika, 35*, 179–197.

Browne, M. W. (1984). Asymptotically distribution-free methods for the analysis of covariance structures. *British Journal of Mathematical and Statistical Psychology, 37*, 62–83.

Cai, L. (2008). SEM of another flavour: Two new applications of the supplemented EM algorithm. *British Journal of Mathematical and Statistical Psychology, 61*, 309–329.

Cai, L., & Lee, T. (2009). Covariance structure model fit testing under missing data: An application of the supplemented EM algorithm. *Multivariate Behavioral Research, 44*, 281–304.

Cai, L., Maydeu-Olivares, A., Coffman, D. L., & Thissen, D. (2006). Limited-information goodness-of-fit testing of item response theory models for sparse 2^p tables. *British Journal of Mathematical and Statistical Psychology, 59*, 173–194.

Dempster, A. P., Laird, N. M., & Rubin, D. B. (1977). Maximum likelihood estimation from incomplete data via the EM algorithm. *Journal of the Royal Statistical Society – Series B, 39*, 1–38.

Enders, C. K., & Peugh, J. L. (2004). Using an EM covariance matrix to estimate structural equation models with missing data: Choosing an adjusted sample size to improve the accuracy of inferences. *Structural Equation Modeling, 11*, 1–19.

Little, R. J. A., & Rubin, D. B. (1987). *Statistical analysis with missing data*. New York: John Wiley & Sons.

Mardia, K. V., Kent, J. T., & Bibby, J. M. (1979). *Multivariate analysis*. London: Academic Press.

Maydeu-Olivares, A., & Joe, H. (2005). Limited and full information estimation and testing in 2^n contingency tables: A unified framework. *Journal of the American Statistical Association, 100*, 1009–1020.

Meng, X.-L., & Rubin, D. B. (1991). Using EM to obtain asymptotic variance–covariance matrices: The SEM algorithm. *Journal of the American Statistical Association, 86*, 899–909.

Mislevy, R. J., Beaton, A. E., Kaplan, B., & Sheehan, K. M. (1992). Population characteristics from sparse matrix samples of item responses. *Journal of Educational Measurement, 29*, 133–161.

Rao, C. R. (1973). *Linear statistical inference and its applications* (2nd ed.). New York: John Wiley & Sons.

Rubin, D. B. (1996). Multiple imputation after 18+ years. *Journal of the American Statistical Association, 91*, 473–489.

Schafer, J. L. (1997). *Analysis of incomplete multivariate data*. New York: Chapman & Hall/CRC.

Thissen, D. (1982). Marginal maximum likelihood estimation for the one-parameter logistic model. *Psychometrika, 47*, 175–186.

Thissen, D., & Wainer, H. (2001). *Test scoring*. Mahwah, NJ: Lawrence Erlbaum Associates.

Yuan, K.-H., & Bentler, P. M. (2000). Three likelihood-based methods for mean and covariance structure analysis with nonnormal missing data. *Sociological Methodology, 30*, 165–200.

Yuan, K.-H., & Jennrich, R. I. (1998). Asymptotics of estimating equations under natural conditions. *Journal of the Multivariate Analysis, 65*, 245–260.

Chapter 8

Nonparametric Bayesian Modeling of Item Response Curves with a Three-Parameter Logistic Prior Mean

Kristin A. Duncan and Steven N. MacEachern

8.1 INTRODUCTION

Item response theory (IRT) models the relationship between a person's observed performance on a test, and a latent trait or ability the test is seeking to measure. IRT is a wide and actively researched field. In this work, we restrict our attention to the case of binary responses where items on a test are coded simply as right or wrong, and also make the assumption that the tests we are dealing with are unidimensional, measuring just one trait. The relationship between values of the trait and the probability of a correct or "1" response is modeled with the item characteristic curve (ICC).

The traditions of IRT within the psychometric community have been heavily influenced by formal measurement theory, leading to models such as the Rasch model. In contrast, we come to the problem from statistics, where traditions are tilted toward empirical modeling. Additionally, our background as faculty members has exposed us to data sets where questions are written for exams in a one-off fashion. These questions do not receive the extensive vetting of questions for exams prepared, for example, by ETS. Consequently, we expect some substantial departures from formal test theory.

A nonparametric Bayesian model for item responses using a Dirichlet process prior on the item characteristic curve was introduced by Qin (1998) and further explored by Duncan and MacEachern (2008). In these works, a two-parameter logistic curve was used as the prior mean for the Dirichlet process. Here we extend the nonparametric Bayesian methodology to use a three-parameter logistic curve as the prior mean. The extension, while natural, is not trivial.

In Section 8.2, we begin by outlining the basic logistic models for IRT along with Bayesian and nonparametric Bayesian versions of these models. In Section 8.3, model nesting is used to describe the relationships between these models. In Section 8.4, we fit a data set from an exam in an introductory statistics class with both parametric and nonparametric methods and compare the results.

8.2 LOGISTIC MODELS FOR IRT

Logistic curves are a natural choice for item response because they are monotonic in the latent trait and they allow the probability of a correct response, conditional on the trait, to vary from zero to one.

Notation-wise, we represent the number of individuals taking the test by N. Individuals are indexed with the letter i, $i = 1 \ldots N$. The number of items in the test is denoted by J. Items are indexed with the letter $j, j = 1 \ldots J$. Y_{ij} is the response of individual i to item j. 1 denotes a correct answer and 0 an incorrect answer. θ is the ability parameter, and the true ability of individual i is denoted θ_i. $P_j(\theta)$ represents the probability an individual with ability θ answers item j correctly; it is the height of the ICC curve for item j at θ.

The three-parameter logistic model (3PLM) is commonly used for multiple-choice problems. The ICC for the 3PLM is specified by

$$P_j(\theta) = c_j + \frac{1 - c_j}{1 + e^{-a_j(\theta - b_j)}}, \tag{8.1}$$

where b_j is a location parameter interpreted as the difficulty of item j, a_j is a scale, or discrimination parameter which must be positive to maintain the increasing monotonicity of the ICC, and c_j is a parameter that accounts for the probability of guessing the correct answer. The two-parameter logistic model (2PLM) is a special case of the 3PLM in which c_j is restricted to be zero. The one-parameter logistic model is the most basic of the logistic models for IRT, and also a special case of the 3PLM in which $c_j = 0$, and $a_j = 1$.

Scoring a test with one of the above IRT models involves implicit estimation of both the parameters of the ICC and the ability levels of the test takers. It is commonly assumed that the abilities follow a Normal(0,1) distribution. Fixing the location and scale parameters of the ability distribution ensures identifiability of the curve parameters. Baker and Kim (2004) give an extensive review of parameter estimation techniques, many of which involve maximum likelihood methods.

8.2.1 Parametric Bayesian IRT Models

Bayesian methods have also been applied to the IRT estimation problem. A hierarchical Bayesian version of the 3PLM may be defined as follows,
Bayesian 3PL Model (3PLM)

$$Y_{ij} \mid \theta_i, a_j, b_j \sim Bernoulli\left(c_j + \frac{1 - c_j}{1 + e^{-a_j(\theta_i - b_j)}}\right) \tag{8.2}$$

$$\theta_i \mid \mu_\theta, \sigma_\theta^2 \sim Normal(\mu_\theta, \sigma_\theta^2) \tag{8.3}$$

$$b_j \mid \mu_b, \sigma_b^2 \sim Normal(\mu_b, \sigma_b^2) \tag{8.4}$$

$$a_j \mid \nu_a, \omega_a \sim Gamma(\nu_a, \omega_a) \tag{8.5}$$

$$c_j \mid \alpha_c, \beta_c \sim Beta(\alpha_c, \beta_c). \tag{8.6}$$

The Bayesian 2PL model (*2PLM*) eliminates the guessing parameter, c_j. Examples of Bayesian IRT models in the literature include Albert (1992), Patz and Junker (1999), and the LSAT example from the OpenBUGS user manual (2006).

8.2.2 Nonparametric Bayesian IRT Models

The one-parameter logistic, two-parameter logistic, and three-parameter logistic, are the most commonly used parametric models for item response because of their interpretability and the ease of working with the logistic function. However, they may not capture the true item functioning, resulting in bias in ability estimates. With the pervasiveness of high-stakes testing today, accurate ability estimates are of great importance. Nonparametric methods do not place restrictions on the form of the ICC. Frequentist methods for modeling the ICC curve nonparametrically, based on kernel smoothing are described by Ramsay (1991). In this subsection, we put forth a nonparametric Bayesian method based on Dirichlet process priors, which allow us to impose strict monotonicity on the ICC while allowing departures from the logistic form.

A Dirichlet process prior is a means of placing a probability distribution on a random variable that is itself a probability distribution. The parameters of the Dirichlet process are the prior mean and the prior mass. The prior mean, also called the base measure, reflects the investigator's prior belief about where the distribution is centered.

The prior mass is a measure of deviation around this prior mean. The Dirichlet process is formulated as

$$x_i \sim F, \quad i = 1 \ldots n$$
$$F \sim DP(MF_0).$$

F is the random probability distribution, $\boldsymbol{x}$ is the observed data from this random distribution, F_0 is the prior mean of the Dirichlet process, and M is the prior mass. M behaves roughly like the bandwidth parameter in kernel smoothing in the sense that larger values of M result in smoother curves, more similar to the base measure. The distinct x values make up a random sample from F_0, and the deviation from the prior mean is evident in the clustering of the observations from the Pólya urn formulation of the Dirichlet process.

In order to model an ICC with a Dirichlet process prior, the ICC is viewed as a cumulative distribution function (cdf) and therefore forced to be monotonically nondecreasing. A two-parameter logistic curve is the cdf for the logistic distribution. However, the three-parameter logistic curve with a left asymptote of c_j rather than 0 is not a cdf. In order to use a 3PL curve in the Dirichlet process, we make the base measure a mixture of the 2PL curve with parameters a_j and b_j and a degenerate distribution with mass c_j at $-\infty$.

Another hurdle to overcome in constructing this model is that we do not observe data from the distribution whose cdf is the item characteristic curve. The solution to this problem is to reparameterize the model by introducing latent variables z_{ij} that allow us to work "directly" with the density F_j. These latent variables are best described in terms of the dose/response problem where z_{ij} is individual i's tolerance to drug (item) j. If the dose (ability) θ_i exceeds the individual's tolerance, the individual responds (answers correctly), whereas if the dose is less than the individual's tolerance he or she does not respond (answers incorrectly). For example, see Finney (1964) or Muthén (1984). Whereas the Y_{ij} values were previously expressed as Bernoulli random variables, in this parameterization the Y_{ij} are completely determined by θ_i and z_{ij}.

The hierarchical model is formulated as follows:

Nonparametric Bayesian 3PL Model (NP3M)

$$\theta_i \mid \mu_\theta, \sigma_\theta^2 \sim N(\mu_\theta, \sigma_\theta^2)$$
$$F_j \mid a_j, b_j, c_j, M_j \sim DP(M_j F_{0j})$$
$$z_{ij} \mid F_j \sim F_j$$
$$Y_{ij} \mid \theta_i, z_{ij} = \begin{cases} 1 & \text{if } z_{ij} \leq \theta_i \\ 0 & \text{if } z_{ij} > \theta_i \end{cases}$$

$$F_{0j} = \begin{cases} \dfrac{1}{1+e^{-a_j(\theta-b_j)}} & w.p. \quad 1-c_j \\ < -\infty > & w.p. \quad c_j \end{cases}$$

$$b_j \mid \mu_b, \sigma_b^2 \sim N(\mu_b, \sigma_b^2)$$

$$a_j \mid \nu_a, \omega_a \sim Gamma(\nu_a, \omega_a)$$

$$c_j \mid \alpha_c, \beta_c \sim Beta(\alpha_c, \beta_c)$$

The the notation $< x >$ indicates a discrete distribution with mass at the point x. F_{0j} is the prior mean of the distribution function from the Dirichlet process for item j. Since F_{0j} varies depending on the values of a_j, b_j, and c_j, this model is a mixture of Dirichlet processes. The nonparametric Bayesian 2PL model (*NP2M*) is a special case of NP3M with $c_j = 0$, giving $F_{0j} = \dfrac{1}{1+e^{-a_j(\theta-b_j)}}$.

8.3 NESTED MODELS

The concept of a nested model appears throughout statistics, underlying both formal procedures, such as the full versus reduced model F-test, and data-analytic procedures, such as model expansion and contraction on the basis of graphical summaries. However, it is difficult to find a formal definition of nested models in the literature. We begin with a common setting.

In regression modeling, the common notion of nested models is clear. A model m_B in the larger family, B, is defined by a set of parameter values, say $\beta_1, \ldots, \beta_p, \sigma^2$, along with the assumption that connects the parameters to the distribution of the data:

$$Y_i \sim N\left(\sum_{j=1}^{p} x_j\beta_j, \sigma^2\right), \qquad i = 1, \ldots, n,$$

where the Y_i are independent, and $N(a, b)$ represents the Gaussian distribution with mean a and variance b. The parameter vector is a point in $R^p \times R^+$. A model m_A in the smaller family, A, is defined by the parameters $\gamma_1, \ldots, \gamma_k, \tau^2$, along with the map from parameters to the distribution of observable data:

$$Y_i \sim N\left(\sum_{j=1}^{k} z_j\gamma_j, \sigma^2\right), \qquad i = 1, \ldots, n,$$

where the Y_i are independent. The parameter vector is allowed to lie in $R^k \times R^+$.

The family of models, A, is nested in the family of models, B, if the "mean space" for A is a subspace of the "mean space" for B. This is seen by taking $\sigma^2 = \tau^2$ and choosing appropriate vectors $\beta = (\beta_1, \ldots, \beta_p)$ and $\gamma = (\gamma_1, \ldots, \gamma_k)$. For any vector γ, there exists a corresponding vector β.

The γ mean space is contained in the β mean space if, for any γ, there is some β for which $Z^t\gamma = X^t\beta$ for all (Z, X). Typically, the relationship between Z and X is of the form $Z = CX$, in which case the relationship becomes $Z^t\gamma = X^tC^t\gamma$ and the question of nesting revolves around $C^t\gamma$ and β. If, for each γ, there exists a β for which $C^t\gamma = \beta$, the models are nested. Since β can lie anywhere in R^p, nesting occurs.

This description includes such things as principal components regression, where X is the original design matrix and Z consists of a (subset) of principal components derived from X. It includes variable selection, where columns of X are dropped when moving to Z. In this latter instance, nesting may be described as one parameter space being contained in the other; in both cases, nesting translates into a linear restriction on the parameters in the larger model. Nesting may be broader than this if, for example, the covariates lie in some appropriately restricted space.

The description of nesting in regression carries over to the ANOVA setting. For fixed effects models, the restriction is on the mean space of the model, exactly as in regression. For random effects models, the restriction to the smaller model occurs when a component of variation is set to 0. For the mixed model, nesting is characterized by a combination of restrictions on variance components and mean spaces.

The pattern extends to many other models, including those that make strong use of latent structures. A particularly interesting class includes the latent monotonic polynomial models for discriminant analysis and item response theory that have been developed by Lorraine Sinnot (1997) and Longjuan Liang (2007) in their work with Michael Browne. In these models, restricting the polynomial to a line (equivalent to setting higher-order coefficients to zero) defines the smaller parameter space, and nesting is immediate.

In an effort to formalize the notion of nesting, we venture the following definition which seems appropriate from the perspective of the classical statistician:

> **Definition 8.3.1: classical nesting.** A family of models, A, is nested in a family of models, B, if, for every $m \in A$, we can obtain m by setting a function of a collection of parameters in B to some fixed value.

This definition places the focus on the parameter space, in keeping with the usual practice when comparing a larger model to a smaller model, where the comparison often reduces to a comparison of the value of a single parameter to some set number. The definition captures the nested models that are the mainstay of applied statistical work, including those described above.

8.3.1 Bayesian Nesting

The following example challenges the traditional definition of nested models. Suppose that family A consists of a single model, namely that the data, $X_1, \ldots, X_n$, are independent and identically distributed Bernoulli trials with probability of success 0.5. Family B is driven by a latent, deterministic, dynamical system. The behavior of the system is entirely specified by its initial condition, say $\theta \in [0, 1]$, and so this initial condition becomes the parameter that defines the model. The system produces a sequence of latent values, $Y_1, Y_2, \ldots$ according to the following rules:

$$\begin{aligned} Y_1 &= \theta \\ Y_{t+1} &= 2Y_t \qquad\qquad \text{if } 0 \leq Y_t < 0.5 \\ &= 2 - 2Y_t \qquad \text{if } 0.5 \leq Y_t \leq 1. \end{aligned}$$

The latent values produce observed values through a measurement equation,

$$X_t = \text{round}(Y_t),$$

leading to $X_1, X_2, \ldots$, a sequence of 0s and 1s.

Following Definition 8.3.1, we consider whether model A is nested in model B or vice-versa. Nesting hinges upon probability calculations under members of the two families. For family (model) A,

$$P_A(x_1, \ldots, x_n) = \prod_{i=1}^{n} (0.5)^{x_i}(0.5)^{1-x_i} = 2^{-n}$$

for each sequence of 0s and 1s. Family B contains many members, indexed by θ. The probability calculation is

$$\begin{aligned} P_B(x_1, \ldots, x_n|\theta) &= 0 \qquad \text{if } \theta \text{ does not lead to } x_1, \ldots, x_n \\ P_B(x_1, \ldots, x_n|\theta) &= 1 \qquad \text{if } \theta \text{ does lead to } x_1, \ldots, x_n. \end{aligned}$$

We recognize immediately that, under Definition 8.3.1, family A is *not* nested in family B, as there is no value of θ for which $P_A(x_1, \ldots, x_n) = P_B(x_1, \ldots, x_n|\theta)$. However, there is a way to produce model A from the family B. The key is to place a probability distribution on the initial condition, θ, thereby moving toward a Bayesian method.

Taking a Bayesian view of the problem, for family A, which consists of a single model, the sole possible prior distribution assigns probability one to the single model in the family—that is, the data are as independent flips of a fair coin. For family B with a prior density $\pi(\theta)$ over $\theta \in [0, 1]$, we have

$$P_{B,\pi}(x_1, \ldots, x_n) = \int_0^1 P_B(x_1, \ldots, x_n|\theta)\pi(\theta)d\theta = \int_{R(x_1,\ldots,x_n)} \pi(\theta)d\theta,$$

where $R(x_1, \ldots, x_n)$ denotes the set of initial conditions which lead to the sequence $x_1, \ldots, x_n$. Notice that, if θ follows the uniform distribution on $[0, 1]$, $P_{B,\pi}(x_1, \ldots, x_n) = 2^{-n}$ for each n-tuple $(x_1, \ldots, x_n)$. The ability to specify a prior distribution over family B that yields the identical distribution for data under any member of family A leads us to conclude that a Bayesian should consider family A to be nested in family B. This suggests the following definition of nesting:

> **Definition 8.3.2: Bayesian nesting.** A family of models, A, is n-nested in a family of models, B, if, for every prior distribution π_A over the family A, there exists some prior distribution π_B over the family B such that
>
> $$P_{A,\pi_A}[(X_1, \ldots, X_n) \in C_n] = P_{B,\pi_B}[(X_1, \ldots, X_n) \in C_n]$$
>
> for all measureable C_n. If A is n-nested in B for all n, A is nested in B.

This definition repackages the concept of nested models. Instead of a comparison of points in the parameter spaces, the comparison is of distributions over observable quantities which are, in turn, driven by distributions over the parameter spaces. We note that the two definitions can be reconciled by recasting the notion of a parameter: The traditional parameter, $\theta \in \Theta$, is replaced by the distribution $\pi(\theta)$, a distribution over Θ. With this change, the Bayesian definition matches the classical definition—for A to be nested in B, for each π_A there must exist some π_B which induces the identical distribution on the data.

8.3.2 The Role of the Prior Distribution

In practical problems, the Bayesian often restricts herself to classes of prior distributions, say a normal distribution for regression coefficients (or abilities in an item response theory model). These restrictions shrink the family of models and can lead to reversals of the usual notion of nesting. The following example illustrates this behavior in the context of a sample of finite size.

Define two families of models. In both, the data X_1, X_2 are, given θ, conditionally independent Bernoulli(θ) trials. Family A has a large parameter space, but a severely restricted space for the prior distribution. The parameter space is the interval [0, 1], and the prior distribution on θ is uniform on the interval. Family B has a small parameter space and a flexible prior distribution. $\Theta = \{0.5 - 1/\sqrt{12}, 0.5 + 1/\sqrt{12}\}$ consists of two points, and any prior distribution supported on these two points is allowed.

The joint distribution of a set of conditionally independent Bernoulli trials, $X_1, \ldots, X_n$, having common parameter θ is characterized by the first n moments of θ. This follows from the exchangability of $X_1, \ldots, X_n$,

$$P\left[\sum_{i=1}^{n} X_i = k|\theta\right] = \int \binom{n}{k}\theta^k(1-\theta)^{n-k}d\pi(\theta),$$

and the fact that the polynomial inside the integral expands to involve only the 0th through nth prior moments of θ.

The first two prior moments under family A are $E[\theta] = 0.5$ and $\text{Var}[\theta] = 1/12$. These moments are exactly those under family B with a prior distribution that assigns mass 0.5 to each possible value of θ. Thus any distribution on (X_1, X_2) that can arise under family A can be reproduced exactly under family B, and we refer to family A as two-nested in family B. Family A is not, however, nested in family B, as shown by examination of the distribution on observables for larger sample sizes. Nor is family B nested in family A, as an asymmetric prior distribution on θ for family B cannot be replicated in family A, even for a single observation.

8.3.3 Subsetted Models

The development of subsetted models has potential application for the computation of the Bayes factor, the usual mechanism for assessing evidence in the comparison of a pair of Bayesian models. The concept of subsetted models is closely related to the idea of nesting. We say that family A of models is subsetted in family B of models if, for every model $m_A \in A$, there exists a model $m_B \in B$ such that we may represent m_B as a non-null mixture of m_A and some other model. That is, if we can write $m_B(x) = pm_A(x) + (1 - p)m_C(x)$, for some $p \in (0, 1]$ and all x. The models we refer to here are the eventual distribution on observable quantities, perhaps induced by a distribution over some parameter space.

Suppose that the models are for discrete observable quantities, as in IRT. The above expression indicates that the probability mass function for model m_B can be written as a convex combination of the probability mass function for model m_A and some other, possibly unusual probability mass function. A similar statement applies in the event that the models represent densities with respect to Lebesgue measure.

The Bayes factor comparing two models is the multiplier that takes one from the prior odds of the models to the posterior odds of the models, provided that the prior odds are neither 0 nor ∞. Alternatively, the Bayes factor can be described as the posterior odds when the prior probabilities of the two models are each 0.5. The Bayes factor comparing model m_B to model m_A is

$$B = m_B(x)/m_A(x).$$

When family A is subsetted in family B, with appropriate choice of m_B, we can write $m_B(x) = pm_A(x) + (1 - p)m_C(x)$ for $p \in (0, 1)$. This results in the expression

$$\begin{aligned}
B &= m_B(x)/m_A(x) \\
&= [(1 - p)m_C(x) + pm_A(x)]/m_A(x) \\
&= p[(1 - p)m_C(x) + pm_A(x)]/[pm_A(x)] \\
&= P_{m_B}(m_A)/P_{m_B}(m_A|x).
\end{aligned}$$

In the last line, $P_{m_B}(m_A)$ represents the prior probability of m_A in the mixture over m_A and m_B, and $P_{m_B}(m_A|x)$ represents its posterior probability.

The above decomposition is determined by p, a parameter that can assume a range of values. One might wonder whether the Bayes factor computed from a given set of data depends on the particular decomposition, indexed by p. This does not happen, for the Bayes factor is always $B = m_B(x)/m_A(x)$, no matter which decomposition is used. With the expression for the Bayes factor established, one can use the expression for computational purposes. The advantage of the formula is that one can perform computations only under model m_B and yet make inference for the comparison of m_B to m_A. The flexibility of the decomposition of m_B into m_A and m_C provides scope for the creation of effective computation of the Bayes factor.

8.3.4 IRT Models

The IRT models of Section 8.2 provide an interesting case study for contrasting the various versions of nesting. For the subsequent discussion, the ability distribution is taken to be standard normal for all models, and so the focus for nesting is on the item characteristic curves.

The support of the item characteristic curves under the 2PL model is the set of logistic distribution functions; for the 3PL model it is the set of logistic distribution functions, rescaled to run from c to 1, with $c \in [0, 1]$. We use the support of the prior distributions to determine the relevant parameter space for the models when studying classical nesting. For these models, family A, the parameter space Θ_A consists of the collection of logistic distributions or rescaled logistic distributions.

The nonparametric Bayesian models rely on the Dirichlet process. The support of the Dirichlet process for the analog of the 2PL model is the collection of distribution functions on the real line (Ferguson, 1973). The support of the NP3 model expands to allow mass at the point $-\infty$, with the item characteristic curve running in a nondecreasing fashion from c to 1, with $c \in [0, 1]$. Thus, we take Θ_B to be this collection of functions and note that $\Theta_A \in \Theta_B$. From a classical perspective, 3PL $\in$ NP3.

The issue of nesting is more difficult from a Bayesian perspective. If the forms of the prior distributions are allowed to vary arbitrarily over Θ_A and Θ_B, then classical nesting (Definition 8.3.1) implies Bayesian nesting (Definition 8.3.2). If, however, the forms of the prior distributions are restricted to those described in Section 8.2, an extended proof demonstrates that the models are not Bayesianly nested (although for small problems they are n-nested—for example, if the exam consists of a single question and an arbitrary number of examinees, the models imply an identical distribution on the binary response vector).

The NP3M does have the ability to mimic the 3PLM very closely, in the sense of Townsend and Ashby (1983). Standard results on the Dirichlet process guarantee that, for any fixed data set, matching all common parameters in the models, and taking the limit as the mass parameters M_j tend to ∞, the distribution of the observable data under NP3 tends to that under 3PL. Consequently, by considering large enough M_j, the difference between these two distributions can be made arbitrarily small.

The 3PLM is also subsetted in the NP3M. One particular representation of the Dirichlet process, known as the Pólya urn scheme (Blackwell & MacQueen, 1973), describes a mechanism by which the latent z_{ij} in the description of NP3M arise. This description states that, with a positive probability, the z_{ij} are independent draws from the F_{0j}. Conditioning on this event, the NP3M and 3PLM are identical, demonstrating the subsetting.

Simulation studies show the ability of nonparametric versions of the 2PLM and 3PLM to provide sound estimates of both item characteristic curves and abilities when the true model generating the data is parametric in form. The success of the models in recovering near-parametric forms from the data appears to be related to subsetting, as the analyses are successful, even for small values of the M_j. See Duncan (2004) for a selection of simulation results.

8.4 STAT 133 DATA STUDY

Our data set consists of results from the final exam for STAT 133, a business statistics class at The Ohio State University. The exam has 28 multiple-choice questions, each with four response options, and we use binary scores in our analysis; 1 for a correct answer and 0 for an incorrect answer. There were 258 students who took this exam with raw scores ranging from 6 to 28. The item scores, or number of individuals answering the item correctly range from 85 to 255.

The BRugs package in R (Spiegelhalter, Thomas, Best, & Gilks, 1995; R Development Core Team, 2006) was used to fit the 2PLM and 3PLM, with $\mu_\theta = 0, \sigma_\theta^2 = 1$, $\mu_b = -.57, \sigma_b^2 = 9, \nu_a = 3.2, \omega_a = 1, \alpha_c = 1.5, \beta_c = 4.5$. The choice of μ_β is based on our prior knowledge that final exam scores in STAT 133 tend to be higher than 50%, and thus the average difficulty of items should be negative. Initial values for θ were based on the normal quantile of the raw score, and initial values for a, b, and c were generated from the priors. A burn in period of 5,000 updates was used before 20,000 updates were collected with a thinning interval of 10.

A custom R package was developed to fit the nonparametric models. We used many of the same prior distributions as for the parametric models and again used the initial values of θ based on the raw score. Initial values for a, b, and c were generated from the priors while initial values for Z were generated according to a Pólya urn scheme. The other additional prior parameter is M, the prior mass for the Dirichlet process. Here we chose to use $M = 5$ based on Qin's (1998) exploration. We also used a burn-in period of 5,000 updates and stored 20,000 iterations with a thinning interval of 10.

To obtain the estimated curves with the parametric models, we used the posterior means for a_j, b_j, and in the case of the 3PL model c_j to plot logistic curves. The estimated curves with the NP models were obtained using the posterior expectation of F_j at a sequence of points from -4 to 4. The posterior probability bands were obtained by taking percentiles of the expected curve height. Abilities were estimated under each model with the posterior mean of θ.

For many items, the estimated curves were similar under the different models, but for a few items large differences were evident. Figure 8.1 gives the plots of both the parametric and nonparametric curves for items 12 and 23, two curves where the parametric and nonparametric results gave different results. In Item 12, we see that the 2PLM and NP2M curves differ substantially, but when the guessing parameter is included in the model, the 3PLM and NP3M curves are quite similar. This indicates that the 2PLM does not adequately describe the item functioning for item 12. The more flexible NP2M captures the lower threshold of the guessing parameter in ability ranges near -1 and 0, but falls back toward the 2PLM prior in lower regions of θ where there is little observed data. In Item 23, we again see the NP2M flattening out in the lower to middle ability region where a guessing parameter would insert a lower threshold. However, for Item 23, differences between the parametric and nonparametric models persist even after the guessing parameter is included. To assess the size of this difference in estimated curves relative to the variability of the estimates, a posterior probability band is included. We see that the NP3M reaches the edge of the 95% posterior probability band for the 3PLM, indicating that there is true disagreement in the models.

Figure 8.2 gives scatterplots of posterior mean abilities under the four models. The correlations are all greater than .988, but some students see substantial differences in estimated ability. For instance, student 217 has an estimated ability of 1.245 in the 3PLM but this drops to 0.759 in the NP3M. Figure 8.3 gives plots of individual ability posterior densities under the 3PLM and NP3M. We see that the nonparametric model usually shows comparable or smaller standard deviation. For student 231, who had a raw score of 20 out of 28 and an ordinary response pattern, the nonnormality of the posterior density resulting from the nonparametric model is notable.

Further examination of Figure 8.2 reveals a more subtle pattern. The dispersion about the line of equal abilities varies differentially with ability. NP2M and NP3M show the greatest differences with their parametric counterparts at the high ability levels, and we attribute this to these models' greater facility at tracking the data. In contrast, the NP3M versus NP2M comparison shows negligible differences in fits at high ability levels and more sizeable differences in fits at lower ability levels. The hook in the lower left corner of this plot (and, to a lesser extent, the 3PLM versus 2PLM plot) shows a reduction in ability estimate when the threshold parameter is added, allowing

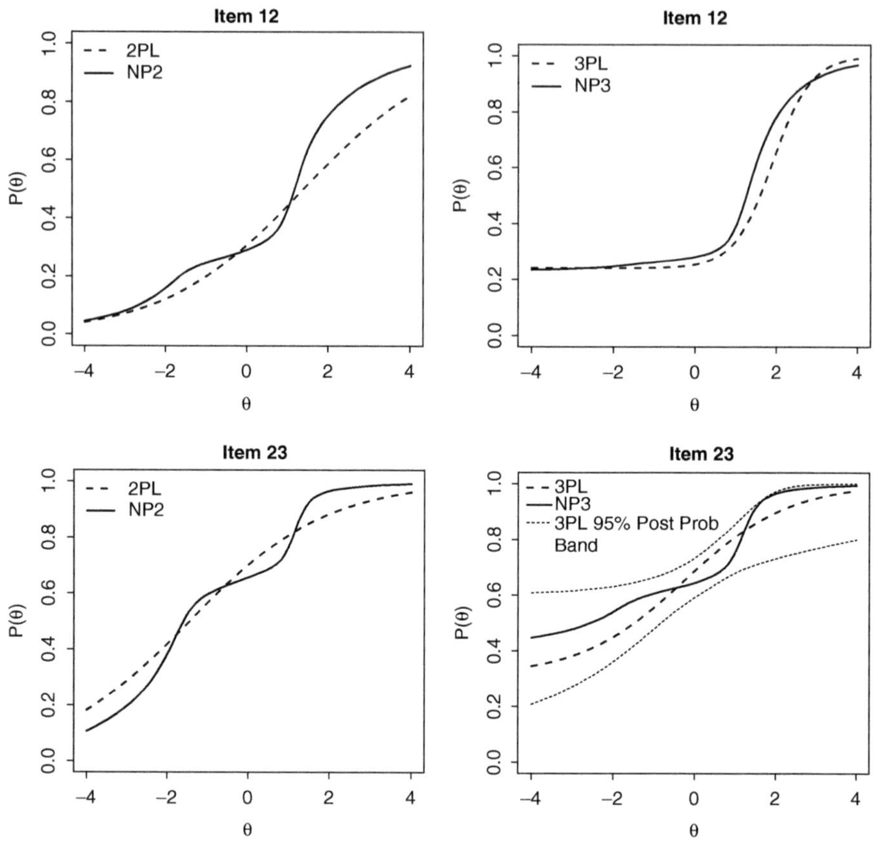

Figure 8.1 Comparison of item characteristic curves for Item 12 (row 1) and Item 23 (row 2). Two-parameter models are compared in column 1 and three-parameter models in column 2. Note that no single transformation of the abilities can reconcile the fits of the 3PLM and NP3M.

for correct responses to be attributed to a source other than ability-based solving of the problem. An examination of the questions driving this difference suggests that some of them may have been memorized as facts. This suggests a breakdown of the unidimensional ability model for this range of abilities.

Figure 8.4 shows a smoothed plot of posterior standard deviation as a function of estimated ability for the 3PLM and NP3M. The precision of ability estimates is related to the information in the items, with more information in regions where the curves are steeper or discriminate more strongly. The "bumpiness" permitted by the nonparametric model yields many curves that are steeper (higher discrimination) at

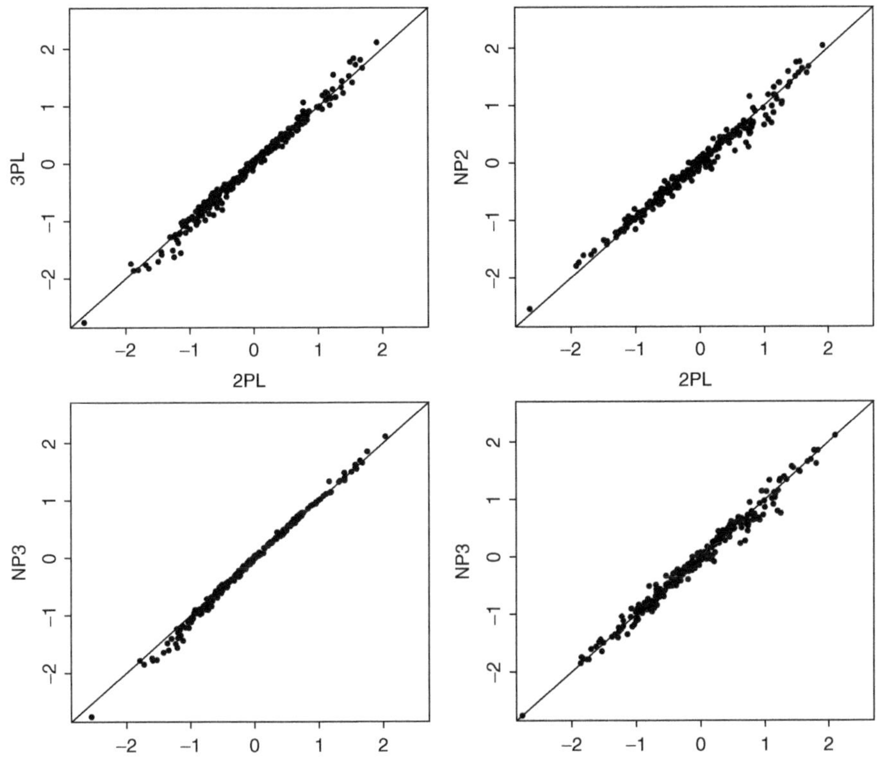

Figure 8.2 Comparison of posterior mean abilities. The hook in the lower corner of the plots in the left column hints at a breakdown of unidimensional ability.

higher ability levels and flatter in the low to mid ability range. This leads to improved accuracy of ability estimates for the upper half of students.

In summary, our inferences about abilities and item functioning are not robust to the choice of model in this example. The STAT 133 dataset illustrates that the 2PLM was insufficient for describing the item functioning for some of the items on the test. The nonparametric Bayesian model with the two-parameter logistic curve as a prior mean for the ICC improved upon the parametric model, particularly by flattening the curve in the lower middle values, where the guessing threshold is missing from the prior and there are enough students with abilities in that range to cause the curve to take that shape. There were fewer differences between the parametric

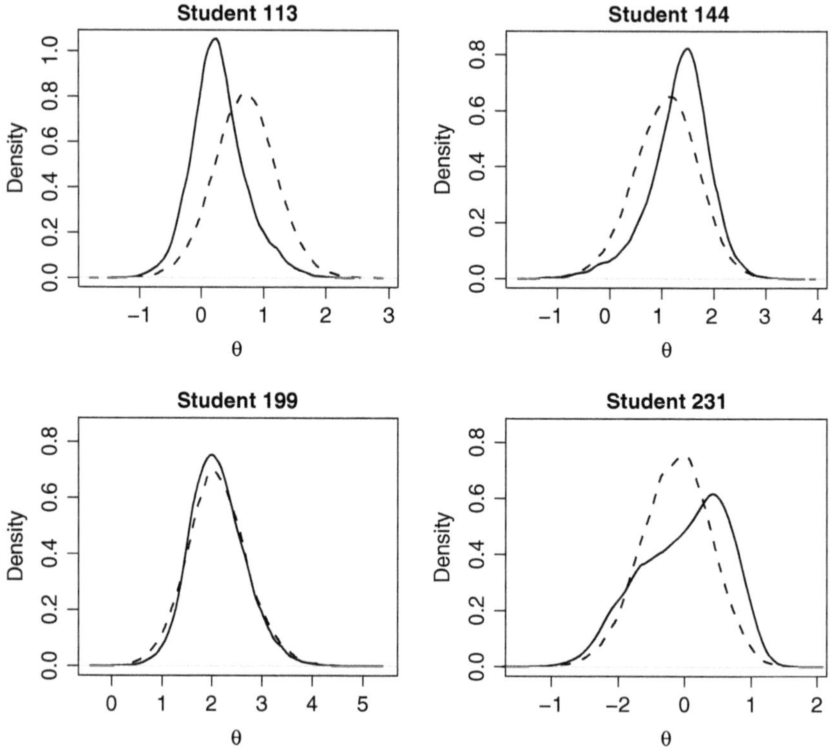

Figure 8.3 Posterior density of ability for selected students. The solid line is NP3M, the dashed line is 3PLM. Many patterns are observed, ranging from little difference between the models (student 199), to sharper inference under the NP3M (students 113, 144), to marked nonnormality of the posterior distribution (student 231).

and nonparametric models based on the three-parameter logistic curve, but some persisted.

While item functioning is of interest to test developers, abilities are of primary interest to end-users of test results because they are used to rank students and assign grades. When considering abilities individually, the differences in point estimates were for the most part small across models, as evidenced by most points falling close to the line on the scatterplots. However, the models did lead to differences in ranks of abilities, statistics that are the basis for grade assignments. The smaller posterior variance of the nonparametric models for individuals in the upper middle ability levels then leads to more accurate grade assignments.

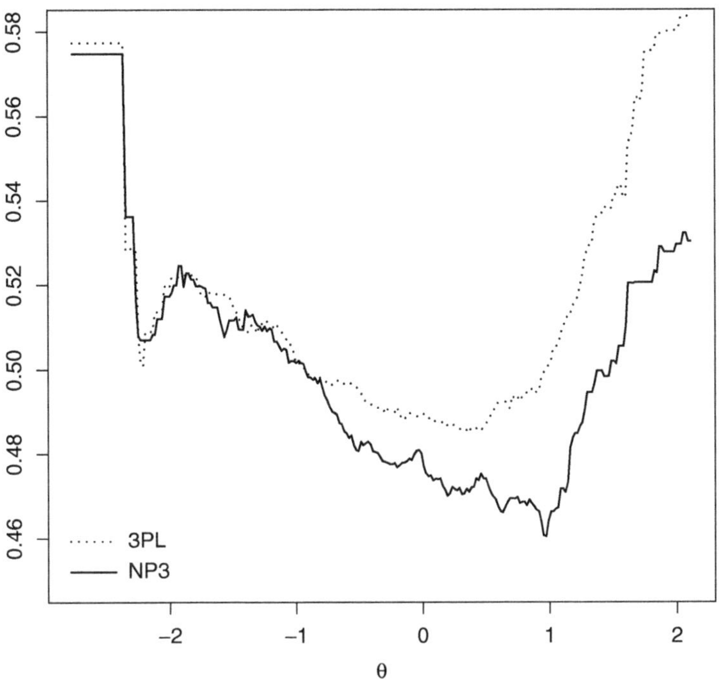

Figure 8.4 Posterior standard deviation as a function of ability. The two models are roughly equivalent on this measure for low ability levels; for high ability levels, NP3M leads to smaller posterior standard deviations.

In this work we coded responses as correct or incorrect and fit the data with the four models described. We acknowledge that there are several other models in the literature for multiple choice responses, such as the partial credit model and Bock's nominal model, which could be fit to this data as well.

8.5 CONCLUSIONS

Parametric logistic ICC curves or item response functions are widely used in both educational assessment and measurement of health outcomes, as evidenced by the NIH Patient Reported Outcomes Measurement Information System (PROMIS). The parametric curves often do a good job of describing item functioning for "good" items, those that satisfy the model assumptions. These parametric curves may be appropriate for high quality data for which the model assumptions are nearly satisfied, and for which IRT was developed. However, as the scope of problems to which IRT is applied expands, there is a growing need to fit flexible models to data sets for which the traditional assumptions are untenable. We believe that nonparametric Bayesian techniques provide a promising means of exploring these difficult problems.

Items for which the unidimensionality assumption does not hold are prone to showing non-logistic ICC curves when only one trait is being considered. We propose that the nonparametric models are good to use for all items, but for the practitioner who prefers a parametric model, the nonparametric models can be a useful tool to detect misfit of the parametric model and perhaps discard problem items. Many measures of model fit have been proposed in the IRT literature, though none has been widely adopted. The joint estimation of abilities and item characteristics, along with the typically large numbers of response patterns make the task of model assessment complicated. Cross-validation (Browne, 2000) is a technique often overlooked due to computation time. However, it is a method that makes good sense to use in the IRT setting for both assessment and comparison (Duncan, 2004).

The STAT 133 data set examined here shows some evidence of mixed dimensionality. While a one-dimensional trait, coupled with a monotone nonparametric ICC may capture high ability performance, low ability students appear to require more than one dimension. Nonparametric Bayesian models allow one to tailor the ability distribution to exhibit differing dimensionality as ability varies.

The concept of nested models is fundamental to much of statistical practice. The definitions we have proposed suggest that the concept, viewed sharply, is Bayesian in nature, rather than classical. The definitions also highlight the role of the prior distribution and the space over which it is allowed to range. The definitions also point in the direction of acknowledging that models may be nearly, but not quite right.

REFERENCES

Albert, J.H. (1992). Bayesian estimation of normal ogive item response curves using Gibbs sampling. *Journal of Educational Statistics, 17*, 251–269.

Baker, F.B., & Kim, S.H. (2004). *Item response theory: Parameter estimation techniques,* 2nd ed. New York: Marcel Dekker.

Blackwell, D., & MacQueen, J.B. (1973). Ferguson distributions via Pólya urn schemes. *The Annals of Statistics, 1*, 353–355.

Browne, M.W. (2000). Cross-validation methods. *Journal of Mathematical Psychology, 44*, 108–132.

Duncan, K.A. (2004). *Case and covariate influence: Implications for model assessment.* PhD Dissertation, The Ohio State University.

Duncan, K.A., & MacEachern, S.N. (2008). Nonparametric Bayesian modeling for item response. *Statistical Modelling, 8*, 41–66.

Ferguson, T.S. (1973). A Bayesian analysis of some nonparametric problems. *The Annals of Statistics, 1*, 209–230.

Finney, D.J. (1964). *Statistical method in biological assay*, second ed. New York: Hafner.

Liang, L. (2007). *A semi-parametric approach to estimating item response functions.* PhD Dissertation, The Ohio State University.

Muthén, B. (1984). A general structural equation model with dichotomous, ordered categorical, and continuous latent variable indicators. *Psychometrika,* 49, 115–132.

Patz, R.J., & Junker, B.W. (1999). A straightforward approach to Markov chain Monte Carlo methods for item response models. *Journal of Educational and Behavioral Statistics, 24*, 146–178.

Qin, L. (1998). *Nonparametric bayesian models for item response data.* PhD Dissertation, The Ohio State University.

R Development Core Team (2006). *R: A language and environment for statistical computing.*

Ramsay, J.O. (1991). Kernel smoothing approaches to nonparametric item characteristic curve estimation. *Psychometrika, 56*, 611–630.

Sinnott, L.T. (1997). *Filtered polynomial density approximations and their application to discriminant analysis.* Masters Thesis, The Ohio State University.

Spiegelhalter, D.J., Thomas, A., Best, N., & Gilks, W.R. (1995). *Bugs examples,* volume 1. Technical Report.

Thomas, A. (2006). *BRugs: OpenBUGS and its R interface BRugs. R package version 0.3-2.*

Townsend, J.T., & Ashby, F.G. (1983). *Stochastic modeling of elementary psychological processes.* Cambridge: Cambridge University Press.

Chapter 9

Exact Solutions for IRT Latent Regression Slopes and Latent Variable Intercorrelations

W. Alan Nicewander

9.1 INTRODUCTION

Using the factor analytic version of item response theory (IRT) models, estimates of the correlations between the latent variables measured by test items and correlations between observed covariates and IRT latent variables are derived. The methods derived here can be applied to dichotomous items, polytomous items, or mixtures of both of these types of item. The estimates are simple in form and entail IRT item a-values (see below) and tetrachoric or biserial correlations—for dichotomous items—and IRT item a-values and polychoric or polyserial correlations for polytomous items or mixed-pairs of polytomous and dichotomous items. Applications are illustrated for:

1. Latent variable correlations underlying pairs of items or pairs of tests; and
2. Estimating slopes in latent variable regression problems.

The estimates are "exact" in the sense that they follow directly from the assumptions and involve no asymptotic or approximate statistical methods.

The normal ogive model for dichotomous items may be written as,

$$p(u_{il} = 1|\theta_l) = c_{il} + (1 - c_{il}) \int_{-\infty}^{-a_{il}(\theta_l - b_{il})} \varphi(t)\,dt = \Phi[-a_{il}(\theta_l - b_{il})], \tag{9.1}$$

where u_{il} is the ith dichotomous, item score, and θ_l is a latent proficiency variable; a_{il} is the item slope parameter, and b_{il} is the item location parameter; c_{il} is a guessing parameter; $\varphi(t)$ is the normal density function; and Φ is the normal cumulative distribution function (CDF).

Another useful version of this model is the so-called factor analytic representation: Let Z_{il} be a latent response variable that is a linear function of θ_l plus error,

$$Z_{il} = \lambda_{il}\theta_l + \epsilon_{il}, \tag{9.2}$$

where λ_{il} is a factor loading and ϵ_{il} is an error variable (see Lord, 1981). It is further assumed that the distribution of θ_l is normal with zero mean and unit variance, and that the conditional distributions of Z_{il} are normal with means $\lambda_{il}\theta_l$ and standard deviations, $\sqrt{1 - \lambda_{il}^2}$. Finally, it is assumed that ϵ_{il} is uncorrelated with θ_l and Y_{il}. Let γ_{il} be a threshold defined so that an item is answered correctly if Z_{il} exceeds γ_{il}; otherwise, it is incorrect. Now define a_{il} and b_{il} in (9.1) as,

$$a_{il} = \frac{\lambda_{il}}{\sqrt{1 - \lambda_{il}^2}}; \quad b_{il} = \frac{\gamma_{il}}{\lambda_{il}}. \tag{9.3}$$

Then (9.1) may be rewritten as,

$$p(u_{il} = 1|\theta_l) = c_{il} + (1 - c_{il})\Phi\left[\frac{\gamma_{il} - \lambda_{il}\theta_l}{\sqrt{1 - \lambda_{il}^2}}\right] = c_{il} + (1 - c_{il})\Phi[-a_{il}(\theta - b_{il})]. \tag{9.4}$$

9.2 COMPUTING CORRELATIONS AMONG THE LATENT VARIABLES MEASURED BY DICHOTOMOUS ITEMS

Within the context of latent variable regression, the correlation matrix for a vector of IRT latent variables is necessary for interpreting solutions. In what follows, solutions are developed for estimating these latent correlations using single items (dichotomous or polytomous) for measuring the latent variables—or using a mix of dichotomous and polytomous items assembled into (unidimensional) tests or subtests designed to measure the latent variables, θ_l and θ_m. Let Z_{il} and Z_{jm} be the latent response variables (see (9.2) above) that underlie two dichotomous items developed to measure θ_l and θ_m. The correlation between the variables Z_{il} and Z_{jm} is, given the assumptions listed above,

$$\rho(Z_{il}, Z_{jm}) = E(\lambda_{il}\theta_l + \epsilon_{il})(\lambda_{jm}\theta_m + \epsilon_{jm}) = \lambda_{il}\lambda_{jm}\rho(\theta_l, \theta_m) \tag{9.5}$$

or,

$$\rho(\theta_l, \theta_m) = \frac{\rho(Z_{il}, Z_{jm})}{\lambda_{il}\lambda_{jm}}. \tag{9.6}$$

We now have the desired correlation between the IRT latent variables expressed as a function of the item a-values (transformed to λs) and $\rho(Z_{il}, Z_{jm})$. At first glance, this does not seem to be a very useful expression for $\rho(\theta_l, \theta_m)$, because it appears to involve another latent correlation, $\rho(Z_{il}, Z_{jm})$. However, given that this correlation coefficient is the correlation *between the normally distributed continua underlying two dichotomous variables*, it is, by definition, the tetrachoric correlation between the observed items scores, u_{il} and u_{jm}—*corrected for guessing*; therefore, (9.6) may be rewritten as,

$$\rho(\theta_l, \theta_m) = \frac{\rho^*_{tet}(u_{il}, u_{jm})}{\lambda_{il}\lambda_{jm}} = \frac{\rho^*_{tet}(u_{il}, u_{jm})\sqrt{1 + a^2_{il}}\sqrt{1 + a^2_{im}}}{a_{il}a_{im}}, \tag{9.7}$$

where, $\rho^*_{tet}(u_{il}, u_{jm})$ is the tetrachoric correlation between two dichotomous items, corrected for guessing using the method due to Carroll (1945). Equation (9.7) has an interesting connection to classical test theory. Since the squares of λ_{il} and λ_{jm} can be interpreted as reliability coefficients for Z_{il} and Z_{jm}, (9.7) may be interpreted as the tetrachoric, item correlation corrected for guessing and other sources of unreliability. Equations (9.5) and (9.7) also suggest an upper-bound to the guessing-corrected tetrachoric correlation between dichotomous test items, $\rho^*_{tet}(u_i, u_j)$. Since the correlation, $\rho(\theta_l, \theta_m)$, is less than or equal to one, an upper-bound to $\rho^*_{tet}(u_{il}, u_{jm})$ is,

$$\rho^*_{tet}(u_{il}, u_{jm}) \leq \lambda_{il}\lambda_{jm} = \frac{a_{il}a_{jm}}{\sqrt{1 + a^2_{il}}\sqrt{1 + a^2_{jm}}}. \tag{9.8}$$

Equation (9.7) can be extended to cover the situation where one wants the correlation between the two latent variables presumed to underlie two *test* scores. Let $\sum_{i=1}^{n_i} u_{il}$ and $\sum_{j=1}^{n_j} u_{jm}$ be the summed item scores for tests measuring two latent variables, θ_l and θ_m. It can be shown that (9.7) generalizes to yield,

$$\rho(\theta_l, \theta_m) = \frac{\sum_{i=1}^{n_i}\sum_{j=1}^{n_j}\rho^*_{tet}(u_{il}, u_{jm})}{\sum_{i=1}^{n_i}\sum_{j=1}^{n_j}\lambda_{il}\lambda_{jm}}, \tag{9.9}$$

where n_i and n_j are the number of dichotomous items in the tests measuring two latent variables, θ_l and θ_m respectively.

9.3 COMPUTING CORRELATIONS BETWEEN OBSERVED COVARIATES AND IRT LATENT VARIABLES UNDERLYING DICHOTOMOUS ITEMS

It is fairly easy to extend the logic above in order to derive a means for computing correlations between observed covariates and IRT latent variables. First define Y_q as a covariate scaled to have zero mean and unit variance. Then use (9.2) in order to compute the correlation between Y_q and the latent response variable, Z_{il}, assumed to be underlying a dichotomous item, u_{il}:

$$\rho\left(Y_q, Z_{il}\right) = E[Y_q\left(\lambda_{il}\theta + \epsilon_{il}\right)] = \lambda_{il}\rho\left(Y_q, \theta_l\right). \tag{9.10}$$

Solving (9.10) for $\rho\left(Y_q, \theta_l\right)$ gives,

$$\rho\left(Y_q, \theta_l\right) = \frac{\rho\left(Y_q, Z_{il}\right)}{\lambda_{il}}. \tag{9.11}$$

Since $\rho\left(Y_q, Z_{il}\right)$ is the correlation between the covariate, Y_q, and Z_{il}, the normally distributed continuum underlying the dichotomous variable, u_{il}, it, by definition, is the biserial correlation between Y_q and Z_{il} (corrected for guessing). Equation (9.11) may now be rewritten as,

$$\rho\left(Y_q, \theta_l\right) = \frac{\rho^*_{\text{bis}}\left(Y_q, u_{il}\right)}{\lambda_{il}}, \tag{9.12}$$

where $\rho^*_{\text{bis}}\left(Y_q, u_{il}\right)$ is the biserial correlation corrected for guessing:

$$\rho^*_{bis}\left(Y_q, u_{il}\right) = \frac{\rho^*_{pb}\sqrt{\pi_{il}(1-\pi_{il})}}{\varphi[\Phi^{-1}\left(\pi_{il}\right)]}, \tag{9.13}$$

where $\pi_{il} = (p_{il} - c_{il})/(1 - c_{il})$ is the item p-value corrected for guessing, ρ^*_{pb} is the guessing-corrected point-biserial correlation between Y_q and u_{il}, and is equal to $\rho_{pb}\sqrt{p_{il}/(p_{il} - c_{il})}$. The quantity $\varphi[\Phi^{-1}\left(\pi_{il}\right)]$ is the height of the ordinate of the normal distribution at the probit of the corrected p-value (see Lord & Novick, 1968, Ch. 15). Let $X_l = \sum_{i=1}^{n} u_{il}$ be a summed score for a unidimensional test composed of dichotomous items; then the correlation of the external variable, Y_q, with the latent variable, θ_l, is given by,

$$\rho\left(Y_q, \theta_l\right) = \frac{\sum_{i=1}^{n}\rho^*_{bis}\left(Y_q, u_{il}\right)}{\sum_{i=1}^{n}\lambda_{il}}. \tag{9.14}$$

So far, it has been assumed that Y_q is a (quasi) continuous variable. If Y_q is a dichotomous, covariate, then the biserial correlation in (9.12) would be replaced by the tetrachoric correlation between Y_q and u_{il}; if it is categorical, then the biserial correlation in (9.12) would be replaced by a polybiserial correlation (Drasgow, 1986).

(If any of the u_{il} are polytomous items, then the relevant correlation must be polyserial or polychoric depending on whether Y_q is continuous or dichotomous—more on this later). It should be noted that, to the best of the author's knowledge there is no published method (or software) for correcting a polychoric correlation for random guessing in the case where one member of an item-pair is dichotomous and multiple choice. The author has derived a guessing-correction for the polychoric correlation that is a generalization of Carroll's tetrachoric correction; however, the derivation is beyond the scope of this chapter.

9.4 EXTENDING THE LATENT-VARIABLE CORRELATIONS TO POLYTOMOUS TEST ITEMS

In order to simplify exposition, only polytomous items having three categories are considered here. Generalization of the methods described below to items with more than three categories is very straightforward, but the notation gets a bit tedious. Let u_{il} be the score for item i, measuring θ_l, and scored in category k ($k = 1, 2, \ldots, m$). Under commonly used scoring rules, a three-category item would be scored 0, 1, or 2. In a way similar to the case of dichotomous items having binary scores, let Z_{il} be the latent response tendency underlying the polytomous item, u_{il}. Then,

$$Z_{il} = \lambda_{il}\theta_l + \epsilon_{il}$$

where λ_{il} is a factor loading. Let γ_{i1l} and γ_{i2l} be two response thresholds defined so that:

$$p(u_{il} = 0|\theta_l) = p(Z_{il} < \gamma_{i1l}|\theta_l), \tag{9.15a}$$

$$p(u_{il} = 1|\theta_l) = p(\gamma_{i1l} \leq Z_{il} < \gamma_{i2l}|\theta_l), \tag{9.15b}$$

$$p(u_{il} = 2|\theta_l) = p(Z_{il} \geq \gamma_{i2l}|\theta_l). \tag{9.15c}$$

This model is Samejima's (1969) Graded Response Model, and is nearly identical to the one for dichotomous items with the exception that it has more than one threshold. In general, this model for polytomous items will have $m - 1$ thresholds. Now define,

$$a_{il} = \frac{\lambda_{il}}{\sqrt{1 - \lambda_{il}^2}}, \text{ and } b_{ikl} = \frac{\gamma_{ikl}}{\lambda_{il}} (k = 1, 2). \tag{9.16}$$

The thresholds are then rescaled into standardized, conditional thresholds for fixed values of theta and converted to IRT notation:

$$L_{i1l} = \frac{\gamma_{i1l} - \lambda_{il}\theta_l}{\sqrt{1 - \lambda_{il}^2}} = a_{il}(\theta_l - b_{i1l}) \tag{9.17a}$$

$$L_{i2l} = \frac{\gamma_{i2l} - \lambda_{il}\theta_l}{\sqrt{1 - \lambda_{il}^2}} = a_{il}(\theta_l - b_{i2l}). \tag{9.17b}$$

The category probabilities given in Equations (9.15a)–(9.15c) may be computed using a normal ogive model and approximated with logistic functions:

$$p\left(u_{ikl}=0|\theta_l\right)=\Phi\left(L_{i1l}\right)=\int_{-\infty}^{L_{i1l}}\varphi\left(t\right)dt\cong 1/([1+\exp(-1.7L_{i1l})]; \tag{9.18a}$$

$$p\left(u_{ikl}=1|\theta\right)=\Phi\left(L_{i2l}\right)-\Phi\left(L_{i1l}\right)=$$

$$\int_{L_{i1l}}^{L_{i2l}}\varphi\left(t\right)dt\cong\frac{\exp\left(-1.7L_{i2l}\right)-\exp(-1.7L_{i1l})}{[(1+\exp\left(-1.7L_{i1l}\right))][(1+\exp\left(-1.7L_{i2l}\right)]} \tag{9.18b}$$

$$p\left(u_{ikl}=2|\theta_l\right)=\Phi\left(-L_{i2l}\right)=\int_{L_{i2l}}^{\infty}\varphi\left(t\right)dt\cong\frac{1}{1+\exp(1.7L_{i2l})}. \tag{9.18c}$$

9.5 COMPUTING THE CORRELATION BETWEEN THE LATENT VARIABLES UNDERLYING TWO POLYTOMOUS ITEMS—OR A DICHOTOMOUS–POLYTOMOUS ITEM-PAIR

First, define the correlation between the two latent response processes that underlie two polytomous items, u_{il} and u_{jm}, (designed to measure θ_l and θ_m) by generalizing (9.5),

$$\rho\left(Z_{il},Z_{jm}\right)=E[(\lambda_{il}\theta_l+\epsilon_{il})(\lambda_{jm}\theta_m+\epsilon_{ij})]=\lambda_{il}\lambda_{jm}\rho(\theta_l\theta_m). \tag{9.19}$$

Then, solve for $\rho\left(\theta_l,\theta_m\right)$,

$$\rho\left(\theta_l,\theta_m\right)=\frac{\rho\left(Z_{il},Z_{jm}\right)}{\lambda_{il}\lambda_{jm}}=\frac{\rho_{polyc}\left(u_{il},u_{jm}\right)}{\lambda_{il}\lambda_{jm}}, \tag{9.20}$$

where $\rho_{polyc}\left(u_{ikl},u_{jkm}\right)$ is the polychoric correlation (Olsson, 1979) for two polytomous items. Given two tests composed of n_i and n_j polytomous items designed to measure θ_l and θ_m, (9.19) can be generalized as,

$$\rho\left(\theta_l,\theta_m\right)=\frac{\sum_{i=1}^{n_i}\sum_{j=1}^{n_j}\rho_{polyc}(u_{il},u_{jm})}{\sum_{i=1}^{n_i}\sum_{j=1}^{n_j}\lambda_{il}\lambda_{jm}}, \tag{9.21}$$

For the case of dichotomous–polytomous item-pairs, (9.20) and (9.21) can be modified by substituting the guessing-corrected polychoric correlation $\rho^*_{polyc}(u_{il},u_{jm})$ for the uncorrected coefficient, $\rho_{polyc}(u_{il},u_{jm})$—since, if one of the two items is dichotomous, almost certainly guessing is possible for that item.

9.6 COMPUTING THE CORRELATION BETWEEN AN OBSERVED COVARIATE AND THE LATENT VARIABLE MEASURED BY A POLYTOMOUS ITEM

From the earlier derivation resulting in (9.20), it is fairly obvious that the correlation between an observed covariate, Y_q, and the latent variable measured by a polytomous item, u_{il}, is,

$$\rho\left(Y_q, \theta_l\right) = \frac{\rho\left(Y_q, Z_{il}\right)}{\lambda_{il}} = \frac{\rho_{polys}\left(Y_k, u_{il}\right)}{\lambda_{il}}, \tag{9.22}$$

where $\rho_{polys}\left(Y_q, u_{il}\right)$ is the polyserial correlation (Drasgow, 1986) between the external variable and the score on the polytomous item. An upper bound for $\rho_{polys}\left(Y_q, u_{il}\right)$ is,

$$\rho_{polys}\left(Y_q, u_{il}\right) \leq \lambda_{il} = \frac{a_{il}}{\sqrt{1 + a_{il}^2}}. \tag{9.23}$$

Finally, in most applications of regressing IRT latent variables on covariates, the latent variable would be measured by a test composed of a number of items, u_{il}. Let $X_l = \sum_{i=1}^{n} u_{il}$ be a summed, unidimensional test score; then the correlation of the covariate, Y_q, with the latent variable, θ_l, is given by,

$$\rho\left(Y_q, \theta_l\right) = \frac{\sum_{i=1}^{n} \rho\left(Y_q, Z_{il}\right)}{\sum_{i=1}^{n} \lambda_{il}} = \frac{\sum_{i=1}^{n} \rho^{\dagger}\left(Y_q, u_{il}\right)}{\sum_{i=1}^{n} \lambda_{il}}, \tag{9.24}$$

where—depending on whether the items are dichotomous or polytomous, and whether Y_q, is continuous, categorical or binary—$\rho^{\dagger}(Y_q, u_{il})$ is the appropriate, correlation coefficient, biserial, polychoric or polyserial. Equation (9.24) provides a summary of the methods for obtaining estimates of the correlations between covariates, Y_q, and latent variables, θ_l, using a mixture of dichotomous and polytomous items. An equivalent, summary equation for estimating the correlations of latent variables using two, summed test scores (with one or both containing a mixture of item types) is,

$$\rho\left(\theta_l, \theta_m\right) = \frac{\sum_{i=1}^{n_i} \sum_{j=1}^{n_j} \rho^{\dagger}\left(u_{il}, u_{jm}\right)}{\sum_{i=1}^{n_i} \sum_{j=1}^{n_j} \lambda_{il}\lambda_{jm}}, \tag{9.25}$$

where, $\rho^{\dagger}\left(u_{il}, u_{jm}\right)$ is the appropriate correlation coefficient—tetrachoric or polychoric—depending on whether u_{il} and u_{jm} are both dichotomous, both polytomous or a mix of the two item types. Equations (9.24) and (9.25) summarize most of the procedures developed in this chapter.

9.7 ESTIMATING LATENT REGRESSION SLOPES AND THE INTERCORRELATIONS FOR A SET OF LATENT VARIABLES

Consider a multivariate multiple regression that involves regressing a p-dimensional vector of IRT latent variables, $\boldsymbol{\theta}$, on an n-dimensional vector of observed covariates, $\mathbf{y}$. The problem is to find the slopes of the latent variables—represented by 'proxy' items that are related in probability to the latent variables—on the observed Ys. The multivariate, multiple regression model may be expressed as,

$$\theta = \boldsymbol{\Gamma}\mathbf{y} + \epsilon, \tag{9.26}$$

where $\boldsymbol{\Gamma}$ is a $(p \times n)$ matrix containing the regression slopes, and ϵ is the $(p \times 1)$ vector of the residuals. A Bayesian solution to this regression problem, including the estimation of the dispersion matrix, $\boldsymbol{\Sigma}_{\theta\theta}$, has been proposed by Mislevy (1985)—more on this later. However, the regression system in (9.26) could be solved for Γ using ordinary least squares (OLS) if the dispersion matrix $\boldsymbol{\Sigma}_{y\theta} = E\,(y\theta')$ were known. With a little reflection, Equation (9.24) could be used to compute estimates of the elements of $\boldsymbol{\Sigma}_{\theta y}$ using a mixture of dichotomous and polytomous items. These in turn could be used to compute least squares estimates of the regression slopes,

$$\hat{\boldsymbol{\Gamma}} = \boldsymbol{\Sigma}_{yy}^{-1}\boldsymbol{\Sigma}_{y\theta} \tag{9.27}$$

and the latent correlation matrix, $\boldsymbol{\Sigma}_{\theta\theta}$, could be estimated using (9.25). Latent variable regressions have been used extensively in National Assessment of Educational Progress (NAEP) testing through the use of covariates, consisting of student background variables, in order to obtain ancillary estimates of the means and variances of the student posterior distribution of θ for some student, a; viz. $E(\theta_a) = \Gamma y_a$, and $\mathrm{var}(\theta) = \boldsymbol{\Sigma}_{\theta\theta}$, a covariance matrix assumed to be constant for all students (von Davier & Sinharay, 2010). Latent variable regression systems also may be useful as a means of obtaining initial estimates of θ in adaptive testing using student background covariates.

9.8 SOME DATA ILLUSTRATING THE USE OF THESE CORRELATIONS

9.8.1 The Correlations Underlying the Latent Variables Measured by the ACT Assessment

This example is from Hulick and Nicewander (1995). The ACT and the SAT are the principal college admissions tests in the U.S. The ACT is composed of four subtests:

English (75 items); Mathematics (60 items); Reading (40 items); and Science (40 items). All ACT items are multiple choice. The ACT test form used here is a form from 1994, and the data are from a random sample of 2,400 students who took the test that year. For each of the four subtests, the three-parameter logistic (3-PL) model was fitted to the items using the SSI© BILOG computer program. Equation (9.8), (with m set equal to l) was applied to all possible pairs of items within an ACT subtest. The correlations between the latent variables measured by items within an ACT subtest should be equal to unity if the items are truly unidimensional. Additionally, (9.25) was applied to all cross-pairs of items from each of six possible pairs of ACT subtests. If the subtests are measuring different latent variables, then one would expect the cross-correlations between the subtest latent variables to be less than unity but certainly not zero. All tetrachoric correlations used in (9.25) were guessing-corrected using Carroll's correction, and were computed using a method due to Digby (1983). Table 9.1 contains the average, estimated correlation between the latent variables measured by all item-pairs within the same ACT subtest (on the diagonal), and the cross-correlations between the latent variables.

The upper section of Table 9.1 provided evidence that the ACT subtests are unidimensional, in that the average, latent-variable correlations, computed from all pairs of items *within* each ACT subtest, were all slightly greater than one (putatively, due to lack of model fit and estimation error). Also, the latent traits for the four ACT subtests were distinct but rather substantially correlated. That the observed, subtest score correlations (corrected for unreliability) were essentially the same as the latent variable correlations, implies that the IRT latent variables function as true scores and are not the deeper common-factor latent variable(s) that are hinted at by the substantial subtest intercorrelations. The latent variable correlations in Table 9.1 are not particularly interesting theoretically, but do indicate that the methods derived here work properly when used with real data—and may be useful in applied situations.

9.8.2 A Miniature Example of Latent Variable Regression

Suppose one had two sets of four items each that measure two latent variables—θ_1 and θ_2—and that for each set of four items, three were dichotomous and one was polytomous (with three categories). Also, suppose that four covariates, Y_1, Y_2, Y_3, and Y_4 were available, and upon which, the latent variables were to be regressed (using four items as proxies for each θ). Item c-values, for the dichotomous items, were set equal to .20. The category boundaries for the polytomous items are labeled b_{i1} and b_{j2}. The items parameters for these two sets of items are listed in Table 9.2; note that both sets of items have the same a-values, but that the b-values and category boundaries indicate that the second item-set is the more difficult. A (6×6) population correlation

Table 9.1 Correlations among the ACT latent variables estimated from item tetrachoric correlations corrected for guessing and adjusted according to Equation (9.25). Tetrachoric correlations were computed for all item-pairs within-and-between the four ACT subtests

	Math	*Reading*	*Science*	*English*
Math	1.01			
Reading	.71	1.01		
Science	.86	.89	1.02	
English	.77	.93	.87	1.02
Diagonal elements above are the within-subtest latent variable correlations; off-diagonals are cross-correlations between the latent variables underlying the pairs of ACT subtests (See Equation (9.25)).				
	Math	*Reading*	*Science*	*English*
Math	1.00			
Reading	.70	1.00		
Science	.88	.91	1.00	
English	.75	.90	.79	1.00
Off-diagonal elements above are the correlations between the observed scores on the ACT subtests, corrected for unreliability.				
Reliability Coefficients	.89	.88	.79	.92

matrix, $\mathbf{\Sigma}$, for the two latent variables, θ_1 and θ_2, and the four covariates was specified, and is shown in Table 9.3.

A (2,000 × 6) matrix, $\boldsymbol{M}$, of independent, random normal deviates was generated and transformed to have the population correlation structure shown in Table 9.2 by using the transformation,

$$M^* = ML, \tag{9.28}$$

Table 9.2 Item parameters for two sets of four items each—three dichotomous and one 3-category polytomous—for measuring two IRT latent variables, θ_1 and θ_2

		Item parameters				
Items	*Latent* θ	*a*	*b*	*c*	b_{11}	b_{21}
u_{11}	θ_1	1	−.35	.20	–	–
u_{21}	θ_1	1.5	0	.20	–	–
u_{31}	θ_1	2	.90	.20	–	–
u_{41}[a]	θ_1	1	–	–	−.35	.35
Items	*Latent* θ	*a*	*b*	*c*	b_{12}	b_{22}
u_{12}	θ_2	1	0	.20	–	–
u_{22}	θ_2	1.5	.80	.20	–	–
u_{32}	θ_2	2	1.35	.20	–	–
u_{42}[a]	θ_2	1	–	–	0	.70

[a]Polytomous item.

Table 9.3 Population and estimated correlations among two latent variables, θ_1 and θ_2, and four covariates, Y_1, Y_2, Y_3, and Y_4[a]

	θ_1	θ_2	Y_1	Y_2	Y_3	Y_4
θ_1	1					
θ_2	**.50** (.491)	1				
Y_1	**.60** (.588)	**.20** (.200)	1			
Y_2	**.60** (.594)	**.20** (.199)	**.40** (.406)	1		
Y_3	**.20** (.209)	**.60** (.602)	**.24** (.227)	**.24** (.250)	1	
Y_4	**.20** (.208)	**.60** (.618)	**.24** (.241)	**.24** (.234)	**.40** (.426)	1

[a]Based on 2,000 simulated observations—bold values are the specified population correlations and values in parenthesis are estimates computed using Equations (9.24) and (9.25)—except for the covariate correlations which are ordinary Pearsonian coefficients. The shaded area delineates the true-and-estimated values of the matrix, $\Sigma_{y\theta}$.

where, $\boldsymbol{L}$ is a Cholesky factor of Σ (or any other Gram factor of Σ) such that $\boldsymbol{\Sigma} = \boldsymbol{LL}^T$ (Anderson, 1984). The first two columns of $\boldsymbol{M}^*$ were taken to be 2,000 simulated observations on θ_1 and θ_2, and the remaining four columns of $\boldsymbol{M}^*$ were the 2,000 simulated observations on the four covariates, Y_1, Y_2, Y_3, and Y_4. The values of θ_1 and θ_2 were used, in conjunction with the two sets of items parameters in Table 9.2, to generate the 2,000 responses to the eight items. Finally, the requisite correlations between the covariates and θ_1 and θ_2 were estimated using (9.24) and (9.25), for which the tetrachoric, biserial, polychoric, and polyserial correlations were computed in the manner described below. For the cross-pairs of dichotomous items u_1, u_2, and u_3 (from the set measuring $\theta_{1)}$, and u_5, u_6, and u_7 (from the set measuring θ_2), the nine guessing-corrected, tetrachoric correlations were computed as follows:

1. SAS's Proc Freq was used to obtain the 2 × 2 tables for item-pairs;
2. These tables were corrected for guessing using Carroll's method; and
3. Using SAS's PROBBNRM function, the bivariate normal distribution was integrated while looping on values of r, the bivariate normal correlation parameter, until the (0,0) orthant was equal to the empirical, corrected value of the (0,0) cell of the 2 × 2 tables.

The values of r leading to the matches between the (0,0) orthants and (0,0) cells were taken as the values of the tetrachoric correlations. The cross-correlations between the polytomous items in each set and the three dichotomous items in the other, were guessing-corrected polychoric correlations computed as follows:

1. SAS's Proc Freq was used to obtain the 2 × 3 tables for the six, dichotomous–polytomous item-pairs;
2. These tables were corrected for guessing using a method derived by the author; and
3. The bivariate normal distribution was integrated while looping on values of r, the bivariate normal correlation parameter, until the (0,0) orthant was equal to the empirical, corrected value of the (0,0) cell of the 2 × 3 tables.

The values of r leading to the matches between the (0,0) orthants and cells were taken as the values of the polychoric correlations. The guessing-corrected biserial correlation between the six dichotomous items and the covariates, Y_1, Y_2, Y_3, and Y_4 were computed using (9.13). Finally, the correlations between the two polytomous items and the four covariates were polyserial correlations that, by definition, were not affected by guessing, and were computed by integrating the bivariate normal distribution for various values of r until the (0, 4) orthants—as approximations to the (0, ∞) orthants—were equal to the frequency of the 0-category scores for the polytomous items.

Table 9.4 True and estimated vectors of regression slopes of θ_1 and θ_2 on four covariates, Y_1, Y_2, Y_3, and Y_4 along with the true and estimated R^2s[a]

		$\Gamma_{\theta_{1},y}$	$\Gamma_{\theta_{2},y}$
Y_1	True	**.430**	**–.005**
	Estimated	.413	.003
Y_2	True	**.430**	**–.005**
	Estimated	.423	–.009
Y_3	True	**–.005**	**.430**
	Estimated	.007	.415
Y_4	True	**–.005**	**.430**
	Estimated	.007	.443
R^2	True	**.514**	**.514**
	Estimated	.500	.522

[a]Each of the two latent variables had 'proxies' of four items each—three dichotomous and one 3-category polytomous (see text). Based on 2,000 simulated observations.

Correlations among the covariates were ordinary Pearsonian coefficients. The results of the latent regression are summarized in Table 9.3 and Table 9.4. Table 9.3 lists in bold type:

1. The true values of the correlation between the two latent variables;
2. The true correlations between covariates and latent variables; and
3. The true covariate intercorrelations—along with the estimates of these correlations (in parentheses).

Note that the estimate of $\rho(\theta_1, \theta_2)$ was .491 which is very close to the true value of .50. The shaded area of Table 9.3 contains $\mathbf{\Sigma}_{y\theta}$ and its estimate, $\hat{\mathbf{\Sigma}}_{y\theta}$; this matrix and its estimate were also in close agreement. The true values and estimates of the lower-triangular part of $\mathbf{\Sigma}_{yy}$ are adjacent to the shaded portion of Table 9.3. Again, there was close agreement between true values of the covariate correlations and their estimates—this is not surprising since these estimates do not involve latent variables. Table 9.4 contains the true regression slopes for the two latent θs on the four covariates and their least squares estimates. Also listed in Table 9.4, are the true-and-estimated squared multiple correlations for these two regression systems. From this table, one could reasonably conclude that these latent slopes and the correlation of the two latent variables were fairly accurate.

9.8.3 A NAEP-Like Latent Variable Regression Example Comparing the Bayesian and Exact Solutions

The final example is from Nicewander and Lain (1994) and compares the exact solutions, proposed here, for IRT latent variable regression slopes—and the accompanying solution for the correlation matrix of the latent variables—with the Bayesian, EM solution. The Bayesian solution is based on the work of Mislevy (1985) and is implemented in a collection of FORTRAN programs written at ETS© and called MGROUP; in particular, the program used in this inquiry was CGROUP (Thomas, 1993; see also von Davier & Sinharay, 2010). In this comparison, the method derived here is referred to as OLS regression—ordinary least squares regression. In order to compare the CGROUP and OLS solutions for latent variable regression, a data set consisting of five θs and six observed covariates was simulated, each with $n = 2{,}000$, and replicated ten times. These data were generated as above using the following population dispersion matrices:

$\mathbf{\Sigma}_{\theta\theta}$ = a (5 × 5) constant-correlation matrix with a constant correlation of $\rho = .9$

$\mathbf{\Sigma}_{yy}$ = a (6 × 6) identity matrix.

$\mathbf{\Sigma}_{y\theta}$ = a (5 × 6) matrix with identical rows = (.4 .2 0 .4 .2 0).

Given these dispersion matrices, the population value of $\mathbf{\Gamma}$ is equal to $\mathbf{\Sigma}_{y\theta}$. These choices for the dispersion matrices were loosely based on the NAEP Mathematics tests which measure five areas of latent proficiency in mathematics that are known to be highly correlated (hence, the constant correlation of .9). The NAEP covariates were the dominant principal components of a large set of student background variables (hence, the zero correlations) between the covariates used here. Using the procedures discussed earlier, ten replications of normal random numbers were generated for 2,000 cases and transformed to have the above dispersion matrices in expectation. In addition to these simulated data, the responses to 50 dichotomous items from the 1992 NAEP 11th Grade Math items (ten per each $\theta_j; j = 1, 2, \ldots, 5$) were simulated using NAEP item-parameter estimates and a binary random-number generation process. These simulated data were then used to generate estimates of the regression slopes in $\mathbf{\Gamma}$ using both the CGROUP solution and the OLS solution (applied to the estimates of $\mathbf{\Sigma}_{\theta\theta}$ and $\mathbf{\Sigma}_{y\theta}$ obtained from (9.24) and (9.25) above). The whole process was repeated ten times, and means and standard deviations of the estimates computed. These summary statistics are shown in Table 9.5.

The estimates in Table 9.5 show a weak trend wherein the OLS regression estimates had somewhat less bias than the CGROUP estimates, but had slightly higher standard errors. Using the estimates of the regression slopes in Table 9.5, five, squared

Table 9.5 True and estimated latent-variable regression slopes using: a) OLS in conjunction with the correlations estimated using Equation (9.25); and b) CGROUP regression[a]

	θ_1	θ_2	θ_3	θ_4	θ_5
True slope	**.40**	**.40**	**.40**	**.40**	**.40**
OLS	.43 (.04)	.40 (.05)	.39 (.03)	.40 (.02)	.38 (.05)
CGROUP	.43 (.03)	.38 (.03)	.42 (.03)	.39 (.03)	.47 (.04)
True slope	**.20**	**.20**	**.20**	**.20**	**.20**
OLS	.21 (.04)	.21 (.04)	.21 (.04)	.21 (.02)	.21 (.04)
CGROUP	.22 (.02)	.21 (.03)	.22 (.04)	.21 (.03)	.24 (.04)
True slope	**0**	**0**	**0**	**0**	**0**
OLS	.01 (.04)	.04 (.05)	.01 (.03)	.02 (.03)	.01 (.05)
CGROUP	.01 (.03)	.03 (.03)	.01 (.02)	.02 (.03)	.01 (.04)
True slope	**.40**	**.40**	**.40**	**.40**	**.40**
OLS	.44 (.02)	.38 (.04)	.39 (.02)	.43 (.02)	.40 (.04)
CGROUP	.43 (.03)	.39 (.03)	.42 (.02)	.41 (.02)	.47 (.03)
True slope	**.20**	**.20**	**.20**	**.20**	**.20**
OLS	.23 (.04)	.20 (.06)	.20 (.02)	.21 (.03)	.20 (.04)
CGROUP	.22 (.02)	.19 (.02)	.22 (.02)	.21 (.02)	.20 (.04)
True slope	**0**	**0**	**0**	**0**	**0**
OLS	.00 (.03)	.01 (.06)	−.01	−.01	.03 (.04)
CGROUP	.00 (.03)	.01 (.04)	−.01 (.03)	−.01 (.02)	.01 (.04)

[a]Tabled estimates are the means of ten replications with $n = 2{,}000$ each. Values in parentheses are estimated standard errors.

multiple Rs were computed for each of the two sets of slope estimates (the true, squared multiple Rs are all .4): For OLS, these estimates had a mean of .41 with s.d. = .02; for the CGROUP estimates, the squared multiple correlations had a mean of .45 with s.d. = .06.

The estimates of the correlation matrix, $\mathbf{\Sigma}_{\theta\theta}$, for the latent variables is presented in Table 9.6. For these estimates of the true correlation matrix (having a constant-correlation of $\rho = .9$), the ten OLS-estimated correlations had a mean of .908 with s.d. = .055; the CGROUP estimates had a mean of .872 with s.d. = .010. Here again, these data are consistent with the OLS estimates being less-biased but having greater standard error relative to the Bayesian estimates.

Table 9.6 Estimates of the IRT latent variable intercorrelations using OLS in conjunction with Equation (9.24) and CGROUP (all true correlations are .90)[a]

Method		θ_1	θ_2	θ_3	θ_4	θ_5
	θ_1	1				
OLS CGROUP	θ_2	.81(.05) .88(.01)	1			
OLS CGROUP	θ_3	.94(.03) .89(.01)	.87(.06) .87(.01)	1		
OLS CGROUP	θ_4	.93(.04) .88(.01)	.84(.04) .86(.01)	.98(.01) .87(.01)	1	
OLS CROUP	θ_5	.93(.04) .88(.04)	.88(.05) .86(.01)	.94(.03) .87(.04)	.96(.04) .86(.01)	1

[a]Tabled estimates are means based on ten, simulated replications, each with $n = 2{,}000$. Values in parentheses are estimated standard errors.

9.9 SUMMARY

The estimates of correlation coefficients involving IRT latent variables developed in this chapter are exact, in that they follow directly from assumptions and involve no asymptotic or approximate statistical methods. The estimates have a very simple form for two basic reasons:

1. They make strong assumptions about data; and
2. The bivariate-normal correlation coefficients, on which these estimates are based—tetrachoric, polychoric, biserial, and polybiserial—do the "heavy lifting", mathematically, because of the complex calculations that are subsumed in their computation.

The correlations proposed here have a variety of possible uses, not the least of which is a solution for the regression coefficients for predicting IRT latent variables from responses to a mixture of dichotomous and polytomous test items. Another potential use of latent variable regression is the use of background variables for estimating the initial values of θ in computerized adaptive testing. As mentioned earlier, a drawback to the exact solutions derived here, is that there is no published software for computing guessing-corrected, polychoric correlation coefficients for a mixture of polytomous and dichotomous, multiple-choice items (for which guessing the correct answer is possible). As mentioned here, there are other solutions for latent regression problems that are Bayesian in form and use an EM algorithm for fitting the models to data (see von Davier & Sinharay, 2010). This inquiry provided some evidence that the "exact" solution developed here may be less biased but have larger standard errors than the Bayesian–EM solutions.

REFERENCES

Anderson, T. W. (1984). *An introduction to multivariate statistical analysis.* New York: Wiley.

Carroll, J. B. (1945). The effect of difficulty and chance success on correlations between items or between tests. *Psychometrika, 10*, 1–19.

Digby, P. G. N. (1983). Approximating the tetrachoic correlation coefficient. *Biometrics, 9*, 753–757.

Drasgow, F. (1986). Polychoric and polyserial correlations. In S. Kotz and N. L. Johnson (Eds.), *The encyclopedia of statistics*, Volume 7 (pp. 68–74). New York: Wiley.

Hulick, P. A., & Nicewander, W. A. (1995). *Estimation of the ACT latent trait intercorrelations.* Final Report, Summer Internship Project, ACT, Inc.

Lord, F. M. (1981). *Applications of item response theory to practical testing problems.* Hillsdale, NJ: Erlbaum.

Lord, F. M., & Novick, M. R. (1968). *Statistical Theories of Mental Test Scores.* Reading, MA: Addison Wesley.

Mislevy, R. J. (1985). Estimation of latent group effects. *Journal of the American Statistical Association, 80*, 993–997.

Nicewander, W. A., & Lain, M. M. (1994). *The correlation of IRT latent variables with one another and with observable variables.* Paper presented at the Psychometric Society, University of Illinois, June, 1994.

Olsson U. (1979). Maximum likelihood estimation of the polychoric correlation coefficient. *Psychometrika, 44*, 443–460.

Samejima, F. (1969). Estimation of latent ability using a response pattern of graded scores. *Psychometric Monograph*, No. 17.

Thomas, N. (1993). Asymptotic corrections for multivariate posterior moments with factored likelihood functions. *Journal of Computational and Graphical Statistics, 2*, 309–322.

von Davier, M., & Sinharay, S. (2010). Stochastic approximation methods for latent regression Item Response Models. *Journal of Educational and Behavioral Statistics, 35*, 174–193.

Chapter 10

Analysis of Structural Equation Models Based on a Mixture of Continuous and Ordinal Random Variables in the Case of Complex Survey Data

Stephen du Toit[1]

10.1 INTRODUCTION

There has been a growing interest in recent years in fitting models to data collected from cross-sectional and longitudinal surveys that use complex sample designs. This interest reflects expansion in requirements by policy makers and researchers for in-depth studies of social processes.

An important feature of software for the analysis of structural equation models (SEMs) is their facility to deal with a wide class of models for the analysis of latent

[1] Research for this chapter was supported by SBIR grant 5R44AA014999-03 from NIAAA to Scientific Software International.

variables (LVs). In the social sciences, and increasingly in biomedical and public health research, LV models have become an indispensible statistical tool.

There are basically three major reasons for the utility of LV models. First, this kind of model can summarize information contained in many response variables by a few latent variables. Second, when properly specified, an LV model can minimize the biasing effects of errors of measurement in estimating treatment effects. Third, LV models investigate effects between primary conceptual variables, rather than between any particular set of ordinary response variables. This means that an LV model is often viewed as more appropriate theoretically than is a simpler model with response variables only.

This chapter deals with the use of design weights to fit SEM models to a mixture of continuous and ordinal manifest variables with or without missing values, with optional specification of stratum and/or cluster variables. It also deals with the issue of robust standard error estimation and the adjustment of the chi-square goodness-of-fit statistic. In what follows, we assume that the population from which the sample data are obtained can be stratified into H strata. Within each stratum h, n_h clusters or primary sampling units (PSUs) are drawn and within the hth stratum and kth cluster, n_{h_k} ultimate sampling units (USUs) are drawn with design weights w_{hkl}, where l denotes the lth USU within the kth cluster, which in turn is nested within stratum h.

In the subsequent sections we briefly discuss parameter estimation and the Taylor linearization method employed in LISREL to produce robust standard error estimates under single stage sampling. Results based on real data are presented.

10.2 A REPARAMETERIZATION OF THE GENERAL LISREL MODEL

In this section, a reparameterization of the general LISREL model (Jöreskog & Sörbom, 2004) is presented. The relationships between the parameters of the model and the LISREL parameter matrices are also described.

Let $\mathbf{y}_i$ be a $(p + q)$ vector of latent variables and consider the general model

$$\mathbf{y}_i = \boldsymbol{\upsilon} + \boldsymbol{\Lambda}(\boldsymbol{\alpha} + \mathbf{u}_i) + \boldsymbol{\varepsilon}_i \tag{10.1}$$

where

$$\boldsymbol{\upsilon} = \begin{pmatrix} \boldsymbol{\tau}_y : (p\text{x}1) \\ \boldsymbol{\tau}_x : (q\text{x}1) \end{pmatrix}, \tag{10.2}$$

$$\boldsymbol{\Lambda} = \begin{pmatrix} \boldsymbol{\Lambda}_y(\mathbf{I}-\mathbf{B})^{-1} & \boldsymbol{\Lambda}_y(\mathbf{I}-\mathbf{B})^{-1}\boldsymbol{\Gamma} \\ 0 & \boldsymbol{\Lambda}_x \end{pmatrix}, \tag{10.3}$$

$$\boldsymbol{\alpha} = \begin{pmatrix} \boldsymbol{\alpha}_y \\ \boldsymbol{\kappa} \end{pmatrix}, \tag{10.4}$$

$$\mathbf{u}_i = \begin{pmatrix} w_i \\ v_i \end{pmatrix}, \tag{10.5}$$

$$\text{Cov}\,(\mathbf{u}_i) = \Psi = \begin{pmatrix} \Psi_y & \mathbf{0} \\ \mathbf{0} & \Phi \end{pmatrix}, \tag{10.6}$$

$$\boldsymbol{\varepsilon}_i = \begin{pmatrix} \boldsymbol{\varepsilon}_y \\ \boldsymbol{\varepsilon}_\delta \end{pmatrix}, \tag{10.7}$$

and

$$\text{Cov}\,(\boldsymbol{\varepsilon}_i) = \Theta = \begin{pmatrix} \Theta_\varepsilon & \Theta_{\varepsilon\delta} \\ \Theta_{\delta\varepsilon} & \Theta_\delta \end{pmatrix}. \tag{10.8}$$

From the definitions above it follows that:

- $\boldsymbol{\Lambda}_y$ of the general LISREL model is part of the first $p \times m$ submatrix of $\boldsymbol{\Lambda}$
- $\boldsymbol{\Lambda}_x$ of the general LISREL model is the final $q \times n$ submatrix of $\boldsymbol{\Lambda}$
- $\mathbf{B}$ of the general LISREL model is an $m \times m$ matrix with zero diagonal elements and is part of the first $p \times m$ and second $p \times n$ submatrices of $\boldsymbol{\Lambda}$
- $\boldsymbol{\Gamma}$ of the general LISREL model is an $m \times n$ submatrix of $\boldsymbol{\Lambda}$
- $\boldsymbol{\tau}_y$ of the general LISREL model consists of the first p elements of $\boldsymbol{\upsilon}$
- $\boldsymbol{\tau}_x$ of the general LISREL model consists of the final q elements of $\boldsymbol{\upsilon}$
- $\boldsymbol{\alpha}_y$ of the general LISREL model consists of the first m elements of $\boldsymbol{\alpha}$
- $\boldsymbol{\kappa}$ of the general LISREL model consists of the last n elements of $\boldsymbol{\alpha}$
- $\boldsymbol{\Theta}_\varepsilon$ of the general LISREL model is the first $p \times p$ submatrix of $\boldsymbol{\Theta}$
- $\boldsymbol{\Theta}_\delta$ of the general LISREL model is the final $q \times q$ submatrix of $\boldsymbol{\Theta}$
- $\boldsymbol{\Theta}_{\varepsilon\delta} = \boldsymbol{\Theta}'_{\delta\varepsilon}$ of the general LISREL model is the second $p \times q$ submatrix of $\boldsymbol{\Theta}$
- $\boldsymbol{\Phi}$ of the general LISREL model is the final $n \times n$ submatrix of $\boldsymbol{\Psi}$
- $\boldsymbol{\Psi}_y$ of the general LISREL model is the first $m \times m$ submatrix of $\boldsymbol{\Psi}$

10.2.1 Mixture of Ordinal and Continuous Dependent Variables

Let

$$\mathbf{y}_i = \begin{pmatrix} \mathbf{y}_{Oi} : p_i \times 1 \\ \mathbf{y}_{Ni} : q_i \times 1 \end{pmatrix}, \qquad i = 1, 2, \ldots, N \tag{10.9}$$

where N denotes the number of cases and where the $(p_i + q_i) \times 1$ vector $\mathbf{y}_i$ of manifest variables is partitioned into a $p_i \times 1$ vector $\mathbf{y}_{Oi}$ of ordinal and a $(q_i \times 1)$ vector $\mathbf{y}_{Ni}$

of continuous manifest variables. It is further assumed that the SEM model has m latent variables $\mathbf{u}_i$, where $\mathbf{u}_1, \mathbf{u}_2, \ldots, \mathbf{u}_N$ are independently and identically distributed $N(\mathbf{0}, \boldsymbol{\Psi})$.

The likelihood function for case i is evaluated as

$$\begin{aligned} f(\mathbf{y}_i) &= \int_{\mathbf{u}_i} f(\mathbf{y}_{0i}, \mathbf{y}_{Ni}, \mathbf{u}_i) d\mathbf{u}_i \\ &= \int_{\mathbf{u}_i} f(\mathbf{y}_{0i}, \mathbf{y}_{Ni} | \mathbf{u}_i) g(\mathbf{u}_i) \, d\mathbf{u}_i. \end{aligned} \tag{10.10}$$

Under the assumption of conditional independence, it follows that

$$f(\mathbf{y}_i) = \int_{\mathbf{u}_i} \prod_{j=1}^{p_i} f(y_{0ij} | \mathbf{u}_i) \cdot \prod_{j=1}^{q_i} f(y_{Nij} | \mathbf{u}_i) g(\mathbf{u}_i) \, d\mathbf{u}_i. \tag{10.11}$$

Hence

$$f(\mathbf{y}_i) = \int_{\mathbf{u}_i} \exp\left\{ \sum_{j=1}^{p_i} \ln f(y_{0ij} | \mathbf{u}_i) + \sum_{j=1}^{p_i} \ln f(y_{Nij} | \mathbf{u}_i) + \ln g(\mathbf{u}_i) \right\} d\mathbf{u}_i. \tag{10.12}$$

In general, a closed-form solution to this integral does not exist. To evaluate integrals of the type described above, we use a direct implementation of Gauss–Hermite quadrature. With this rule, an integral of the form

$$I(t) = \int f(t) \exp\left[-t^2\right] dt$$

is approximated by the sum

$$I(t) \approx \sum_{u=1}^{Q} w_u f(z_u),$$

where w_u and z_u are weights and nodes of the Hermite polynomial of degree Q.

Adaptive quadrature generally requires fewer points and weights to yield estimates of the model parameters and standard errors that are as accurate as would be obtained with more points and weights in nonadaptive quadrature. The reason for that is that the adaptive quadrature procedure uses the empirical Bayes means and covariances, updated at each iteration to essentially shift and scale the quadrature locations of each case (subject) in order to place them under the peak of the corresponding integral.

10.2.2 The Conditional Distribution for Ordinal Variables

Suppose that y_{0ij} has C categories, then

$$P\left(y_{0ij} = c\right) = P\left(y_{0ij} \leq c\right) - P\left(y_{0ij} \leq c-1\right), \quad c = 1, 2, \ldots, C-1$$

where $P\left(y_{0ij} = 0\right) = 0$ and $P\left(y_{0ij} = c\right) = 1 - P\left(y_{0ij} \leq c-1\right)$.

In LISREL the logit, probit, log-log and cumulative log-log functions are available. For the logistic link function, for example,

$$P\left(y_{0ij} \leq c\right) = 1/\left(1 + \exp\left(-\eta_{ij}\right)\right),$$

where $\eta_{ij} = \tau_{ic} + \boldsymbol{\lambda}_{\mathbf{ij}}^{'}\left(\boldsymbol{\alpha} + \mathbf{u}_i\right)$.

In the case of ordinal variables it is assumed that the corresponding subsets of $\boldsymbol{\upsilon}$ and $\boldsymbol{\Theta}$ are set to zero means and unit variances. The parameter τ_{ic} is the so-called threshold parameter.

10.2.3 The Conditional Distribution for Normal Variables

Suppose that $\mathbf{y}_{Ni}$ has typical element y_{Nij}, then

$$f\left(y_{Nij} | \mathbf{u}_i\right) = \left(2\pi\theta_{jj}\right)^{-1/2} \exp - \frac{1}{2\theta_{jj}} \left(y_{Nij} - \mu_{ij}\right)^2,$$

where $\mu_{ij} = v_j + \boldsymbol{\lambda}_i^{'}\left(\boldsymbol{\alpha} + \mathbf{u}_i\right)$. The parameter v_j is the so-called intercept parameter and θ_{jj} the residual variance.

10.2.4 The Special Case of Uncorrelated Latent Variable Subsets

Suppose that $\mathbf{y}_{0i}$ depends only on $\mathbf{u}_{0i} : m_0 \times 1$, and $\mathbf{y}_{Ni}$ depends only on $\mathbf{u}_{Ni} : m_N \times 1$, where $\mathbf{u}_i : (m \times 1) = \begin{pmatrix} \mathbf{u}_{0i} \\ \mathbf{u}_{Ni} \end{pmatrix}$ and $m = m_0 + m_N$.

$$f\left(\mathbf{y}_{0i}, \mathbf{y}_{Ni}\right) = \int_{\mathbf{u}_0} \int_{\mathbf{u}_N} f\left(\mathbf{y}_{0i}, \mathbf{y}_{Ni}, \mathbf{u}_{0i}, \mathbf{u}_{Ni}\right) d\mathbf{u}_{Ni} d\mathbf{u}_{0i}$$

$$= \int_{\mathbf{u}_0} \left\{ \int_{\mathbf{u}_i} \prod_{j=1}^{q_i} f\left(y_{Nij} | \mathbf{u}_N\right) \cdot g\left(\mathbf{u}_N | \mathbf{u}_0\right) d\mathbf{u}_N \right\} \times \prod_{j=1}^{p_i} f\left(y_{0ij} | \mathbf{u}_0\right) \cdot g\left(\mathbf{u}_0\right) d\mathbf{u}_0.$$

Hence

$$f(\mathbf{y}_i) = \int\limits_{\mathbf{u}_0} \exp\left\{\ln f(y_{Ni}|\mathbf{u}_0) + \sum_{j=1}^{p_i} \ln f\left(y_{0ij}|\mathbf{u}_0\right) \ln g(\mathbf{u}_0)\right\} d\mathbf{u}_0. \tag{10.13}$$

10.2.5 The Conditional Likelihood Function of the Set of Normal Variables

Since $\mathbf{y}_{Ni} \sim N(\boldsymbol{\mu}_i, \boldsymbol{\Sigma}_i)$ where $\mathbf{y}_{Ni} = \mathbf{v} + \boldsymbol{\Lambda}_{Ni}\boldsymbol{\alpha} + \boldsymbol{\Lambda}_{Ni}\mathbf{u}_{Ni}$, hence $\boldsymbol{\mu}_i = \mathbf{v} + \boldsymbol{\Lambda}_{Ni}\boldsymbol{\alpha}$, $\boldsymbol{\Sigma}_i = \boldsymbol{\Lambda}_{Ni}\boldsymbol{\Psi}_{22}\boldsymbol{\Lambda}'_{Ni} + \mathbf{D}_\theta$ and $\mathbf{u}_{0i} \sim N(\alpha, \Psi_{11})$, it follows from well-known results for normal conditional distributions that

$$\mathbf{y}_{Ni} \sim N\left(\boldsymbol{\mu}_{y\cdot 0}, \boldsymbol{\Sigma}_{y\cdot 0}\right)$$

where

$$\boldsymbol{\mu}_{y\cdot 0} = \upsilon + \boldsymbol{\Lambda}_{Ni}\boldsymbol{\alpha} + \boldsymbol{\Lambda}_{Ni}\boldsymbol{\Psi}_{21}\boldsymbol{\Psi}_{11}^{-1}(\mathbf{u}_{0i} - \boldsymbol{\alpha}) \tag{10.14}$$

and

$$\boldsymbol{\Sigma}_{y\cdot 0} = \boldsymbol{\Sigma}_i - \boldsymbol{\Lambda}_{N1}\boldsymbol{\Psi}_{21}\boldsymbol{\Psi}_{11}^{-1}\boldsymbol{\Psi}_{12}\boldsymbol{\Lambda}'_{N1}. \tag{10.15}$$

Therefore

$$\ln f(\mathbf{y}_{Ni}|\mathbf{u}_0) = -\frac{p_i}{2}\ln 2\pi - \frac{1}{2}\ln\left|\boldsymbol{\Sigma}_{y\cdot 0}\right|^{-1/2} - \frac{1}{2}\left(\mathbf{y}_{Ni} - \mathbf{u}_{y\cdot 0}\right)' \boldsymbol{\Sigma}_{y\cdot 0}^{-1}\left(\mathbf{y}_{Ni} - \mathbf{u}_{y\cdot 0}\right) \tag{10.16}$$

10.3 PARAMETER ESTIMATION

Parameter estimation is relatively straightforward and can be summarized by the following two steps.

Step 1

Calculate the natural logarithm of the likelihood function, $\ln L$, where

$$\ln L = \sum_{h=1}^{H}\sum_{k=1}^{n_h}\sum_{l=1}^{n_{hk}} w_{hkl} \ln f(\mathbf{y}_{hkl}|\boldsymbol{\gamma}). \tag{10.17}$$

Step 2

Obtain an estimate $\hat{\boldsymbol{\gamma}}$ of $\boldsymbol{\gamma}$ by solving the set of simultaneous equations

$$\left.\frac{\partial \ln L}{\partial \boldsymbol{\gamma}}\right|_{\boldsymbol{\gamma}=\hat{\boldsymbol{\gamma}}} = \mathbf{0}. \tag{10.18}$$

In general, no closed-form solution to the set of equations (10.18) exists, and therefore parameter estimates are obtained interactively using the Fisher scoring algorithm:

$$\hat{\boldsymbol{\gamma}}^{(t+1)} = \hat{\boldsymbol{\gamma}}^{(t)} + \mathbf{I}_n^{-1}\left(\hat{\boldsymbol{\gamma}}^{(t)}\right) \mathrm{g}\left(\hat{\boldsymbol{\gamma}}^{(t)}\right) \tag{10.19}$$

where $\hat{\boldsymbol{\gamma}}^{(t)}$ denotes the parameter values at iteration t, $t = 1, 2, \ldots,$ and $\mathrm{g}(\cdot)$ denotes the gradient vector and where $\mathbf{I}_n(\cdot)$ denotes the information matrix.

Suppose that the elements of $\boldsymbol{\gamma} : (k \times 1)$ are functions of the elements of $\boldsymbol{\gamma}^* : (m \times 1)$, where

$$\boldsymbol{\gamma}^* = vec\begin{bmatrix} \boldsymbol{\Lambda}_y(\mathbf{I}-\mathbf{B})^{-1} & \boldsymbol{\Lambda}_y(\mathbf{I}-\mathbf{B})^{-1}\boldsymbol{\Gamma} \\ \mathbf{0} & \boldsymbol{\Lambda}_x \end{bmatrix}.$$

Suppose further that it is possible to evaluate $\frac{\partial \ln L}{\partial \boldsymbol{\gamma}^*} = \mathrm{g}(\boldsymbol{\gamma}^*)$ and $\frac{\partial^2 \ln L}{\partial \boldsymbol{\gamma}'^* \partial \boldsymbol{\gamma}^*} = \mathbf{H}(\boldsymbol{\gamma}^*)$.

Proposition 1

$$\mathrm{g}(\boldsymbol{\gamma}) = \mathbf{C}' \, \mathrm{g}(\boldsymbol{\gamma}^*)$$

where

$$\mathbf{C} : m \times k \text{ has typical column } \mathbf{C}_i = \frac{\partial \boldsymbol{\gamma}^*}{\partial \boldsymbol{\gamma}_i}, \quad i = 1, 2, \ldots, k$$

Proof:

From the chain rule for first order derivatives it follows that

$$\begin{aligned} g(\boldsymbol{\gamma}) &= \frac{\partial \ln L}{\partial \boldsymbol{\gamma}} = \frac{\partial \ln L}{\partial \boldsymbol{\gamma}'^*} \frac{\partial \boldsymbol{\gamma}^*}{\partial \boldsymbol{\gamma}_i} \\ &= g'(\boldsymbol{\gamma}^*)\mathbf{C}_i \end{aligned}$$

or, alternatively,

$$\mathrm{g}_i(\boldsymbol{\gamma}) = \mathbf{C}_i' \mathrm{g}(\boldsymbol{\gamma}^*).$$

Hence $\mathrm{g}(\boldsymbol{\gamma}) = \mathbf{C}' \mathrm{g}(\boldsymbol{\gamma}^*)$.

Proposition 2

Let $\mathbf{I}_n(\boldsymbol{\gamma}) = -E\left[\mathbf{H}(\boldsymbol{\gamma})\right]$, where

$$[\mathbf{H}(\boldsymbol{\gamma})]_{ij} = \frac{\partial^2 \ln L}{\partial \boldsymbol{\gamma}_i \partial \boldsymbol{\gamma}_j}.$$

Then

$$vecs\left[\mathbf{H}(\boldsymbol{\gamma})\right] = \mathbf{S}\mathrm{g}(\boldsymbol{\gamma}^*) + vecs\left[\mathbf{C}^{'}\mathbf{H}(\boldsymbol{\gamma}^*)\mathbf{C}\right], \tag{10.20}$$

where $S : (k^* \times m)$ has typical row $\frac{\partial^2 \gamma^*}{\partial \gamma_i \partial \gamma_j}$, $i = 1, 2, \ldots, k$; $j = 1, 2, \ldots, i$ and $k^* = k(k+1)/2$.

Proof:

From (10.18) we have

$$\frac{\partial \ln L}{\partial \boldsymbol{\gamma}_j} = \mathbf{C}_j^{'}\mathrm{g}\left(\boldsymbol{\gamma}^*\right).$$

Therefore, using the product rule for matrix differentiation

$$\begin{aligned}
\frac{\partial^2 \ln L}{\partial \gamma_i \partial \gamma_j} &= \left[\mathbf{H}\left[\boldsymbol{\gamma}\right]\right]_{i,j} \\
&= \left[\frac{\partial}{\partial \gamma_i}\mathbf{C}_j^{'}\right]\mathrm{g}\left(\boldsymbol{\gamma}^*\right) + \mathbf{C}_j^{'}\left[\frac{\partial}{\partial \gamma_i}\mathrm{g}\left(\boldsymbol{\gamma}^*\right)\right] \\
&= \mathbf{s}_{ij}^{'}\mathrm{g}\left(\boldsymbol{\gamma}^*\right) + \mathbf{C}_j^{'}\left[\frac{\partial^2 \ln L}{\partial \gamma_i \partial \boldsymbol{\gamma}^*}\right].
\end{aligned} \tag{10.21}$$

Applying the chain rule to the second term in (10.21) it follows that

$$\begin{aligned}
\mathbf{C}_j^{'}\left[\frac{\partial^2 \ln L}{\partial \gamma_i \partial \boldsymbol{\gamma}^*}\right] &= \mathbf{C}_j^{'}\frac{\partial^2 \ln L}{\partial \boldsymbol{\gamma}^* \partial \boldsymbol{\gamma}^{*'}}\frac{\partial}{\partial \gamma_i} \\
&= \mathbf{C}_j^{'}\mathbf{H}\left(\boldsymbol{\gamma}^*\right)\mathbf{C}_i.
\end{aligned} \tag{10.22}$$

Iterations are continued until $\left|\hat{\gamma}_i^{(t+1)} - \hat{\gamma}_i^{(t)}\right| \prec \varepsilon \; \forall i = 1, 2, \cdots, nfree$ where ε is a small scalar value, e.g. 10^{-6}.

10.3.1 Approximate Covariance Matrix of Estimators

There are a number of basic methods available (see, e.g., Rao, 1975) for estimation of the sampling variances and covariances of the estimated parameters. The method used in LISREL is based on the linearization methods described by Wolter (1985) and

Binder (1983). An approximate expression for the asymptotic covariance matrix of $\hat{\boldsymbol{\gamma}}$ is given by

$$Cov(\hat{\boldsymbol{\gamma}}) \approx \mathrm{I}_n^{-1}(\boldsymbol{\gamma})\,\mathbf{G}\mathrm{I}_n^{-1}(\boldsymbol{\gamma}) \tag{10.23}$$

Using results derived by Binder (1983) and Fuller (1975), it follows that, under single stage sampling with replacement (WR) or without replacement (WOR)

$$\mathbf{G} = \sum_{h=1}^{H} \frac{n_h(1-f_h)}{n_h - 1} \sum_{i=1}^{n_h} \left(\mathbf{t}_{h_{i.}} - \bar{\mathbf{t}}_{h_{..}}\right)\left(\mathbf{t}_{h_{i.}} - \bar{\mathbf{t}}_{h_{..}}\right)' \tag{10.24}$$

where

- $n_h = \sum_{j=1}^{n_{h_i}} m_{h_{ij}}$, with $m_{h_{ij}}$ the number of cases within stratum h, cluster i, and USU j.
- $f_h = \frac{n_h}{N}$, the sampling rate for stratum h.
- $\mathbf{t}_{h_{ij}} = \mathbf{g}_{h_{ij}}(\hat{\boldsymbol{\gamma}})$, where $\mathbf{g}_{h_{ij}}(\hat{\boldsymbol{\gamma}})$ is the h_{ij}th contribution to the gradient vector $\mathbf{g}(\boldsymbol{\gamma})$.
- $\mathbf{t}_{h_{i.}} = \sum_{j=1}^{m_{h_{ij}}} \mathbf{t}_{h_{ij}}$
- $\bar{\mathbf{t}}_{h_{..}} = \frac{1}{n_h} \sum_{i=1}^{n_h} \mathbf{t}_{h_{i.}}$

In practice, we assume a zero contribution to $\mathbf{G}$ for strata that contain a single PSU (cluster). Additionally, if there is no variable to define clusters, the observations within each stratum are treated as being the primary sampling units.

10.3.2 Adjustment to the Chi-Square Goodness-of-Fit Statistics

Simulation studies indicated that the χ^2_{LR}-statistic based on the difference between two deviance statistics in general yields a too high rejection rate. The chi-square adjustment described below is based on contributions by Satorra and Bentler (1994, 2010).

Let

$$d_1 = tr\left(\mathbf{I}_n(\hat{\boldsymbol{\gamma}}_1)\,Cov(\hat{\boldsymbol{\gamma}}_1)\right),\ d_2 = tr\left(\mathbf{I}_n(\hat{\boldsymbol{\gamma}}_2)\,Cov(\hat{\boldsymbol{\gamma}}_2)\right) \tag{10.25}$$

where $\mathbf{I}_n\left(\hat{\gamma}_s\right)$, $s = 1, 2$ denotes the information matrix under H_1 and H_2.

A correction to the χ^2-statistic for testing the difference in two deviance statistics is given by

$$\chi^2_{robust} = c \times \chi^2_{LR} \tag{10.26}$$

where $(nfree_2 > nfree_1)$

$$c = \frac{nfree_2 - nfree_1}{abs(d_1 - d_2)}, \tag{10.27}$$

and where $nfree_1$ and $nfree_2$ respectively denote the total number of parameters to be estimated under the H_1 and H_2 models.

10.4 EXAMPLES

10.4.1 Confirmatory Factor Analysis Model

The data set forms part of the data library of the Alcohol and Drug Services Study (ADSS).[2] The ADSS is a national study of substance abuse treatment facilities and clients. Background data and data on the substance abuse of a sample of 1,752 clients were obtained. The sample was stratified by census region and within each stratum a sample was obtained for each of three facility treatment types within each of the four census regions.

The following variables included in the .PSF (Figure 10.1) were selected from the survey data:

- CENREG: This variable indicates the census region and has four categories, these being "Northeast," "Midwest," "South," and "West" respectively.
- FACTYPE: The facility treatment type has four categories, too, representing facilities with "residential treatment," "outpatient methadone treatment," "outpatient nonmethadone treatment," and "more than one type of treatment" respectively.
- COCEU: An indicator variable with value "1" if the respondent has ever used cocaine, and "0" otherwise.
- MAREU: An indicator variable with value "1" if the respondent has ever used marijuana, and "0" otherwise.
- DEPR: This indicator variable is coded "1" if the respondent is depressed, and "0" otherwise.

[2] ICPSR Study no. 3088, 1996-1999. Ann Arbor, MI: Inter-university Consortium for Political and Social Research [distributor]

drug_select.PSF

	CENREG	FACTYPE	A2TWA0	COCEU	MAREU	JAILR	DEPR	EDU	NUMTE
1	2.00	4.00	80.50	0.00	1.00	1.00	0.00	3.00	3.00
2	2.00	4.00	615.60	0.00	1.00	0.00	0.00	3.00	0.00
3	4.00	4.00	61.20	1.00	1.00	1.00	0.00	4.00	0.00
4	3.00	4.00	1584.40	0.00	0.00	0.00	1.00	2.00	0.00
5	4.00	4.00	54.30	0.00	1.00	0.00	0.00	4.00	0.00
6	3.00	4.00	105.80	1.00	1.00	0.00	1.00	2.00	1.00
7	3.00	2.00	51.40	1.00	1.00	1.00	1.00	4.00	2.00
8	2.00	4.00	125.40	0.00	1.00	0.00	0.00	4.00	0.00
9	4.00	5.00	65.00	1.00	1.00	0.00	1.00	2.00	2.00
10	2.00	4.00	80.50	0.00	1.00	1.00	0.00	3.00	1.00
11	1.00	5.00	495.90	0.00	0.00	1.00	0.00	2.00	1.00
12	4.00	4.00	166.70	1.00	1.00	0.00	0.00	2.00	0.00

Figure 10.1 First 12 records of a subset of the ADSS data.

- EDU: A categorical variable representing the respondent's level of education at admission. It has five categories, these being (from 1 to 5) "less than 8 years," "8–11 years or less than High School graduate," "High School graduate/GED," "some college," and "college graduate / postgraduate."
- JAILR: This indicator variable indicates whether the respondent had a prison or jail record prior to admission.
- NUMTE: A count variable, indicating the total number of treatment episodes prior to admission.

From the main menu bar, select the **Data, Survey Design ...** option (Figure 10.2).

Add the survey design variables to the appropriate text boxes (Figure 10.3).

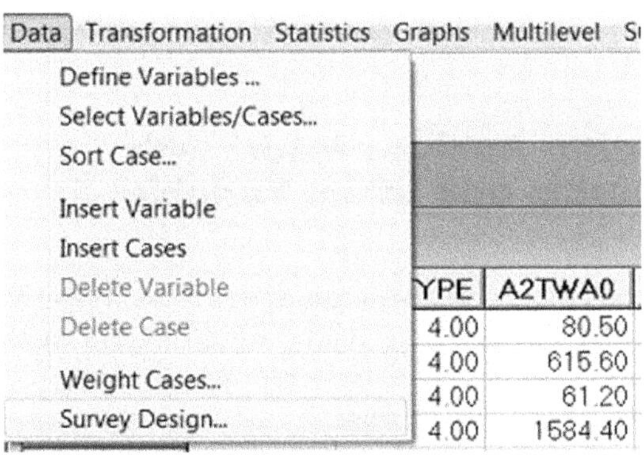

Figure 10.2 Data menu, survey design option.

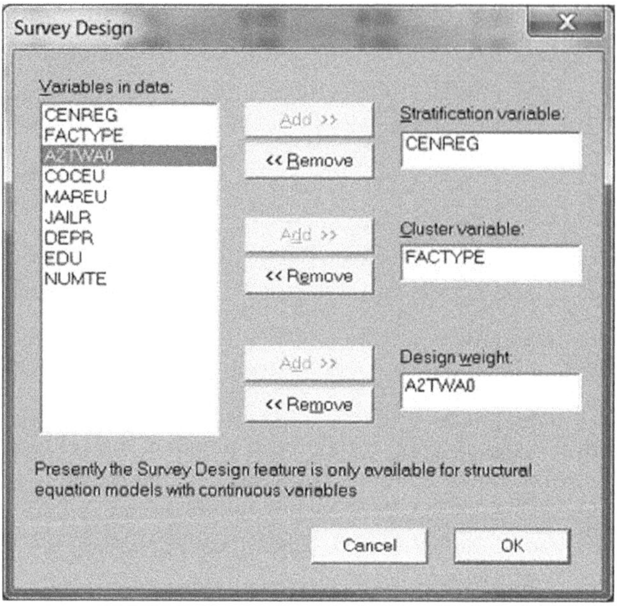

Figure 10.3 Survey design dialog.

The SIMPLIS syntax file is shown in Figure 10.4. The only difference between this and the usual SIMPLIS syntax is the addition of the paragraph

$ADAPQ(12) CLL

Other options are PROBIT, LOGLOG and LOGIT.

A portion of the LISREL output is shown in Figures 10.5 and 10.6.

Fit statistics and threshold values are shown in Figure 10.7.

10.4.2 Confirmatory Factor Analysis Model with Latent Variable Relationship and Latent Variable Means

This example is based on six political efficacy measurements as described in Aish & Jöreskog (1990). The dataset EFFICACY.PSF (Figure 10.8) consists of 1,719 cases obtained in a U.S. sample where the number 9 denotes a missing value.

The data are the responses to the following statements:

- NOSAY: “People like me have no say in what the government does.”

```
drug_use2.spl
Raw Data from file 'Drug_Select.PSF'
$ADAPQ(12) CLL
!------------------------------------------------------------
! Complex survey design example
! CENREC = Stratification variable
! FACTYPE = PSU's (Clusters)
! A2TWAO = Respondent design weights
! Complimentary log-log link function, 12 quadrature points
!------------------------------------------------------------
Latent Variables  DrugUse Environm
 COCEU = DrugUse
 MAREU = DrugUse
 JAILR = DrugUse
 DEPR = Environm
 EDU =  Environm
 NUMTE = Environm
LISREL OUTPUT: ND=3
Path Diagram
End of Problem
```

Figure 10.4 SIMPLIS syntax file for confirmatory factor analysis (ADSS example).

```
drug_use2.OUT

 LISREL Estimates (Maximum Likelihood)

         LAMBDA-X

             DrugUse    Environm
            --------    --------
   COCEU      1.494        - -
            (0.543)
              2.752

   MAREU      1.447        - -
            (0.150)
              9.667

   JAILR      0.754        - -
            (0.188)
              3.998
```

Figure 10.5 Maximum likelihood parameter estimates (ADSS example).

drug_use2.OUT

```
     DEPR          - -        0.341
                             (0.102)
                              3.350

      EDU          - -       -0.133
                             (0.082)
                             -1.626

    NUMTE          - -        4.323
                             (1.402)
                              3.084


         PHI

              DrugUse    Environm
             --------    --------
  DrugUse       1.000

 Environm       0.487       1.000
              (0.073)
                6.634
```

Figure 10.6 Maximum likelihood parameter estimates (ADSS example), continued.

drug_use2.OUT

```
 Fit Statistic Scale Factor =       0.25240
 Number of Strata    =                4
 Number of Clusters =                12

 Number of quadrature points =                12
 Number of free parameters =                  17
 Number of iterations used =                  13

 -2lnL (deviance statistic) =       12685.15371
 Akaike Information Criterion       12719.15371
 Schwarz Criterion                  12806.79545

      Threshold estimates and standard deviations
      -------------------------------------------
      Threshhold               Estimates          S.E.        Est./S.E.

      TH1_COCEU                -0.71273       0.27535        -2.58846
      TH1_MAREU                -1.89351       0.19717        -9.60337
      TH1_JAILR                -0.26160       0.13898        -1.88223
      TH1_DEPR                 -0.01710       0.08917        -0.19178
      TH1_EDU                  -2.85333       0.13439       -21.23097
      TH2_EDU                  -1.22371       0.07418       -16.49622
      TH3_EDU                   0.07611       0.05628         1.35250
      TH1_NUMTE                -2.17086       0.83776        -2.59127
      TH2_NUMTE                 0.73476       0.17195         4.27303
      TH3_NUMTE                 3.03918       0.56226         5.40529
```

Figure 10.7 Fit statistics and threshold estimates (ADSS example).

EFFICACY.PSF

	NOSAY	VOTING	COMPLEX	NOCARE	TOUCH	INTEREST
1	2.00	2.00	1.00	1.00	1.00	1.00
2	2.00	3.00	3.00	3.00	2.00	3.00
3	3.00	2.00	2.00	3.00	3.00	3.00
4	3.00	3.00	2.00	3.00	2.00	3.00
5	2.00	2.00	1.00	2.00	2.00	2.00
6	2.00	2.00	1.00	1.00	2.00	1.00
7	3.00	2.00	2.00	3.00	3.00	3.00
8	2.00	2.00	2.00	2.00	1.00	2.00
9	3.00	1.00	2.00	2.00	2.00	2.00
10	2.00	1.00	2.00	2.00	1.00	1.00
11	2.00	2.00	1.00	2.00	2.00	2.00
12	1.00	1.00	1.00	1.00	2.00	9.00
13	1.00	9.00	1.00	1.00	1.00	1.00
14	3.00	3.00	2.00	3.00	3.00	3.00

Figure 10.8 First 14 records of the political efficacy data.

- VOTING: "Voting is the only way that people like me can have any say about how the government runs things."
- COMPLEX: "Sometimes politics and government seem so complicated that a person like me cannot really understand what is going on."
- NOCARE: "I don't think that public officials care much about what people like me think."
- TOUCH: "Generally speaking, those who are elected to Congress in Washington lose touch with the people pretty quickly."
- INTEREST: "Parties are only interested in people's votes but not in their opinions."

The ordered categories are:

- 1: agree strongly
- 2: agree
- 3: disagree
- 4: disagree strongly

It is hypothesized that the six variables are indicators of two factors, where Factor 1 represents Internal Efficacy measured by NOSAY, VOTING, and COMPLEX. Factor 2 represents External Efficacy measured by NOCARE, TOUCH, and INTEREST.

eff1.ls8

```
TI  Efficacy Data
DA NI=9 NO=0
RA FI = 'EFFICACY.PSF'
$ADAPQ(6) PROBIT
!-----------------------------------------
! Probit link function, 8 quadrature points
! Six ordinal variables
! Two factors, PHI unconstrained
! Percentage missing values = 2.32%
!-----------------------------------------
MO NY=6 NE=2 TE=SY BE=FU AL=FR PS=SY
LE
f1 f2
FR LY(2,1) LY(3,1)
FR LY(5,2) LY(6,2)  BE(2,1)
VA 1 LY(1,1)
VA 1 LY(4,2)
PD
```

Figure 10.9 LISREL syntax file for the political efficacy example.

It is further hypothesized that Factor 2 depends on Factor 1. The LISREL syntax for fitting the model is shown in Figure 10.9. Six quadrature points are selected for each of the two dimensions using the probit link function.

A portion of the output is shown in Figure 10.10, being the parameter specifications where values greater than zero correspond to parameters that are estimated, while zero values correspond to parameters that are fixed at zero or the values entered in the "VA" paragraphs.

The path diagram representation shown in Figure 10.11 provides the values of the estimated parameters for factor means.

The path diagram for the basic model showing the parameter estimates is given in Figure 10.12.

10.5 DISCUSSION

This chapter demonstrates that marginal maximum likelihood estimation of structural equation models with a mixture of ordinal and continuous manifest variables can be undertaken by a direct implementation of the Gauss–Hermite formulae for numerical

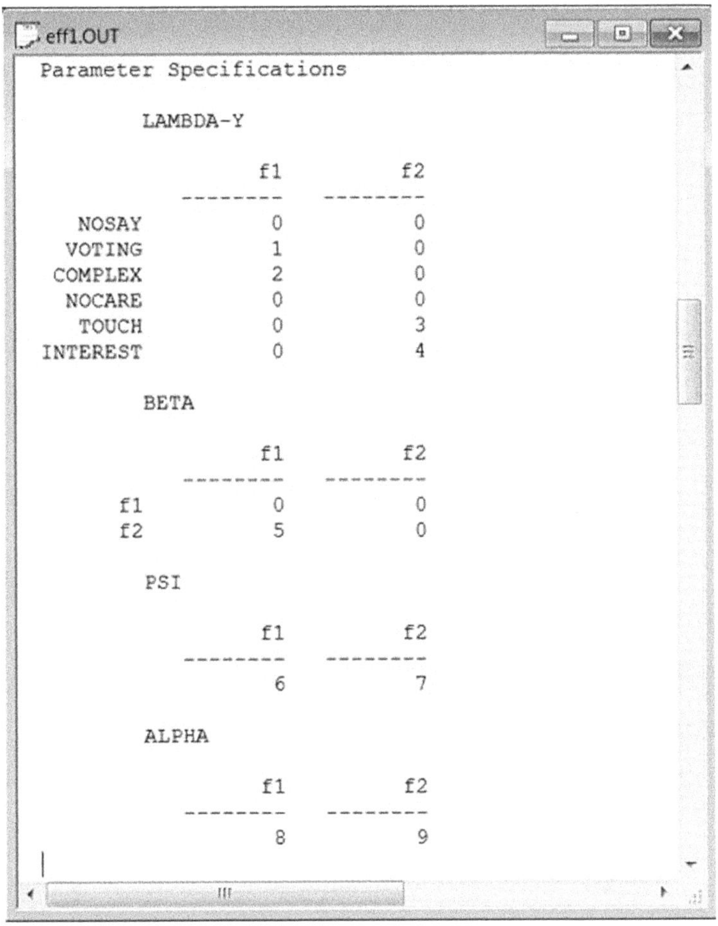

Figure 10.10 Parameter specifications for the political efficacy example.

integration. Accuracy of the parameter estimates and the computing time are dependent on the number of latent variables and quadrature points used. Computation time can be particularly burdensome when six or more latent variables are included in the model. Hence, research has been conducted to investigate alternative ways to fit this type of model. One approach that seems promising is the Metropolis–Hasting Robinson—Monro estimation procedure proposed by Cai (2010a, 2010b). However, the approach described in this chapter can be considered a benchmark for the evaluation of alternative methods as, with enough quadrature points, the parameter estimates, estimated standard errors, and fit statistics will be precise.

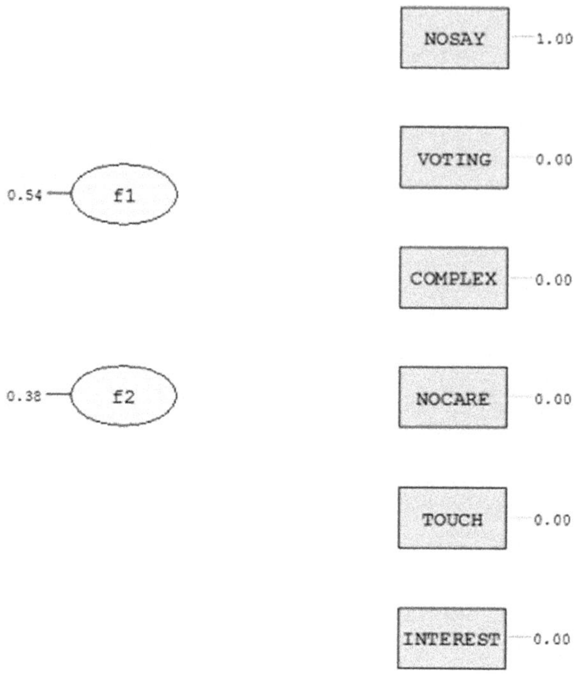

Figure 10.11 Path diagram for the means part of the model (political efficacy example).

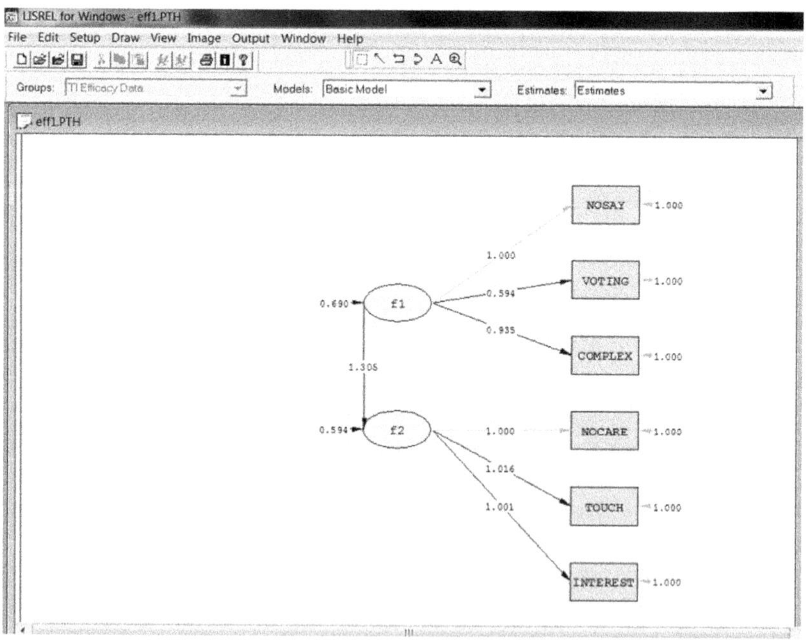

Figure 10.12 Path diagram for the basic model (political efficacy example).

REFERENCES AND FURTHER READING

Aish, A. M., & Jöreskog, K. G. (1990). A panel model for political efficacy and responsiveness: An application of LISREL 7 with weighted least squares. *Quality and Quantity, 24*, 405–426.

An, A. B. (2003). Performing Logistic Regression on Survey Data with the New SURVEYLOGISTIC. Paper 258-27 presented at SUGI 27, held on April 14–17, 2003. Orlando, Florida.

Binder, D. A. (1983). On the variances of asymptotically normal estimators from complex surveys. *International Statistical Review, 51*, 279–292.

Binder, D. A., & Patak, Z. (1994). Use of estimating functions for individual estimation from complex surveys, *Journal of the American Statistical Association, 83*, 1035–1043.

Cai, L. (2010a). High-dimensional exploratory item factor analysis by a Metropolis–Hastings Robbins–Monro algorithm. *Psychometrika, 75*, 33–57.

Cai, L. (2010b). Metropolis–Hastings Robbins–Monro algorithm for confirmatory item factor analysis. *Journal of Educational and Behavioral Statistics, 35*, 307–335.

Chambers, R., & Skinner, C. J. (Eds.) (2003). *Analysis of survey data.* New York: Wiley.

Feder, M., Nathan, G., & Pfefferman, D. (2000). Multilevel modeling of complex survey longitudinal data with time varying random effects. *Survey Methodology, 26*, 53–65, Statistics Canada.

Fuller, W. A. (1975). Regression analysis for sample survey. *Sankhya, Series C, 37*, 117–132.

Graubard, B. I., & Korn, E. L. (1993). Hypothesis testing with complex survey data, *Journal of the American Statistical Association, 88*, 629–641.

Horvitz, D. G., & Thompson, D. J. (1952). A generalization of sampling without replacement from a finite universe. *Journal of the American Statistical Association, 47*, 663–685.

Jöreskog, K. G., & Moustaki, I. (2001). Factor analysis of ordinal variables: A comparison of three approaches. *Multivariate Behavioral Research, 36*, 347–387.

Jöreskog, K. G., & Sörbom, D. (2004). LISREL 8.70 for Windows [Computer Software]. Lincolnwood, IL: Scientific Software International, Inc.

Kish, L. (1965). *Survey sampling,* New York: Wiley.

Kish, L., & Frankel, M. R. (1974). Inference from complex samples. *Journal of Royal Statistical Society (B), 36*, 1–37.

Muthén, B. O. (1984). A general structural equation model with dichotomous, ordinal, categorical, and continuous latent variable indicators. *Psychometrika, 49*, 115–132.

Muthén, B. O., & Satorra, A. (1995). Complex sample data in structural equation modeling. In P. Marsden (Ed.) *Sociological Methodology, 59*, 216–316.

Nathan, G. (1988). Inference based on data from complex sample designs. In P. R. Krishnaiah & C. R. Rao (Eds.) *Handbook of statistics* (pp. 247–266). Amsterdam: Elsevier.

Rao, J. N. K. (1975). Unbiased variance estimation for multistage designs, *Sankhya, Series C, 37*, 133–139.

Rao, J. N. K. (1994). Estimation of totals and distributing functions using auxiliary information at the estimation stage. *Journal of Official Statististics, 10*, 153–165.

Rao, J. N. K., & Scott, A. J. (1981). The analysis of categorical data from complex sample surveys: Chi-squared tests for goodness-of-fit and independence in two-way tables. *Journal of the American Statistical Association, 76*, 221–230.

Rao, J. N. K., Scott, A. J., & Skinner, C. J. (1998). Quasi-score test with survey data. *Statistica Sinica, 8*, 1059–1070.

Sarndal, C. E., Swensson, B., & Wretman, J. (1992). *Model assisted survey sampling.* New York: Springer.

SAS Institute, Inc. (2004). SAS/STAT®: *User's Guide.* Cary, NC: SAS Institute, Inc.

Satorra, A., & Bentler, P. M. (1994). Corrections to test statistics and standard errors in covariance structure analysis. In A. van Eye & C.C. Clogg (Eds.), *Latent variable analysis in developmental research* (pp. 285–305). Thousand Oaks, CA: Sage.

Satorra, A., & Bentler, P. M. (2010). Ensuring positiveness of the scaled difference chi-square test statistic. *Psychometrika, 75*, 243–248.

Shapiro, G. M., & Bateman, D.V. (1978). A better alternative to the collapsed stratum variance estimate. *Proceedings of the Survey Research Methods Section*, American Statistical Association, 451–456.

Skinner, C. J. (1989). Domain means, regression and multivariate analysis. In C. J. Skinner, D. Holt, & T. M. F. Smith (Eds.), *Analysis of complex surveys* (pp. 59–87). Chichester: Wiley.

Skinner, C. J., Holt, D., & Smith, T. M. F. (1989). *Analysis of complex surveys*, Chichester: Wiley.

Smith, T., & Holmes, D. (1989) Multivariate analysis. In C. J. Skinner, D. Holt, & T. M. F. Smith (Eds.) *Analysis of complex surveys* (pp. 165–190). Chichester: Wiley.

SUDAAN User Manual Release 8.0, Second Edition (2002). Research Triangle Institute.

Traat, I., Meister, K., & Sostra, K. (2001). Statistical inference in sampling theory. *Theory of Stochastic Processes, 7*, 301–316.

Wolter, K. M. (1985). *Introduction to variance estimation.* New York: Springer-Verlag.

Yates, F., & Grundy, P. M. (1953). Selection without replacement from within strata with probability proportional to size. *Journal of the Royal Statistical Society (B), 15*, 253–261.

PART 2
Drawing Meaning from Latent Variable Models

Chapter 11

A Simulation Paradigm for Evaluating Approximate Fit

Roger E. Millsap

In preparation for the 2010 conference held in honor of Michael Browne, I checked online to find out which paper is most often cited from among the many contributions he has made to the quantitative methods literature. I was initially surprised at the result: The chapter Browne and Cudeck (1993) had well over 6,000 citations. This book chapter describes the properties of the root mean square error of approximation (RMSEA) fit index, gives advice on how the index might be used, and illustrates its use. The chapter is the source for the widely used (.05, .08) RMSEA cut-points for evaluating model fit in structural equation modeling. In retrospect, the high citation rate for this publication should not have been surprising. The number of fit indices used in structural equation modeling has multiplied, starting with the chi-square test of exact fit, and expanding to the several dozen fit indices now available in standard software. The growth in the number of fit indices has led to a corresponding growth in the demand for standards for judging model fit. If we are not going to depend solely on the chi-square test of exact fit, which indices should be used, and what values of these indices should define acceptable fit?

The RMSEA index described in Browne and Cudeck (1993), and first proposed by Steiger and Lind (1980), is of particular interest because, under standard assumptions, the sampling distribution of the RMSEA is known for large samples, leading to confidence intervals and test statistics. Browne and Cudeck (1993) provide tentative advice for what values of the RMSEA are consistent with good or acceptable fit. Their recommended values were not intended to be absolute cut-points, adopted regardless of the model or context. Unfortunately, the (.05, .08) cut-points have become just that: absolute rules applied by researchers in many contexts and across all models. This situation has evolved despite the clear advice of Browne and Cudeck (1993) that the RMSEA and other indices should not be used in a knee-jerk fashion to decide questions of fit.

Much research has now accumulated to undercut the use of fixed cut-points for the RMSEA in distinguishing acceptable versus unacceptable fit (Beauducel & Wittmann, 2005; Fan & Sivo, 2005; Marsh, Hau, & Wen, 2004; Yuan, 2005). The general trend in this research has been to show that the RMSEA is affected by model and data conditions that are not directly a function of fit, such as model size or complexity, the number of measured variables, the type of model, distributional conditions, and sample size. Depending on the configuration of such conditions, a model that has a known, important misspecification may still often yield a sample RMSEA value less than .05, indicating good fit under the usual rule. The implication of this research is that it is not possible to formulate cut-points for the RMSEA that will distinguish acceptable and unacceptable models across all conditions. This implication is not unique to the RMSEA. Cut-points for other global fit indices can be criticized on similar grounds.

11.1 APPROXIMATE FIT: WHAT IS THE INTERPRETATION?

The cut-point problem is part of a more general problem with global descriptive measures of fit, whether considered from sample or population perspectives. An approximate fit index is supposed to gauge the degree to which a model is misspecified, accepting the fact that the model is not properly specified. Given that the models under study are multivariate, a global fit index represents a many-to-one mapping of varied local misspecifications onto a single scale. The problem is that for any given fit index value, it is difficult to say exactly which types of misspecifications are consistent with that value. Suppose that one is fitting a two-factor model with each indicator loading on a single factor, with a nonzero correlation between the two factors. If the RMSEA = .03 in the population for this model, what types of misspecification are possible under this RMSEA value? Could a third factor be consistent with RMSEA = .03? Could there be nonzero correlations among several unique factors? The answers to such questions are unclear. As a result, the nature of "approximate fit" is unclear when it is gauged by one of the global fit indices. It is also difficult to decide on an appropriate cut-point for dividing acceptable and unacceptable levels of model misspecification.

This ambiguity in the interpretation of global fit indices has produced varied reactions. Some researchers have rejected the use of global indices of approximate fit (Barrett, 2007). Other researchers have placed emphasis on the use of local fit information (McDonald & Ho, 2002; Saris, Satorra, & van der Veld, 2009). Local fit information is certainly valuable even when global fit indices are considered as well. Here we will consider a different approach that retains the global fit index, but dispenses with *a priori* cut-points (Millsap, 2007). These cut-points are replaced with a simulation-based decision process that seeks evidence for the interpretation of any global lack of fit.

11.2 A SIMULATION-BASED FIT EVALUATION METHOD

Suppose that one is fitting a model M0 to sample data in a sample of size N, and that a statistically significant chi-square test has shown lack of fit for M0. Suppose also that the RMSEA index has a value of RMSEA $= .04$ in this sample. Without further scrutiny, conventional rules of thumb would suggest that M0 be retained as providing adequate approximate fit. In adopting this position, one is saying that while M0 may be the wrong model, the right model should be close to M0 in some sense. Model M0 thus provides a good approximation to the right model. What do we mean when we say that M0 provides a good approximation? One way of answering this question is to construct an alternative model M1 that is close enough to M0 that we would regard the two models as equally good from a practical perspective. For example, suppose that M0 is a single-factor model for a 15-item questionnaire. The alternative model M1 might be a single-factor model that permits nonzero covariances between the unique factors for several pairs of items, with these nonzero covariances fixed to small values (e.g., producing unique factor correlations $< .10$). Here we would be arguing that while M1 is the correct model, the sizes of the unique factor correlations are small enough to ignore. As another example, suppose that M0 is a path model for seven measured variables, with some pairs of variables having no direct paths between them. The alternative model M1 may be identical to M0 except for the inclusion of two additional nonzero direct paths, with the two new path coefficients fixed at small values (e.g., producing increments to R-squared $< .01$). Here we again argue that while M1 is the better model, the additional paths are small enough to be ignored. More examples can be constructed. The main point is that if we wish to advocate M0 as an adequate approximation, we should be able to specify what we mean by an "approximation" in exact terms.

The specification of M1 requires that not only must M0 be respecified to include new parameters, we must also assign values to all of the model parameters. As shown below, this step is required to enable the generation of data from M1. Once values of all model parameters are specified, the covariance matrix (and mean vector, if needed) can be constructed under the model. With the covariance matrix and mean vector in hand, we can generate data under multivariate normality from M1. The values to be assigned to the model parameters can come from a variety of sources, including the data at hand if real data are available. In some cases, parameter values are explicitly selected to represent "small" effects, as in the above two examples. In other cases, the investigator may simply assign values to parameters that are reasonable choices given previous use of M1 or models similar to M1.

11.2.1 Steps in the Analyses

Once M0 is available and M1 has been specified, the analysis proceeds in four steps that can be replicated as needed for different alternative models. It is assumed here that M0 has been fit to real data, and that statistically significant lack of fit has been found. The real data analyses have also generated real sample values of the studied fit index, such as the RMSEA.

Step one Generate q repeated random independent samples of size N from the M1 model under multivariate normality assumptions for the latent and/or measured variables. In each of the q generated samples, fit the original model M0 to the data and record the resulting value of the studied fit index. The RMSEA will be used here as the main focus, but any other fit statistic of interest could be used. Note here that we are fitting the "wrong" model to the generated data, given that M1 is the generating model.

Step two Using the q sample values of the RMSEA as obtained in Step one, create a frequency distribution of the RMSEA values.

Step three Let F0 be the sample RMSEA value for M0 in the real data with sample size N. Let F1 be a random variable that is the RMSEA value for M0 when fit to the simulated data generated in Step one, generating q values for F1. Using the distribution of F1 values assembled in Step two, evaluate P1 = P(F1 > F0) as the proportion of F1 values across q replications that exceed F0 in value.

Step four We now make a decision about M0 based on our choice of M1 and the value of P1 in Step three. If P1 is small enough (e.g., below .10), we conclude that the misfit in M0 in the real data is much larger than we would expect if M1 was the model underlying the real data. This result casts doubt on the idea that we can accept M0 because it is a good approximation to M1, and M1 is the real model. We therefore reject M0. On the other hand, suppose that P1 is not small. This result is consistent with the idea that our sample value of F0 could have arisen due to M1 being the real model. It is also possible that the fit of M0 in the real data is much better than its fit in data generated by M1 (e.g., P1 is large). The evidence then supports the idea that M0 is misspecified to a degree that is ignorable, under the assumption that M1 is the true model.

The decision process in step four does not need to consider the value of F0 in relation to traditional cut-points for the global fit index under study. The process depends instead on the empirical probability estimate P1 in step three for the chosen M1. As a result, one may actually reject M0 if F0 has a sufficiently low probability estimate, even if the F0 value would have been acceptable under conventional cut-points for the global index. The converse situation is also possible: we may retain M0 if the estimate P1 is high, even if F0 would have led to rejection under conventional cut-points.

Once the four-step process has been completed for one choice of M1, it is possible to repeat the process with a new alternative model M2. The new model M2 might be a different parameterization of M1, or it could be a qualitatively different model. The same defining characteristic should apply however: M2 should be a model that is sufficiently "close" to M0 so that they are good approximations of each other in practical terms. The use of multiple alternative models is useful in showing that M0 can be supported (or rejected) from multiple perspectives. One possible result however is that the multiple alternative models lead to different outcomes, sometimes suggesting rejection of M0 and sometimes suggesting retention. We will return to this problem below.

The next section presents two illustrations of the above simulation approach in real data. All analyses were conducted using the MONTECARLO procedure in Mplus 5.21 (Muthén & Muthén, 1998–2009). This Mplus procedure is convenient for the proposed simulation approach because it permits the user to generate repeated random samples from M1 while fitting M0 in each sample, generating a distribution of fit statistics that can be saved for later analyses.

11.3 ILLUSTRATIONS

The first example uses data from the WAIS-R standardization sample in a multiple-group confirmatory factor analysis. Males ($N = 940$) and females ($N = 940$) are the two groups to be compared. A subset of four of the 11 subtests of the WAIS-R is used here: Information, Vocabulary, Comprehension, and Arithmetic. The model M0 of interest is a single-factor model with invariance constraints on the factor loadings. The model is shown in Figure 11.1. This model represents metric invariance (Horn &

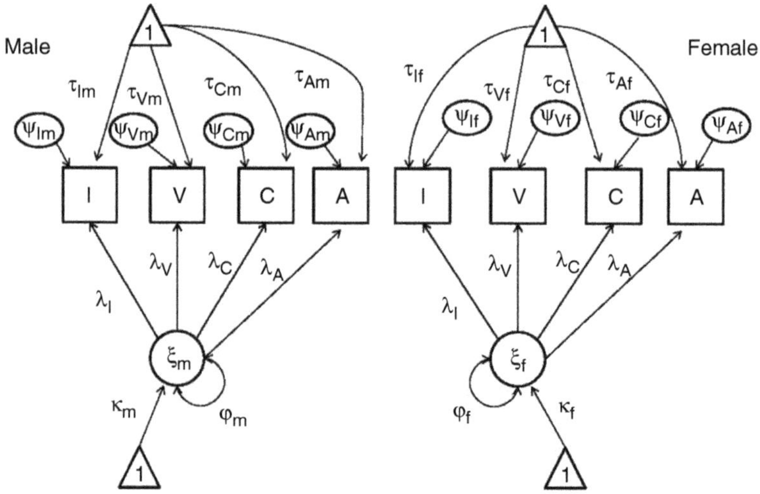

Figure 11.1 M0 single-factor model with metric invariance.

McArdle, 1992; Thurstone, 1947) in the conventional model sequence for multiple-group invariance studies. When fit to the data, M0 produces a significant chi-square value ($\chi^2(7) = 18.959, p < .01$) and an RMSEA estimate of .043. We can conclude that the specified model is incorrect in some way. Is it reasonable to argue that M0, while incorrect, offers a good approximation to the true model? To address this question, we first evaluate M0 under an alternative model M1 that retains the single-factor structure but permits group differences in three of the four factor loadings. The differences are specified to be small at .10 for each loading. Specification of M1 requires that all parameters be given values. These parameter values are given in Table 11.1.

The MONTECARLO facility in Mplus was used to generate $q = 1{,}000$ samples under M1, each sample involving 940 observations in each of two independent groups. Figure 11.2 shows the histogram of the resulting RMSEA values when M0 was fit to each sample. Across the 1,000 samples, the average RMSEA is .048, with an SD of .014. The Mplus code for this simulation and all subsequent simulation examples is given in the Appendix. The RMSEA value of .043 from the real data lies in the center of the distribution in Figure 11.2. This result suggests that the lack of fit for M0 in the real data is consistent with the results obtained when M1 is the actual model generating the data. In other words, the results support the contention that the RMSEA value found in the real data could have arisen from misspecification that is ignorable in practical terms.

The M1 model used here is not the only type of misspecification that can be considered for M0. As a second alternative, a model M2 was created that retains

Table 11.1 M1 model parameters for the single-factor WAIS-R example

Parameter	*Group 1 value*	*Group 2 value*
τ_I	9.82	8.98
τ_V	9.41	9.25
τ_C	9.61	9.33
τ_A	9.50	9.20
λ_I	.90	.80
λ_V	1.00	1.00
λ_C	.88	.78
λ_A	.70	.60
κ	0	0
φ	8.4	7.8
θ_I	2.0	1.8
θ_V	1.0	1.2
θ_C	4.0	4.0
θ_A	3.0	3.0

the single-factor structure with invariant loadings, but permits a nonzero covariance between the unique factors for Information and Comprehension. The covariance was specified in each group to produce a correlation between the unique factors of about .20. All other parameter values were fixed as in Table 11.1, except that the Group 2 factor loadings were fixed at the Group 1 values to achieve metric invariance. The model M2 was used to simulate data for q = 1,000 samples, each with two independent groups of sample size 940 as before. Figure 11.3 shows the results for the RMSEA across the 1000 samples. The average RMSEA value in these simulated samples is .065 (SD = .014), with the original RMSEA value of .043 from the real data falling at around the 6th percentile. This result suggests that the fit of the M0 model in the real data is better than we would expect to see if M2 was the true model. Again, this evidence argues for retention of M0 as a reasonable approximation for the data-generating model.

The next example uses data presented in Preacher, Wichman, MacCallum and Briggs (2008) involving five waves of measurement of mother's closeness to her child in a sample of N=851 mothers. The M0 model is a linear latent growth model with error variances constrained to be invariant across time. Both intercepts and slopes are assumed to be random, with a nonzero correlation between slopes and intercepts. Figure 11.4 presents the model. This model led to a significant chi-square ($\chi^2(14) = 75.996$, $p < .01$) and an RMSEA estimate of .072 in the real data. The chi-square suggests that M0 is the wrong model, with the RMSEA reflecting a substantial degree of misfit by conventional standards.

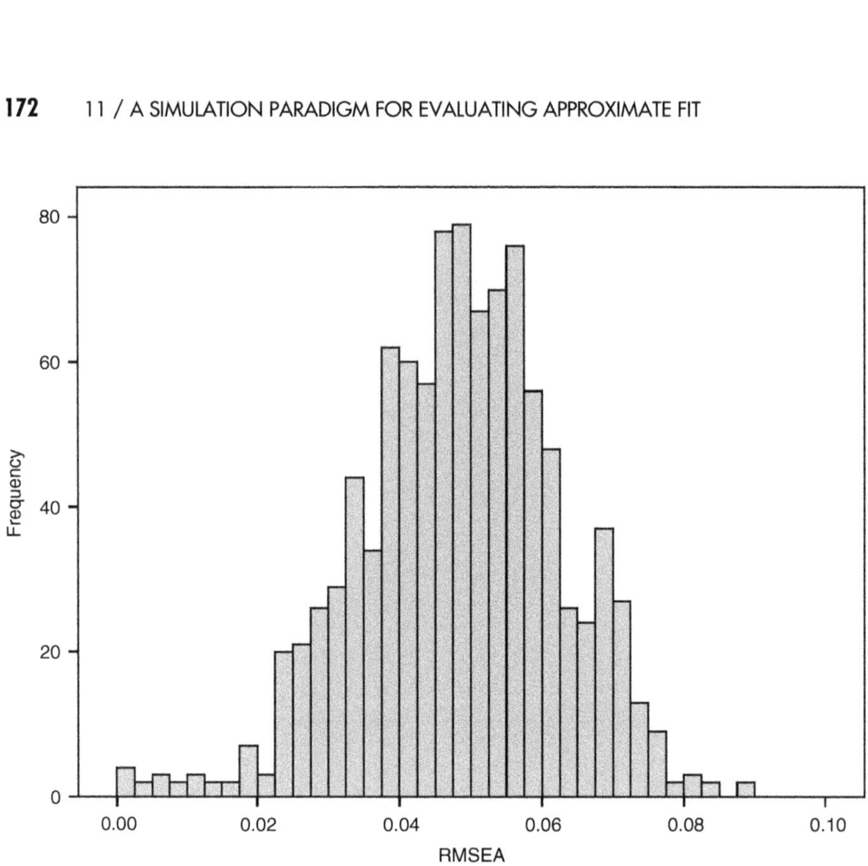

Figure 11.2 Frequency histogram for RMSEA over 1000 simulated samples from WAIS-R single-factor M1 model.

To examine this conclusion further, an alternative model M1 was created that permits the error variances to vary over time, while retaining all other model features in M0. The magnitudes of the changes in error variance were limited to 10% of the value found in the real data under M0. The factor loadings for this model are given in Figure 11.4. The other parameter values for this model were:

$$\kappa = \begin{bmatrix} \kappa_{\mathrm{I}} \\ \kappa_{\mathrm{L}} \end{bmatrix} = \begin{bmatrix} 38.000 \\ -.358 \end{bmatrix}, \tag{11.1}$$

$$\Theta = \begin{bmatrix} \varphi_{\mathrm{I}} & \varphi_{\mathrm{IL}} \\ \varphi_{\mathrm{IL}} & \varphi_{\mathrm{L}} \end{bmatrix} = \begin{bmatrix} 2.977 & .253 \\ .253 & .136 \end{bmatrix}, \tag{11.2}$$

$$Diag\,(\Psi) = [3.3 \quad 4.1 \quad 4.1 \quad 3.3 \quad 4.1]\,. \tag{11.3}$$

The model M1 was used to generate $q = 1{,}000$ samples, each with $N = 851$, and M0 was fit in each sample. Figure 11.5 displays the histogram of RMSEA values across

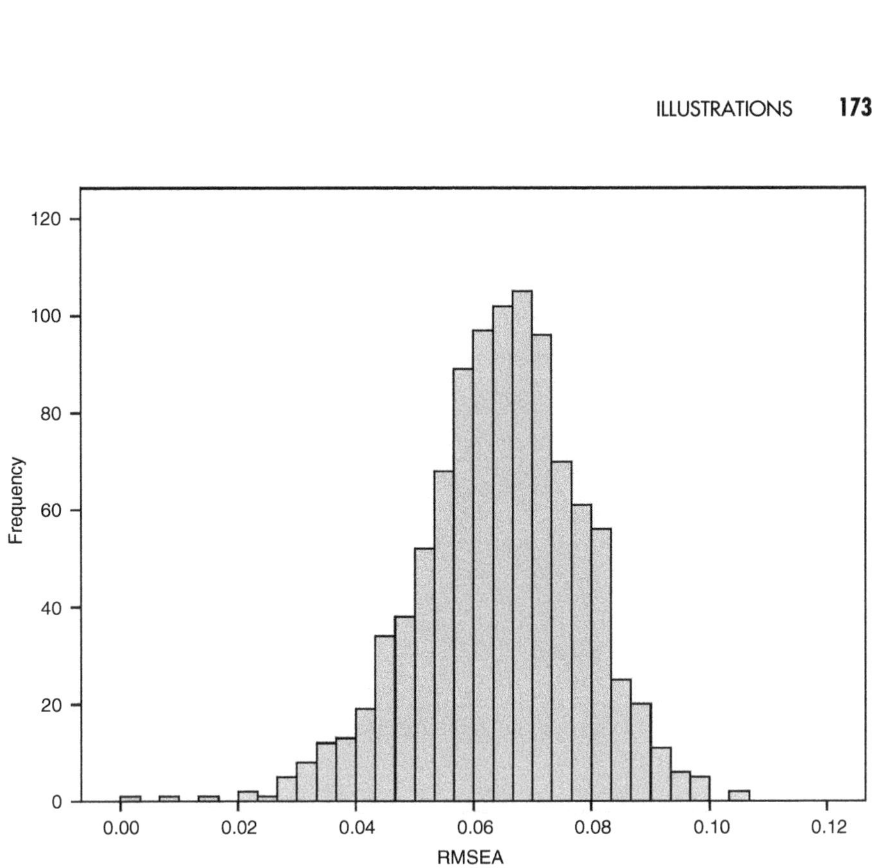

Figure 11.3 Frequency histogram for RMSEA over 1000 simulated samples from WAIS-R single-factor M2 model.

the 1,000 samples, with a mean RMSEA of .029 (SD = .013). In this distribution, the sample RMSEA estimate found in the real data is in the extreme right tail of the histogram, suggesting that the lack of fit found for M0 in the real data cannot be taken as consistent with M1 as the model underlying the real data. On this evidence, we would reject M0.

To evaluate M0 from a different perspective, a second alternative model M2 was created. Model M2 retains the invariant error variances (all fixed to 3.696), but introduces a small quadratic effect in addition to the linear effect in M0. The variance of the quadratic effect is fixed to .01, in contrast to the estimates of the variances of the intercept (2.977) and linear (.136) effects. This small variance was selected to limit the amount of variance explained by the quadratic effect, relative to the intercept and linear effects. Also, the covariances between the quadratic effect and the intercept and linear effects were limited at .008. This M2 model was used to generate data for $q = 1{,}000$ samples, with $N = 851$ per sample. The original model M0 was fit to each sample, generating the histogram of RMSEA values in Figure 11.6. The mean RMSEA value

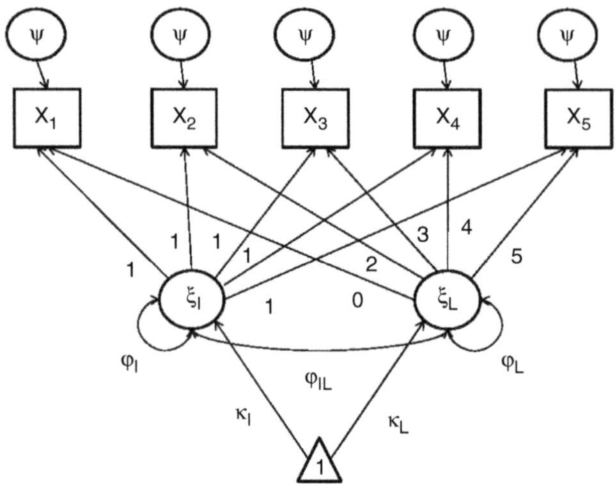

Figure 11.4 M0 linear latent growth model with invariant error variance.

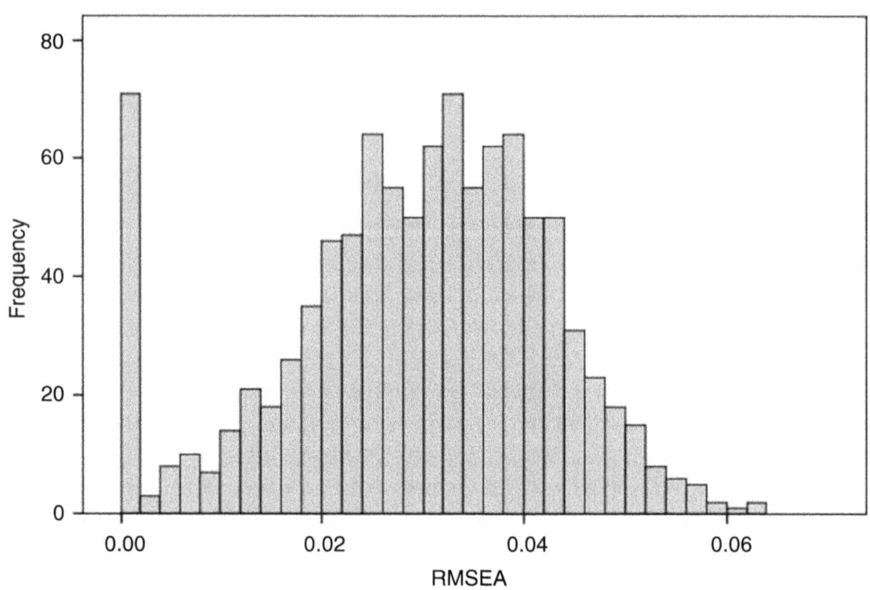

Figure 11.5 Frequency histogram for RMSEA over 1000 simulated samples from linear latent growth M1 model.

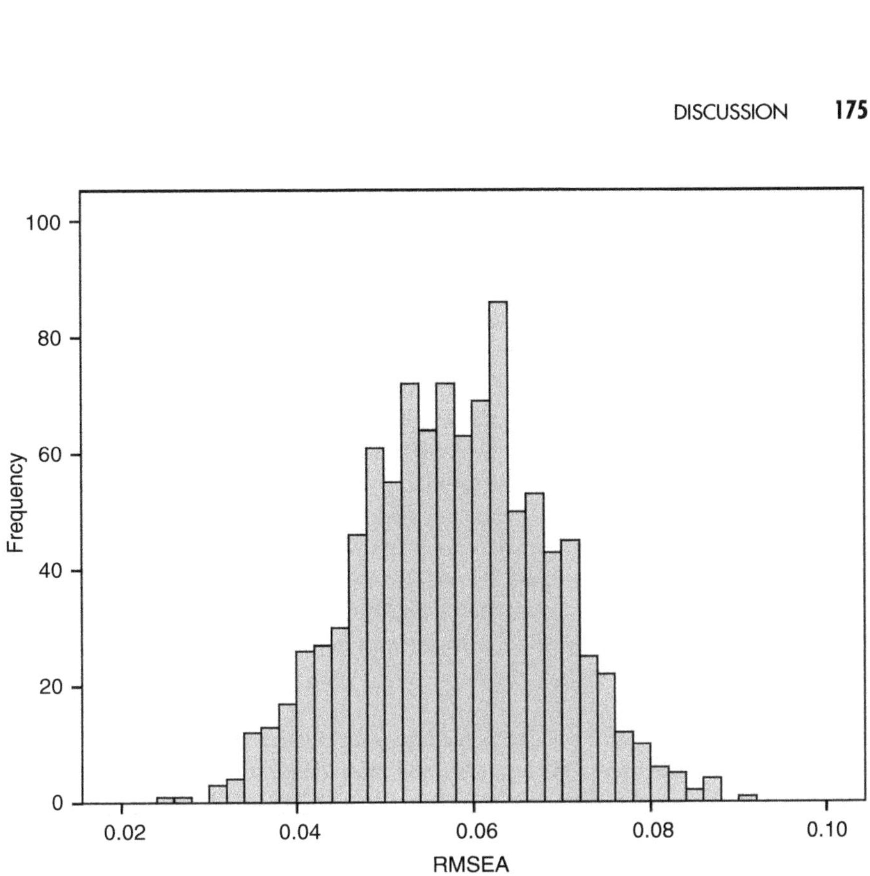

Figure 11.6 Frequency histogram for RMSEA over 1000 simulated samples from quadratic latent growth M2 model.

here is .058 (SD = .011), and our original RMSEA estimate from the real data falls at the 91st percentile in the distribution in Figure 11.6 (P2 = .09). While not as extreme as the results found with M1, the results for M2 again suggest that the lack of fit for M0 in the real data is larger than it should have been if M2 had been the model underlying the real data. We are again compelled to reject M0.

11.4 DISCUSSION

The fit evaluation method presented here is intended as a way of closely examining the idea of approximate fit in SEM. The method forces the researcher to explicitly define approximate fit by creating an alternative model M1. Judged against this alternative model M1, the model of interest M0 is an adequate approximation. The models (M0, M1) are exchangeable in this sense. The model M1 is fully specified, with all parameters fixed to values chosen by the researcher. Finally, the fit of M0 in the real data at hand is evaluated in the context of the fit of M0 in simulated data generated by M1. Here "fit" is evaluated using the researcher's preferred index of global fit (the RMSEA was chosen in the illustrations presented here). The model M0 is rejected if the fit in the simulated data is much better than the fit of M0 in the real data. Otherwise, M0 is retained.

One objection that can be raised with regard to this fit evaluation method lies in the choice of M1. What is to prevent a researcher from "cheating" by choosing an M1 model that is, in fact, quite different from M0? Such a model would presumably lead to substantial lack of fit for M0 in the simulated data, suggesting that M0 be retained unless the fit of M0 in the real data was very poor. The answer to this objection is that any researcher who uses this fit evaluation method should be required to report both M0 and M1 in detail, along with the fit results in the real and simulated data. Readers, reviewers, and journal editors must then judge whether the M0 and M1 models can really be considered "exchangeable." Any attempt to cheat by strategic choice of M1 would then need to first convince reviewers and journal editors that the choice of M1 is permissible.

A second objection to the above procedure is that there are many possible choices for alternative models M1, M2,..., MP, and they need not all lead to the same conclusion with regard to M0. We might find that the fit of M0 in relation to M1 is comparable to the fit of M0 in the real data. We might also find that the fit of M0 in relation to M2 is much better than the fit in the real data. Given such contrasting results, what decision should be made? We should first note that this situation is not really a weakness in the proposed method, but is rather an intrinsic difficulty in fit evaluation generally. If the fit of M0 in the real data is not perfect, different alternative models can make the lack of fit appear large or small depending on the context. The practice of evaluating fit against fixed cut-points hides this problem by failing to be specific about the "true model" to which M0 is to be considered an approximation. The proposed method simply makes clear what has always been true in any case. Faced with contrasting results across alternatives, the researcher can pursue any one of several approaches. One approach is to order the alternatives with regard to their theoretical importance or likelihood, and then follow the conclusion suggested by the top alternative model. Another approach is to adopt a conservative position and reject M0 if any of the alternatives suggest rejection. This approach would retain M0 only if it passes all hurdles. In general, the researcher must decide which approach is suitable given the goals of the research.

A third objection to the proposed method is that the simulated data in the examples given here are generated under assumptions of multivariate normality. These assumptions may not hold in the real data, and this fact may lead to inconsistencies between the outcomes of the analyses in real versus simulated data. This objection is certainly a potential problem, but it is one that can be addressed in a number of ways.

First, the researcher could simulate data from any theoretical distribution deemed appropriate. The use of normal distributions in the examples was not intended to be prescriptive. In practice, however, the researcher may not have a good alternative distribution to use in place of the normal. In this case, a different approach is to use a bootstrap procedure that would resample the real data as a basis for generating data. Prior to resampling, the researcher would need to impose the desired covariance and mean structure on the data using the structures implied by M1. This can be done using procedures developed by Bollen and Stine (1993).

To describe the Bollen–Stine procedure, let X be the $N \times p$ sample data matrix for N participants on p measured variables. Let S and $\overline{X}$ be the sample covariance matrix and mean vector, respectively, for X. Finally, let Σ_1 and μ_1 be the covariance matrix and mean vector, respectively, implied by M1. We can then use a triangular factoring $\Sigma_1 = \mathrm{T}_1\mathrm{T}_1'$ and $\mathrm{S} = \mathrm{AA}'$ to compute

$$\mathrm{X}^* = \left(\mathrm{X} - 1\overline{\mathrm{X}}'\right)\mathrm{A}'^{-1}\mathrm{T}_1' + 1\mu_1'. \tag{11.4}$$

Here 1 is an $N \times 1$ unit vector and X^* is an $N \times p$ matrix of transformed data. The transformed data will have a sample covariance matrix of Σ_1 and a mean vector of μ_1. The bootstrap procedure can then proceed using bootstrap samples of size N from the transformed data. If a new alternative model M2 is chosen, the data would need to be transformed once again using Σ_2 and μ_2 to reflect M2. This bootstrap procedure eliminates any prior assumptions about distributional form, apart from the usual bootstrap assumption that the sample distribution is representative of the population.

The widespread use of global indices of approximate fit has meant that many models are retained in spite of less-than-perfect fit (as indexed by chi-square), because researchers believe that in such models, the lack of fit is small enough to ignore for practical purposes. If we accept this practice, it is fair to ask the researcher to describe an alternative model that is similar enough to the researcher's model of interest so that the two models are exchangeable for all practical purposes. The fit evaluation method proposed here takes this exercise one step further. We ask whether the lack of fit found for the original model in real data could have arisen in a scenario in which the real data were generated by the specified alternative model. This question can be investigated by actually simulating data from the alternative model, evaluating the fit of the original model in this simulated data, and comparing the fits of the original model in the real and simulated data. If the fit in the real data is much worse than the fit in the simulated data, we reject the idea that the lack of fit is ignorable in the above sense. On the other hand, if the fits in the real and simulated data are comparable, this result lends support to the researcher's position that the lack of fit found in the real data can be ignored. The key element in this fit evaluation method is an attempt to make visible the reasoning process that many researchers covertly pursue when retaining a model on the basis of approximate fit. In making the process visible, we hope to force researchers to think more carefully about the nature of approximate fit, and whether the approximation argument is truly plausible for the data at hand.

REFERENCES

Barrett, P. (2007). Structural equation modeling: Adjudging model fit. *Personality and Individual Differences, 42*, 812–824.

Beauducel, A., & Wittmann, W. (2005). Simulation study on fit indices in confirmatory factor analysis based on data with slightly distorted simple structure. *Structural Equation Modeling, 12*, 41–75.

Bollen, K. A., & Stine, R. A. (1993). Bootstrapping goodness-of-fit measures in structural equation modeling. In K. A. Bollen and J. S. Long (Eds.) *Testing structural equation models* (pp. 111–135). Newbury Park, CA: Sage.

Browne, M. W., & Cudeck, R. (1993). Alternative ways of assessing model fit. In K. A. Bollen and J. S. Long (Eds.) *Testing structural equation models* (pp. 136–162). Newbury Park, CA: Sage.

Fan, X., & Sivo, S. A. (2005). Sensitivity of fit indices to misspecified structural or measurement model components: Rationale of the two-index strategy revisited. *Structural Equation Modeling, 12*, 343–367.

Horn, J. L., & McArdle, J. J. (1992). A practical guide to measurement invariance in research on aging. *Experimental Aging Research, 18*, 117–144.

Marsh, H. W., Hau, Kit-Tai, & Wen, Z. (2004). In search of Golden Rules: Comment on hypothesis-testing approaches to setting cutoff values for fit indexes and dangers in overgeneralizing Hu and Bentler's (1999) findings. *Structural Equation Modeling, 11*, 320–341.

McDonald, R. P., & Ho, M. H. (2002). Principles and practice in reporting structural equation analyses. *Psychological Methods, 7*, 64–82.

Millsap, R. E. (2007). Structural equation modeling made difficult. *Personality and Individual Differences, 42*, 875–881.

Muthén, L. K., & Muthén, B. O. (1998–2009). *Mplus user's guide, fourth edition.* Los Angeles, CA: Muthén & Muthén.

Preacher, K. J., Wichman, A. L., MacCallum, R. C., & Briggs, N. E. (2008). *Latent growth curve modeling.* Thousand Oaks, CA: Sage Publications.

Saris, W. E., Satorra, A., & van der Veld, W. (2009). Testing structural equation models, or detection of misspecification? *Structural Equation Modeling,16* , 561–582.

Steiger, J. H., & Lind, J. M. (1980). Statistically based tests for the number of common factors. Paper presented at the annual meeting of the Psychometric Society, Iowa City, IA.

Thurstone, L. L. (1947). *Multiple factor analysis.* Chicago: University of Chicago Press.

Yuan, K. H. (2005). Fit indices versus test statistics. *Multivariate Behavioral Research, 40*, 115–148.

APPENDIX

Mplus code used to simulate data from M1 and M2 in the WAIS-R and latent growth model examples is given below.

1) For M1 in the WAIS-R analyses

```
TITLE: simulations for two-grp WAIS verbal model unequal loadings .1 diff
MONTECARLO:
   NAMES ARE info voca comp arit;
   NGROUPS = 2;
   NOBSERVATIONS = 940 940;
   NREPS = 1000;
   SEED = 878;
   RESULTS = waismonte22.dat;
MODEL POPULATION:
   f1 BY info@.9 voca@1.0 comp@.88 arit@.7;
   [f1@0];
   [info@9.82 voca@9.41 comp@9.61 arit@9.5];
   f1@8.4;
   info@2.0 voca@1.0 comp@4.0 arit@3.0;
MODEL : f1 BY voca@1.0 info* comp* arit*;
   [voca* info* comp* arit*]; [f1@0];
   f1*;
MODEL POPULATION-G2:
   f1 BY info@.8 voca@1.0 comp@.78 arit@.6;
   [f1@0];
   [info@8.98 voca@9.25 comp@9.33 arit@9.2];
   f1@7.8;
   info@1.8 voca@1.2 comp@4.0 arit@3.0;
MODEL G2: [voca* info* comp* arit*];
   f1*;
OUTPUT: TECH9;
```

2) For M2 in the WAIS-R analyses

```
TITLE: simulations for two-grp WAIS verbal model equal loadings, nonzero uniq corr
MONTECARLO:
   NAMES ARE info voca comp arit;
   NGROUPS = 2;
```

```
   NOBSERVATIONS = 940 940;
   NREPS = 1000;
   SEED = 555;
   RESULTS = waismonte21.dat;
MODEL POPULATION:
   f1 BY info@.9 voca@1.0 comp@.88 arit@.7;
   [f1@0];
   [info@9.82 voca@9.41 comp@9.61 arit@9.5];
   f1@8.4;
   info@2.0 voca@1.0 comp@4.0 arit@3.0;
   info with comp@.6;
MODEL : f1 BY voca@1.0 info* comp* arit*;
   [voca* info* comp* arit*]; [f1@0];
   f1*;
MODEL POPULATION-G2:
   f1 BY info@.9 voca@1.0 comp@.88 arit@.7;
   [f1@0];
   [info@8.98 voca@9.25 comp@9.33 arit@9.2];
   f1@7.8;
   info@1.8 voca@1.2 comp@4.0 arit@3.0;
   info with comp@.5;
MODEL G2: [voca* info* comp* arit*];
   f1*;
OUTPUT: TECH9;
```

3) For M1 in the latent growth model analyses

```
TITLE: simulations for lgc mother-child model
MONTECARLO:
   NAMES ARE c1 c2 c3 c4 c5;
   NOBSERVATIONS = 851;
   NREPS = 1000;
   SEED = 547;
   RESULTS = lgcmonte1.dat

MODEL POPULATION:
   f1 BY c1@1 c2@1 c3@1 c4@1 c5@1;
   [f1@38.00];
   f2 BY c1@0 c2@2 c3@3 c4@4 c5@5;
   [f2@-.358]; f1 with f2@.253;
```

```
   f1@2.977; f2@.136;
   [c1@0 c2@0 c3@0 c4@0 c5@0];
   c1@3.3 c2@4.1 c3@4.1 c4@3.3 c5@4.1;
MODEL: f1 by c1@1 c2@1 c3@1 c4@1 c5@1;
   f2 by c1@0 c2@2 c3@3 c4@4 c5@5;
   [C1@0 C2@0 C3@0 C4@0 C5@0];
   F1*; F2*; [F1* F2*];
   F1 with F2*;
   C1*(1);
   C2*(1);
   C3*(1);
   C4*(1);
   C5*(1);
OUTPUT: TECH9;
```

4) For M2 in the latent growth model analyses

```
TITLE: simulations for lgc mother-child model quadratic case
MONTECARLO:
   NAMES ARE c1 c2 c3 c4 c5;
   NOBSERVATIONS = 851;
   NREPS = 1000;
   SEED = 780;
   RESULTS = lgcmonte21.dat
MODEL POPULATION:
   f1 BY c1@1 c2@1 c3@1 c4@1 c5@1;
   [f1@38.00];
   f2 BY c1@0 c2@2 c3@3 c4@4 c5@5;
   f3 by c1@0 c2@4 c3@9 c4@16 c5@25;
   [f3@–.01]; f1 with f3@.008;
   f2 with f3@.008;
   [f2@–.358]; f1 with f2@.253;
   f3@.01;
   f1@2.977; f2@.136;

   [c1@0 c2@0 c3@0 c4@0 c5@0];
   c1@3.696 c2@3.696 c3@3.696 c4@3.696 c5@3.696;
MODEL: f1 by c1@1 c2@1 c3@1 c4@1 c5@1;
   f2 by c1@0 c2@2 c3@3 c4@4 c5@5;
   [C1@0 C2@0 C3@0 C4@0 C5@0];
```

```
  F1*; F2*; [F1* F2*];
  F1 with F2*;
  C1*(1);
  C2*(1);
  C3*(1);
  C4*(1);
  C5*(1);
OUTPUT: TECH9;
```

Chapter 12
Fungible Parameter Estimates in Latent Curve Models

Robert C. MacCallum, Taehun Lee, and
Michael W. Browne

This chapter explores a phenomenon that occurs to some degree in the process of fitting virtually any statistical model to empirical data. The phenomenon involves a kind of uncertainty that is inherent in parameter estimates, uncertainty arising from the sensitivity of the relationship between those parameter estimates and the fit of the model to the data. If there is low sensitivity in this relationship, then small changes in model fit would be associated with large changes in parameter estimates. Under such circumstances, very different parameter estimates will be associated with nearly identical model fit, calling into question an investigator's substantive interpretation of the optimal parameter estimates. On the other hand, high sensitivity would mean that small changes in fit would be associated with small changes in parameter estimates, thus supporting more rigorous substantive interpretation of parameter estimates.

We will explore this phenomenon in the context of linear latent curve modeling (LCM). LCM is a special case of structural equation modeling (SEM) widely used in psychology to model change over time. In this modeling context we will define, quantify, and illustrate the phenomenon just described. Our findings illustrate that this type of uncertainty in LCM parameters can cast considerable doubt on the interpretation of model parameter estimates. The work reported here represents one aspect in a series of collaborative projects being carried out by the authors.

12.1 FUNGIBLE PARAMETER ESTIMATES

Consider the general context of fitting a statistical model to empirical data. The formal expression of the model includes parameters to be estimated; let $\boldsymbol{\theta}$ be a vector containing q such parameters. The model is fit to the data by optimizing a fitting function, F, which is a function of the model, the parameters in $\boldsymbol{\theta}$, and the data. Typical fitting functions include maximum likelihood and least squares. Without loss of generality, let us assume that F is to be minimized. A vector of parameter estimates, $\hat{\boldsymbol{\theta}}$, yields the minimized value of the discrepancy function, $\hat{F}$. Assuming all parameters in $\boldsymbol{\theta}$ are identified, any perturbation of $\hat{\boldsymbol{\theta}}$ will result in an increase in $\hat{F}$.

Given a value of $\hat{F}$, suppose we choose another value of this function that is just slightly larger than $\hat{F}$; call it $\hat{F}^*$. The perturbed $\hat{F}^*$ is chosen so that its difference from $\hat{F}$ is so small as to not be of practical importance with respect to model fit. We then wish to obtain parameter estimates that would yield this slightly suboptimal level of model fit represented by the value $\hat{F}^*$. Let any such solution be designated $\hat{\boldsymbol{\theta}}^*$. We will show in the next section how such values can be computed. For now, we note that in all but trivial situations there will be an infinite number of solutions for $\hat{\boldsymbol{\theta}}^*$. The set of such solutions has been referred to as a *fungible* set (Waller, 2008) because all of these solutions for $\hat{\boldsymbol{\theta}}^*$ are fungible, or exchangeable, with respect to the level of model fit provided. And because the difference between $\hat{F}$ and $\hat{F}^*$ is not of practical significance, all of the solutions for $\hat{\boldsymbol{\theta}}^*$ explain the observed data virtually as well as the optimal solution, $\hat{\boldsymbol{\theta}}$.

Examination of such a fungible set of solutions for $\hat{\boldsymbol{\theta}}^*$ relative to $\hat{\boldsymbol{\theta}}$ can be highly informative and can have important implications for interpretation of the optimal estimates in $\hat{\boldsymbol{\theta}}$. If some of the solutions in $\hat{\boldsymbol{\theta}}^*$ are very different from each other and very different from $\hat{\boldsymbol{\theta}}$, the implication is that there exist very different parameter estimates that provide nearly equivalent fit of the model to the data. Such a circumstance would call into question any substantive interpretation specific to $\hat{\boldsymbol{\theta}}$. On the other hand, if the fungible solutions are not very different from each other or from $\hat{\boldsymbol{\theta}}$, then reasonably rigorous interpretation of estimates in $\hat{\boldsymbol{\theta}}$ would be more justifiable.

Waller (2008) investigated this phenomenon in the context of the linear regression model. In ordinary least squares (OLS) regression with k independent variables we obtain a $k \times 1$ vector of regression coefficients, $\hat{\boldsymbol{\beta}}$, and a squared multiple correlation, R^2. Then choosing a perturbed value R^{*2} that is just slightly smaller than R^2, Waller showed how to compute linear weights, $\hat{\boldsymbol{\beta}}^*$, that yield that slightly suboptimal R^{*2}; he called such solutions "fungible regression weights." Waller also showed that all members of such a fungible set of regression weights yield the same sum of squared residuals. In an application to empirical data, Waller illustrated how solutions for $\hat{\boldsymbol{\beta}}^*$ can be very different from $\hat{\boldsymbol{\beta}}$ and from each other, and he discussed the considerable implications for interpretation of relative magnitude of regression coefficients. In short, the existence of an infinite number of alternative solutions that, for practical purposes, account for essentially the same amount of variance in the dependent variable as does the optimal OLS solution, along with the observed high variation among those alternative solutions, creates substantial uncertainty for the substantive interpretation of the OLS weights.

In this chapter we will explore this same phenomenon in the context of LCM, demonstrating how the existence and nature of fungible parameter estimates can inject a kind of uncertainty into interpretations about individual differences in patterns of change over time. Because LCM is a special case of SEM, we will first provide a general framework for defining and computing fungible parameter estimates in SEM. We will then implement this approach in an application of LCM to empirical data and discuss the results and implications.

12.2 FUNGIBLE PARAMETER ESTIMATES IN SEM

We present here a general framework for defining and computing fungible parameter estimates in SEM. The developments described in this section represent ongoing work by the authors and have been presented at several conferences (MacCallum, Browne, & Lee, 2009; MacCallum, Lee, & Browne, 2010a); several papers are in preparation.

A structural equation model implies a structure for the covariances among the measured variables under study. That covariance structure can be represented as

$$\boldsymbol{\Sigma} = \boldsymbol{\Sigma}(\boldsymbol{\theta}), \tag{12.1}$$

where $\boldsymbol{\Sigma}$ is the population covariance matrix for the p manifest variables and $\boldsymbol{\theta}$ is a $q \times 1$ vector of model parameters. The model represents $\boldsymbol{\Sigma}$ as a matrix-valued function of the parameters in $\boldsymbol{\theta}$. Given a sample covariance matrix, $\mathbf{S}$, the model is fit to the data by obtaining parameter estimates, $\hat{\boldsymbol{\theta}}$, that minimize a measure of discrepancy between $\hat{\boldsymbol{\Sigma}} = \boldsymbol{\Sigma}(\hat{\boldsymbol{\theta}})$ and $\mathbf{S}$. For present purposes we consider the maximum likelihood (ML) discrepancy function:

$$F(\boldsymbol{\Sigma}, \mathbf{S}) = \ln|\Sigma| - \ln|\mathbf{S}| + tr(\mathbf{S}\boldsymbol{\Sigma}^{-1}) - p. \tag{12.2}$$

Substituting $\hat{\boldsymbol{\Sigma}}$ into this expression yields the minimized value of the discrepancy function, $\hat{F} = F(\hat{\boldsymbol{\Sigma}}, \mathbf{S})$.

A common extension of this model involves the addition of a mean structure; i.e., a model specifying the means of the measured variables as a function of model parameters. This extension introduces additional parameters into the full vector, $\boldsymbol{\theta}$, of model parameters and extends the formula for the discrepancy function in (12.2) to include terms associated with means. The approach described in the following paragraphs for defining and computing fungible parameter estimates is applicable to an SEM with or without a mean structure. For simplicity of presentation, we use the context of the covariance structure and the discrepancy function as shown in (12.2).

Assuming all parameters in $\boldsymbol{\theta}$ are identified, any perturbation of $\hat{\boldsymbol{\theta}}$ would cause the value of $\hat{F}$ to increase. The first step in defining fungible parameter estimates is then to specify a value of the discrepancy function that is just slightly larger than $\hat{F}$. This perturbed value will be denoted $\hat{F}^*$, and is chosen so that the difference between $\hat{F}^*$ and $\hat{F}$ is small enough so as not to be considered of practical significance. One approach to specifying $\hat{F}^*$ will be discussed later in the context of an empirical demonstration. Given a specified $\hat{F}^*$, the problem then is to obtain parameter estimates, $\hat{\boldsymbol{\theta}}^*$, that would yield this slightly suboptimal level of model fit. That is, we wish to find $\hat{\boldsymbol{\theta}}^*$ such that

$$F(\boldsymbol{\Sigma}(\hat{\boldsymbol{\theta}}^*), \mathbf{S}) = \hat{F}^*. \tag{12.3}$$

This problem is approached by perturbing the ML estimates in $\hat{\boldsymbol{\theta}}$ so that $\hat{F}$ increases to a value of $\hat{F}^*$.

The desired perturbation of $\hat{\boldsymbol{\theta}}$ is accomplished by first selecting any unit length vector, $\mathbf{d}_n$ of order $q \times 1$. A perturbed parameter vector is then defined as

$$\hat{\boldsymbol{\theta}}_n^* = \hat{\boldsymbol{\theta}} + \kappa_n \mathbf{d}_n, \tag{12.4}$$

where κ_n is an unknown scaling constant that defines the degree of perturbation of $\hat{\boldsymbol{\theta}}$. Substitution from (12.4) into (12.3) yields

$$F(\Sigma(\hat{\boldsymbol{\theta}} + \kappa_n \mathbf{d}_n), \mathbf{S}) = \hat{F}^* \tag{12.5}$$

or equivalently

$$F(\Sigma(\hat{\boldsymbol{\theta}} + \kappa_n \mathbf{d}_n), \mathbf{S}) - \hat{F}^* = 0. \tag{12.6}$$

Given $\hat{\boldsymbol{\theta}}$, $\mathbf{d}_n$, $\mathbf{S}$, and $\hat{F}^*$, the computational problem is to solve for the scaling constant κ_n. We have developed a computational approach based on a general root-solving algorithm suggested by Brent and described in detail in Section 9.2 of Press, Teukolsky, Vetterling, & Flannery (1992). We used this same algorithm for a similar purpose in a study involving the identification of classes of alternative hypotheses that yield identical power for tests of model fit in SEM (MacCallum, Lee, & Browne, 2010b). The algorithm is described in the appendix of that article. Using this computational algorithm, for any specified unit vector, $\mathbf{d}_n$, a solution for κ_n can be obtained, in turn yielding a solution for a perturbed parameter vector $\hat{\boldsymbol{\theta}}_n^* = \hat{\boldsymbol{\theta}} + \kappa_n \mathbf{d}_n$ that yields slightly suboptimal model fit as specified by $\hat{F}^*$. There will be an infinite number of such solutions corresponding to the infinite set of unit vectors, $\mathbf{d}_n$. In our application of this procedure we have selected a large number of vectors $\mathbf{d}_n$ to systematically span the space of the parameters of interest. If we are computing fungible estimates for two parameters, then vectors $\mathbf{d}_n$ are selected to systematically sweep a two-dimensional space, etc. As the number of parameters increases, this procedure quickly becomes computationally burdensome. One approach for addressing this problem will be discussed in the context of our example involving LCM.

The solutions for $\hat{\boldsymbol{\theta}}^*$ produced by this approach represent a set of fungible parameter estimates. Each such solution produces a value of the discrepancy function, $\hat{F}^*$, that is just slightly larger than the optimized value, $\hat{F}$. In a sense, then, these fungible estimates represent a set of alternative explanations of the data that are different from the optimal solution in $\hat{\boldsymbol{\theta}}$, with the alternative solutions being as good as the optimal solution for practical purposes.

12.3 FUNGIBLE PARAMETER ESTIMATES IN LCM

12.3.1 Overview of LCM

LCM is a special case of SEM, essentially consisting of a restricted factor analysis model with an accompanying mean structure. Bollen and Curran (2006) and Preacher, Wichman, MacCallum, & Briggs (2008) provide excellent treatments of basic LCM and its many variants and extensions. In a basic linear LCM there is one outcome variable measured on m occasions. The data model can be represented as

$$\mathbf{y} = \boldsymbol{\tau} + \boldsymbol{\Lambda}\boldsymbol{\eta} + \boldsymbol{\delta}, \tag{12.7}$$

where $\mathbf{y}$ is an $m \times 1$ vector of scores on the m repeated measures of the outcome variable; $\boldsymbol{\tau}$ is an $m \times 1$ vector of intercepts; $\boldsymbol{\Lambda}$ is an $m \times r$ matrix of loadings for the m repeated measures on r latent curve factors; $\boldsymbol{\eta}$ is an $r \times 1$ vector of scores on the r latent curves; and $\boldsymbol{\delta}$ is an $m \times 1$ vector of residuals. For present purposes, and as is typical in many applications, the intercept vector, $\boldsymbol{\tau}$, can be set to zero. Thus, the model specifies that a vector of scores for a measured variable on m occasions is a linear combination of latent curves, plus residuals.

The model in (12.7) implies a covariance structure for the repeated measures:

$$\boldsymbol{\Sigma}_{\mathrm{yy}} = \boldsymbol{\Lambda}\boldsymbol{\Phi}\boldsymbol{\Lambda}' + \boldsymbol{\Psi}, \tag{12.8}$$

where $\boldsymbol{\Phi} = E(\boldsymbol{\eta}\boldsymbol{\eta}')$, the covariance matrix for the latent curves; and $\boldsymbol{\Psi} = E(\boldsymbol{\delta}\boldsymbol{\delta}')$, the covariance matrix for the error terms. The model in (12.7) also implies a mean structure given by

$$\boldsymbol{\mu}_{\mathrm{y}} = \boldsymbol{\Lambda}\boldsymbol{\alpha}, \tag{12.9}$$

where $\boldsymbol{\mu}_{\mathrm{y}}$ is an $m \times 1$ vector of means on the m repeated measures, and $\boldsymbol{\alpha} = E(\boldsymbol{\eta})$, containing the means on the latent curves across the individuals.

The model parameters are thus the latent curve loadings in $\mathbf{\Lambda}$, the latent curve means in $\boldsymbol{\alpha}$, the latent curve variances and covariances in $\mathbf{\Phi}$, and the residual variances in the diagonals of $\mathbf{\Psi}$. The parameter vector, $\boldsymbol{\theta}$, defined above as containing all of the model parameters, would contain parameters in these matrices of the LCM framework. In typical applications, the latent curve loadings are fixed by the investigator and the remaining parameters are estimated. Figure 12.1 shows a path diagram for such a model, with five repeated measures of an outcome variable, and two latent curves. The first latent curve is an intercept factor, with all loadings fixed at 1, and the second latent curve is a slope factor with loadings fixed to represent linear change over time. In fitting such a model to data, one would obtain estimates of the means of the intercepts and slopes (α_1 and α_2), the variances of the intercepts and slopes (φ_{11} and φ_{22}), the covariance of intercepts and slopes (φ_{21}), and the error variance at each occasion (ψ_1, ψ_2, ψ_3, ψ_4, ψ_5) (often constrained to be invariant across occasions).

There are many ways to extend this model. One extension that will be employed in the illustration to follow involves introducing an additional measured variable to serve as a predictor of intercepts and slopes. The full range of linear LCM can be fit easily using standard SEM software.

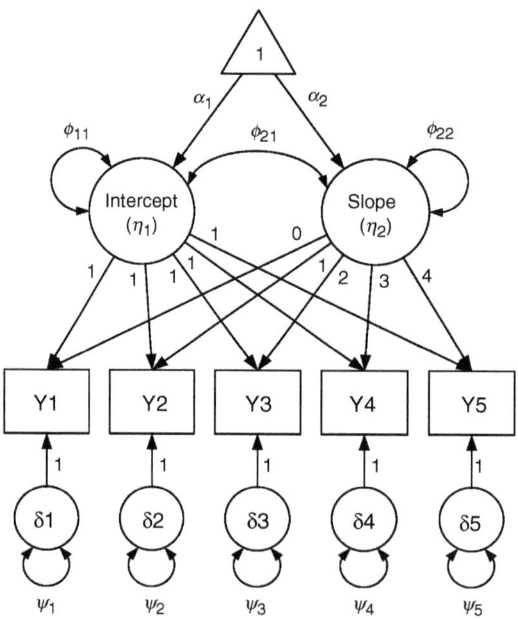

Figure 12.1 Path diagram of latent curve model.

12.3.2 Empirical Example of LCM

Our example makes use of data gathered in the NICHD sponsored Study of Early Child Care (NICHD Early Child Care Research Network, 2006) and used previously by Preacher et al. (2008) in their volume on basic LCM. The outcome variable is a measure of closeness between mother and child obtained from the mother's responses on a 15-item scale. Responses were obtained from each mother in the sample on five occasions, when the child was in grades 1, 3, 4, 5, and 6. The sample consisted of $N = 851$ mother–child pairs. The means across the five occasions were 37.95, 37.28, 37.05, 36.57, and 36.14. The slight downward trend indicates slowly decreasing perceived closeness between mother and child. An LCM as shown in Figure 12.1 was fit to the data so as to model individual differences in slope and intercept. (Note that in specifying this model the loadings for the slope factor were set at 0, 2, 3, 4, 5 so that the intercept would be estimated in terms of level at grade 1, and to represent the fact that no measurement of the outcome was obtained at grade 2.)

The following measures of model fit were obtained: $\chi^2_{14} = 75.90$ ($p < .001$), RMSEA $= 0.072$, NNFI $= 0.96$. (For future reference we also note that $\hat{F} = .089$.) Model fit in this case would be considered fair to good. Parameter estimates are given in Table 12.1. These estimates indicate a mean initial level of the closeness rating of 38.00 with a mean decline of about 0.36 points per year. Standard deviations of intercepts and slopes are estimated to be about 1.72 and 0.37, respectively. Standard errors of all parameter estimates are shown in Table 12.1; resulting tests of statistical significance show all parameter estimates to be significantly different from zero.

12.3.3 Fungible Estimates of Intercept and Slope Parameters

The parameter estimates in Table 12.1 serve to minimize the ML discrepancy function, yielding a value of $\hat{F} = .089$. Any perturbation of the estimates in Table 12.1 would result in an increase in that value of $\hat{F}$. To explore fungible parameter estimates in

Table 12.1 Parameter estimates and standard errors for latent curve model of mother–child closeness

Intercept mean	38.00 (.08)
Intercept variance	2.98 (.29)
Slope mean	−.36 (.02)
Slope variance	.14 (.02)
Intercept–slope covariance	.25 (.06)
Error variance	3.70 (.10)

this example, we choose a value of the discrepancy function just slightly larger than $\hat{F} = .089$, then compute parameter estimates that produce that level of fit. This choice is difficult if we attempt to make it using the scale of F directly. The choice can be made more meaningfully if we employ a fit index that is a direct function of F. One such index is the RMSEA (Browne & Cudeck, 1993; Steiger & Lind, 1980). The relationship between the sample value of RMSEA, designated here as $\hat{\varepsilon}$, and the discrepancy function value, $\hat{F}$, is given by

$$\hat{\varepsilon} = \sqrt{\frac{\hat{F} - d/n}{d}}, \tag{12.10}$$

where d is degrees of freedom for the model, and $n = (N - 1)$. (Note: $\hat{\varepsilon}$ is defined as zero if $(\hat{F} - d/n) < 0$.) In the present case, $\hat{F} = .089$, $d = 14$, $n = 850$, and $\hat{\varepsilon} = .072$. Rather than perturb $\hat{F}$ directly, we suggest perturbing RMSEA. Let us set a perturbed RMSEA as $\hat{\varepsilon}^* = .077$, a value only .005 larger than the observed $\hat{\varepsilon}$. We believe that most researchers working in SEM would consider this difference to be very small in practical terms. Using (12.10) we can determine that a corresponding perturbed value of the discrepancy function would be $\hat{F}^* = .100$.

The task at this point is to implement the procedure described earlier for computing fungible parameter estimates. That is, we wish to find estimates of the model parameters, meaning alternative values to those listed in Table 12.1, that would yield a level of model fit represented by $\hat{F}^* = .100$, rather than a minimized discrepancy function value of $\hat{F} = .089$ that reflects the estimates shown in Table 12.1. Implementation of the computational procedure described earlier becomes problematic with more than a few parameters. In the present case, it would be difficult to obtain fungible estimates for all six model parameters simultaneously. To work around this problem, we adopted a procedure of choosing two focal parameters and obtaining fungible estimates of those while holding all other parameters, which we will call nuisance parameters, at the ML estimates shown in Table 12.1. We acknowledge that this approach is not ideal. Most importantly, this approach ignores the fact that the focal parameters are almost certainly correlated with the nuisance parameters. We will discuss this point further later in this chapter. In any event, we implemented this approach with attention to two different sets of focal parameters: first, we obtained fungible estimates of the mean and variance of the intercepts; second, we obtained fungible estimates of the mean and variance of the slopes.

Figure 12.2 shows a contour of fungible estimates of the intercepts, holding all other parameters at their ML estimates. Results are shown for mean and standard deviation of intercepts. The point in the center of the contour shows the ML estimates of intercept mean and standard deviation, which are associated with a discrepancy function value of $\hat{F} = .089$. Each point on the contour represents an alternative solution, all of which are associated with a slightly suboptimal fit, $\hat{F}^* = .100$. Evaluation of this figure should focus on the range of parameter estimates represented

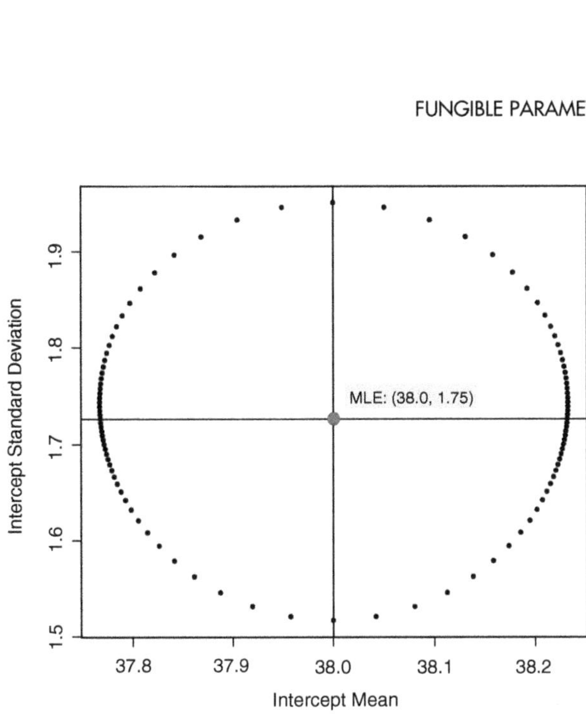

Figure 12.2 Fungible estimates of mean and standard deviation of intercepts for latent curve model of mother-child closeness.

by the contour, particularly how different they are from each other, and how different they are from the ML estimates. In this instance, such an inspection probably would not raise concern. Intercept means and standard deviations represented by points around the contour do not vary radically from each other and are not highly discrepant from the ML estimates. Fungible estimates of the mean intercept range from about 37.77 to 38.23, and fungible estimates of the standard deviation of the intercepts range from about 1.52 to 1.94. These ranges are rather narrow, such that solutions in these ranges would not alter substantive interpretation.

Figure 12.3 shows a contour of fungible estimates for the mean and standard deviation of slopes, again with all other parameters held fixed at their ML estimates, and with the ML estimates for mean and standard deviation of slopes shown in the center of the contour. As was the case with the fungible contour for intercept mean and standard deviation, the contour for slope parameters is relatively compact. Points on the contour represent alternative solutions for mean and standard deviation of slopes that yield slightly suboptimal fit, but the range in these solutions is not particularly large, and they are not highly deviant from the ML estimates. Fungible estimates for the mean slope range from about −.30 to −.42, and fungible estimates for the standard deviation of slopes range from about .30 to .44. Adoption of any of these alternative solutions would not substantially change interpretation regarding the slope parameters. This finding lends more confidence to the substantive interpretation of the ML estimates of slope mean and variance in Table 12.1.

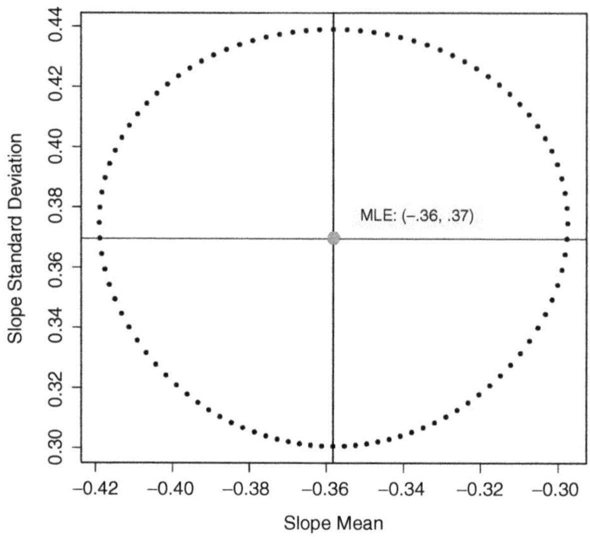

Figure 12.3 Fungible estimates of mean and standard deviation of slopes for latent curve model of mother-child closeness.

12.3.4 Fungible Estimates of Covariate Effects

We extend the example now by introducing a covariate into the model. Specifically, we introduce child's gender as a predictor of intercept and slope, with gender coded as 1 for boys and 0 for girls. This extension allows us to investigate whether mothers' initial level or rate of change in perceived closeness to the child depends on the gender of the child. Formally, this extended model contains two additional parameters, weights representing the effect of child's gender on intercept and on slope, respectively.

When fit to the sample data this model yielded $\chi^2_{17} = 81.69$ ($p < .0001$), $\hat{\varepsilon} = .067$, NNFI = .96, and $\hat{F} = .096$, again indicating adequate-to-good model fit. ML parameter estimates for this model are given in Table 12.2. Comparing the entries in Table 12.2 to those for the unconditional model in Table 12.1, one can see that estimates of parameters that are common to the two models are very similar. Focusing on the effects of child's gender in Table 12.2, we see an estimated effect on intercept of $-.29$ and an estimated effect on slope of $-.05$, meaning that mothers' ratings of closeness to boys tend to start a bit lower and decrease slightly faster in comparison to ratings of closeness to girls. Considering standard errors of these estimates, however, we see that neither effect is significantly different from zero. The 95% confidence interval for the parameter representing the effect of gender on intercept would have bounds of

Table 12.2 Parameter estimates and standard errors for latent curve model of mother–child closeness with child's gender as covariate

Intercept mean	38.14 (.11)
Intercept variance	2.96 (.28)
Slope mean	−.33 (.03)
Slope variance	.14 (.02)
Intercept-slope covariance	.25 (.06)
Error variance	3.70 (.10)
Gender effect on slope	−.05 (.04)
Gender effect on intercept	−.29 (.16)

approximately (−.61, +.03), and the confidence interval for the parameter representing the effect of gender on slope would have bounds of approximately (−.13, +.03). Despite the fact that both intervals contain zero, implying statistical nonsignificance, it is still informative to compute and examine fungible values of these parameters; it could happen that fungible estimates might be considerably larger and very different from these ML estimates.

To obtain fungible estimates for effects of gender on intercept and slope of mothers' ratings of closeness, we first must specify a small perturbation in the fit of the model just considered. We chose to set the perturbed RMSEA at $\hat{\varepsilon}^* = .072$, which is larger than the observed RMSEA by .005. This perturbed RMSEA corresponds to a perturbed discrepancy function value of $\hat{F}^* = .108$. Holding all other parameter estimates at their ML values, we then applied the computational procedure described earlier to obtain fungible estimates of the two parameters representing effects of child's gender on intercept and slope that would yield this slightly suboptimal level of model fit.

Figure 12.4 shows the resulting contour of fungible effects, with the ML estimates represented by the point in the center. Inspection of this contour reveals a much larger range of values compared to that seen in Figures 12.2 and 12.3 for means and standard deviations of intercepts and slopes. In Figure 12.4 we see that fungible estimates of the effect of child's gender on intercept range from about −.81 to +.22, and fungible estimates for the effect of child's gender on slope range from about −.18 to +.08. These ranges reflect considerable variation in substantive interpretation of these effects. To illustrate further, we show two selected points in the extreme ends of the elliptical contour. These points represent two alternative solutions, both of which yield equal fit to the data ($\hat{\varepsilon}^* = .072$), but which are quite different from each other and from the ML estimates. The point at the lower right represents a solution where the effect of child's

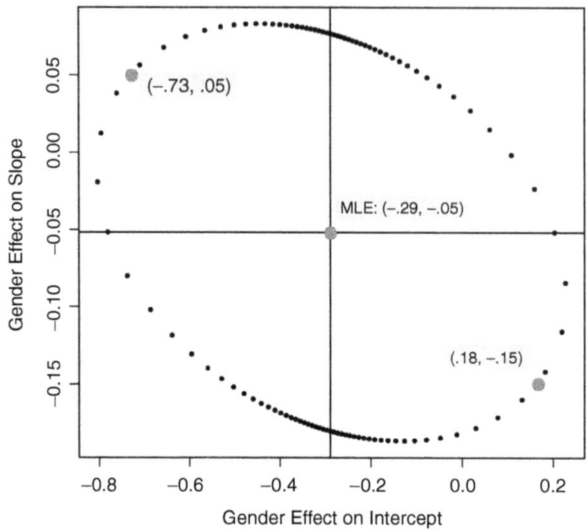

Figure 12.4 Fungible estimates of effect of child's gender on intercepts and slopes for latent curve model of mother-child closeness.

gender on intercept is .18 and the effect on slope is −.15. That is, according to these fungible estimates, mothers' ratings of closeness to boys, as compared to girls, would have an initial level (Grade 1) about .18 higher, but would decrease faster by about .15 per grade. The point at the upper left represents a very different pattern, where mothers' ratings of closeness to boys, as compared to girls, would have an initial level about .73 *lower*, but would decrease more slowly by about .05 per year. Obviously, these two fungible solutions result in very different substantive interpretations, even though they provide equivalent fit to the data; these solutions are also quite different from the ML solution in terms of interpretation of effects. These results would call into question any substantive interpretation of the ML estimates of the covariate effects in Table 12.2.

Given that the ML estimates for the gender effects were not statistically significant, it is tempting to ask whether the fungible estimates of these effects would be significantly different from zero. Unfortunately, at the current stage of development of these methods, it is not feasible to obtain standard errors or conduct significance tests for fungible parameter estimates. In fact we are not sure whether that will ever be possible. Evaluation of fungible estimates is more subjective at this point. Nevertheless, in the current example we believe that such subjective evaluation serves the purpose sufficiently well. The contour in Figure 12.4 shows a range of solutions that reflects a high degree of uncertainty inherent in the ML estimates of the gender effects and in any associated interpretation.

12.4 DISCUSSION

In this chapter we have introduced the concept of fungible parameter estimates in SEM and have outlined a computational procedure for obtaining such estimates. We have also demonstrated this approach in an applied example involving LCM. Our results show how the computation and evaluation of fungible parameter estimates can strengthen or weaken the support for substantive interpretation of optimal parameter estimates. That is, when fungible estimates exhibit a rather compact contour around the optimal estimates, then support for substantive interpretation is strengthened. But when fungible estimates vary widely from the optimal estimates and from each other, then interpretation of the optimal estimates is compromised and weakened by such uncertainty. Such a pattern of results demonstrates the existence of alternative explanations that account for the data virtually as well as the optimal solution, yet are very different from that optimal solution and from each other.

It is of interest to consider the relationship between this kind of uncertainty and a more familiar type that is commonly taken into account in evaluation of parameter estimates. Researchers in the behavioral sciences routinely recognize the potential impact of sampling variability on their parameter estimates and take it into account by obtaining standard errors and confidence intervals for parameter estimates, as well as carrying out tests of statistical significance of those estimates. The underlying issue in that context involves precision of estimation. That is, the parameter estimates carry some uncertainty in the form of imprecision as estimates of population parameters.

The present chapter considers a different kind of uncertainty inherent in those same parameter estimates. In particular, how much would those estimates change if the fit of the model were perturbed to a very small degree? Large changes introduce the notion of alternative explanations of the data that are very different from the optimal solution, but which fit the data to virtually the same degree. This phenomenon represents a different kind of uncertainty in parameter estimates than that arising from sampling variability and imprecision. Interestingly, it is entirely possible for an analysis of these two kinds of uncertainty to show different patterns; e.g., high precision of estimation, reflected by small standard errors and narrow confidence intervals, accompanied by rather large contours of fungible estimates. In fact, such a pattern occurred to some degree in our example involving effects of child's gender on intercept and slope of mothers' ratings of closeness to the child.

We suggest that it is informative to consider both types of uncertainty in applied research, in that both impact on the confidence an investigator can have in the interpretation of parameter estimates. We note also that these two types of uncertainty are not independent. We are currently conducting analytical and computational studies of their relationship.

As noted in our presentation of a computational method and in our example, there are some limitations to the computational approach implemented in our LCM example. Specifically, as the number of parameters increases, implementation of a method for computing fungible estimates of all parameters becomes extremely burdensome. This burden arises due to the need to define a set of vectors that sweep a space of dimensionality equal to the number of parameters, and then solving for the scaling constant associated with each such vector. As noted earlier, we have circumvented this obstacle in our present work by limiting the number of parameters under consideration. We choose two or three focal parameters and obtain fungible estimates of those parameters, holding all other parameters (nuisance parameters) at their ML estimates. The resulting contours are, we believe, conservative approximations of the contours that would be observed for those focal parameters if the computational method could be implemented for the full set of parameters. Our approach in effect ignores the fact that the focal parameters will almost certainly be correlated with the nuisance parameters. A more comprehensive computational method that would take these correlations into account would generally result in larger fungible contours. We do not consider the current approach to be the best possible approach, but rather a method that can be implemented effectively in the present early stages of research on the nature of fungible parameter estimates. We are pursuing the development of improved computational methods, including a method under which we identify the parameter, or linear combination of parameters, characterized by the largest range of fungible estimates.

The development of better computational approaches is only one of the challenges we recognize with regard to future study of fungible parameter estimates. Of primary interest to us is the study of factors that influence the magnitude of fungible contours. What factors inherent in data or models will tend to make fungible contours get larger or smaller? We are pursuing studies of factors such as model fit, magnitude of correlations among measured variables, etc. Results of such investigations may aid applied researchers in designing models and studies that would reduce the impact of this source of uncertainty in their results.

Finally, we wish to close by reminding the reader of the generality of the issue of fungible parameter estimates. It is a phenomenon, a source of uncertainty, that is relevant in virtually any application of statistical modeling. Whenever we are estimating model parameters so as to optimize some measure of model fit, we can define and potentially compute fungible parameter estimates. And we can evaluate those fungible estimates to determine the impact of this type of uncertainty on substantive interpretation of parameter estimates. There exist many sources of uncertainty that impact on results of statistical modeling. In order to fully understand the utility of a model, it is important to represent and quantify all sources of uncertainty to the fullest extent possible.

REFERENCES

Bollen, K. A., & Curran, P. J. (2006). *Latent curve models: A structural equation perspective.* Hoboken, NJ: Wiley.

Browne, M. W., & Cudeck, R. (1993). Alternative ways of assessing model fit. In K. A. Bollen & J. S. Long (Eds.), *Testing structural equation models* (pp. 136–162). Newbury Park, CA: Sage.

MacCallum, R. C., Browne, M. W., & Lee, T. (2009). *Fungible parameter estimates in structural equation modeling.* Paper presented at the annual meeting of the Society of Multivariate Experimental Psychology, Salishan Resort, OR.

MacCallum, R. C., Lee, T., & Browne, M. W. (2010a). *Fungible mediation effects.* Paper presented at the annual meeting of the Society of Multivariate Experimental Psychology, Atlanta, GA.

MacCallum, R. C., Lee, T., & Browne, M. W. (2010b). The phenomenon of isopower in power analysis for tests of structural equation models. *Structural Equation Modeling, 17*, 23–41.

NICHD Early Child Care Research Network (2006). Child-care effect sizes for the NICHD study of early child care and youth development. *American Psychologist, 61*, 99–116.

Preacher, K. J., Wichman, A. L., MacCallum, R. C., & Briggs, N. E. (2008). *Latent growth curve modeling.* Los Angeles: Sage.

Press, W. H., Teukolsky, S. A., Vetterling, W. T., & Flannery, B. P. (1992). *Numerical recipes in Fortran: The art of scientific computing* (2nd ed.). Cambridge, UK: Cambridge University Press.

Steiger, J. H., & Lind, J. M. (1980). *Statistically based tests for the number of factors.* Paper presented at the annual meeting of the Psychometric Society, Iowa City, IA.

Waller, N. G. (2008). Fungible weights in multiple regression. *Psychometrika, 73*, 691–703.

Chapter 13
Statistical Inference of Covariance Structures

Alexander Shapiro[1]

13.1 INTRODUCTION

Let $\boldsymbol{X}$ be a $p \times 1$ random vector representing a certain population of interest. Suppose that we hypothesize that the components $X_1, \ldots, X_p$ of $\boldsymbol{X}$ are uncorrelated with each other. We can write this as $\boldsymbol{\Sigma} = \text{diag}(\theta_1, \ldots, \theta_p)$, where $\boldsymbol{\Sigma}$ is the covariance matrix of $\boldsymbol{X}$ and $\boldsymbol{\theta} = (\theta_1, \ldots, \theta_p)'$ is vector of unknown parameters representing respective variances of the components of $\boldsymbol{X}$. Since these variances should be nonnegative, the parameter vector $\boldsymbol{\theta}$ is restricted to the set (nonnegative orthant) $\mathbb{R}^p_+ = \{\boldsymbol{\theta} : \theta_i \geq 0,\ i = 1, \ldots, p\}$. This gives a relatively simple example of a covariance structural model

$$\boldsymbol{\Sigma} = \boldsymbol{\Sigma}(\boldsymbol{\theta}), \quad \boldsymbol{\theta} \in \Theta, \tag{13.1}$$

[0] This research was partly supported by the NSF award DMS-0914785 and ONR award N000140811104.

where $\boldsymbol{\Sigma}(\boldsymbol{\theta})$ is a symmetric matrix valued function of the parameter vector $\boldsymbol{\theta}$ varying in the corresponding parameter space Θ. In order for the matrix $\boldsymbol{\Sigma}(\boldsymbol{\theta})$ to be nonsingular, we need to enforce constraints $\theta_i > 0$, $i = 1, \ldots, p$, that is to use the parameter space $\mathbb{R}^p_{++} = \{\boldsymbol{\theta} : \theta_i > 0,\ i = 1, \ldots, p\}$.

The case considered above of $\boldsymbol{\Sigma}$ being diagonal gives an example of a linear model and can be written as

$$\boldsymbol{\Sigma} = \theta_1 \boldsymbol{G}_1 + \ldots + \theta_p \boldsymbol{G}_p, \quad \boldsymbol{\theta} \in \Theta, \tag{13.2}$$

where $\Theta = \mathbb{R}^p_{++}$ and $\boldsymbol{G}_i = \text{diag}(0, \ldots, 0, 1, 0, \ldots, 0), i = 1, \ldots, p$, are the corresponding diagonal matrices. A well known example of a nonlinear covariance structural model is factor analysis. That is,

$$\boldsymbol{\Sigma} = \boldsymbol{\Lambda}\boldsymbol{\Lambda}' + \boldsymbol{\Psi}, \tag{13.3}$$

which relates the $p \times p$ covariance $\boldsymbol{\Sigma}$ to the $p \times k$ matrix $\boldsymbol{\Lambda}$ of factor loadings and the $p \times p$ diagonal matrix $\boldsymbol{\Psi}$ of the residual variances. The corresponding parameter vector $\boldsymbol{\theta}$ is composed of the elements of matrix $\boldsymbol{\Lambda}$ and diagonal elements of $\boldsymbol{\Psi}$. In that model the diagonal elements of the matrix $\boldsymbol{\Psi}$ should be nonnegative, while there are no restrictions on the elements of $\boldsymbol{\Lambda}$.

Given a random sample $\boldsymbol{X}_1, \ldots, \boldsymbol{X}_N$ of N observations of vector $\boldsymbol{X}$, one faces two basic questions of statistical inference. That is, testing a fit of the model to the observed data and estimation of the unknown (population) value of the parameter vector. In this article we discuss a large samples statistical inference of covariance structural models of the form (13.1). We assume throughout that the parameter space Θ is a subset of $\mathbb{R}^q$, i.e., that the model involves q parameters, and that the function (mapping) $\boldsymbol{\theta} \mapsto \boldsymbol{\Sigma}(\boldsymbol{\theta})$ is continuously differentiable. These are mild assumptions which are satisfied in all interesting models.

We denote by $\mathbb{S}^{p\times p}$ the space of $p \times p$ symmetric matrices, by $\mathbb{S}^{p\times p}_+$ the set of $p \times p$ positive semidefinite symmetric matrices, and by $\mathbb{S}^{p\times p}_{++}$ the set of $p \times p$ positive definite symmetric matrices. Since any covariance matrix is positive semidefinite, it should hold that $\boldsymbol{\Sigma}(\boldsymbol{\theta}) \in \mathbb{S}^{p\times p}_+$ for any $\boldsymbol{\theta} \in \Theta$. We often would need to enforce the condition that $\boldsymbol{\Sigma}(\boldsymbol{\theta})$ is nonsingular, and hence $\boldsymbol{\Sigma}(\boldsymbol{\theta}) \in \mathbb{S}^{p\times p}_{++}$ for any $\boldsymbol{\theta} \in \Theta$. By $vec(\boldsymbol{\Sigma})$ we denote the $p^2 \times 1$ vector formed by stacking elements of matrix $\boldsymbol{\Sigma}$. Of course, since matrix $\boldsymbol{\Sigma}$ is symmetric, vector $vec(\boldsymbol{\Sigma})$ has duplicated elements. Therefore an alternative is to consider $vecs(\boldsymbol{\Sigma})$, where $vecs(\boldsymbol{\Sigma})$ denotes the $p(p+1)/2 \times 1$ vector formed by stacking (nonduplicated) elements of $\boldsymbol{\Sigma}$ above and including the diagonal.

From a theoretical point of view an advantage of working with the $vec(\cdot)$, rather than $vecs(\cdot)$, operator are the following algebraic identities

$$vec(\boldsymbol{ABC}) = (\boldsymbol{C}' \otimes \boldsymbol{A})vec(\boldsymbol{B}), \tag{13.4}$$

$$tr(\boldsymbol{AB}'\boldsymbol{CB}) = vec(\boldsymbol{B})'(\boldsymbol{A} \otimes \boldsymbol{C})vec(\boldsymbol{B}), \tag{13.5}$$

where $A \otimes B$ denotes the Kronecker product of matrices A and B and $tr(A)$ denotes the trace of (square) matrix A. Throughout this article we denote by σ, s, etc., the respective vectors $vec(\Sigma)$, $vec(S)$ of matrices $\Sigma, S \in \mathbb{S}^{p\times p}$.

It should be noted that the covariance structural model (13.1) can be written in the following equivalent form

$$\Sigma \in \Omega, \tag{13.6}$$

where $\Omega = \{\Sigma : \Sigma = \Sigma(\theta),\ \theta \in \Theta\}$. We refer to the set $\Omega \subset \mathbb{S}^{p\times p}$ as the *image* space of the model. The parametric model (13.1) represents a particular parameterization of the image set Ω. For example, the diagonal model $\Sigma = \text{diag}(\theta_1, \ldots, \theta_p)$ can be written as $\Sigma = \text{diag}(\theta_1^2, \ldots, \theta_p^2)$, with the corresponding parameter space $\Theta = \mathbb{R}^p$. In both parameterizations the image space is the same. We refer to the model $\Sigma \in \mathbb{S}^{p\times p}_{++}$ as the *saturated* model. It also should be noted that there is a certain confusion in the notation here, sometimes model parameters, like Σ, θ etc., are viewed as variables and sometimes as population values. In order to distinguish between these two meanings we sometimes write Σ_0, θ_0 for the respective population values.

With the function (mapping) $\sigma(\theta)$ we associate its $p^2 \times q$ Jacobian matrix $\partial\sigma(\theta)/\partial\theta'$ of partial derivatives and use notation $\Delta(\theta) = \partial\sigma(\theta)/\partial\theta'$. We also use the following notation and terminology throughout the article: A' denotes the transpose of matrix (vector) A; $\|\theta\| = \sqrt{\theta'\theta}$ denotes the Euclidean norm of vector $\theta \in \mathbb{R}^q$; notation $o(x)$ means that $o(x)/x$ tends to 0 as $x \to 0$; notations $X_i = o_p(Y_i)$ and $X_i = O_p(Y_i)$ mean, respectively, that X_i/Y_i tends in probability to 0 and that X_i/Y_i is bounded in probability, in particular $X_i = o_p(1)$ means that the sequence X_i of random variables tends in probability to 0; $\text{dist}(\theta, \Theta) = \min_{\zeta\in\Theta} \|\theta - \zeta\|$ denotes distance from vector θ to set Θ (by " min " we actually mean the " inf " operator if the corresponding minimum is not attained); it is said that $C \subset \mathbb{R}^q$ is a *cone* if $\zeta \in C$ implies that $t\zeta \in C$ for any $t \geq 0$; *polar* of a cone $C \subset \mathbb{R}^q$ is defined as the cone $\{\zeta \in \mathbb{R}^q : \zeta'\theta \leq 0,\ \forall\theta \in C\}$; $|\Sigma|$ denotes the determinant of matrix Σ; I_N denotes the $N \times N$ identity matrix; Prob(A) denotes probability of event A; $\mathbb{E}[Z]$ denotes mean (expected value) of random vector Z; $\mathcal{N}(\mu, \Sigma)$ denotes (multivariate) normal distribution with mean vector μ and covariance matrix Σ; notation $Z_N \Rightarrow Z$ stands for convergence in distribution of random vectors Z_N to Z as $N \to \infty$.

The remainder of this chapter is organized as follows. In the next section we discuss the Maximum Likelihood (ML) approach to testing of fit and estimation of parameters of covariance structural models under the assumption of a normally distributed population. In particular, we show that the ML method can be formulated in a framework of the Minimum Discrepancy Function (MDF) analysis. In the following section we investigate algebraic structure of discrepancy functions and show that locally every discrepancy function can be approximated by a quadratic function. In the section after that we discuss statistical inference of the MDF method. In particular, we single out conditions required for the MDF test statistic to be asymptotically chi-squared. Finally, in the last section we consider the situation when the true value of the parameter vector lies on the boundary of the parameter space. This can happen when the parameter space is defined by the inequality constrained. In such cases asymptotics of the MDF test statistics are described by mixtures of chi-square distributions.

13.2 MINIMUM DISCREPANCY APPROACH

The classical approach to testing and estimation of covariance structures is the Maximum Likelihood (ML) method based on the assumption of normally distributed population. That is, let $\boldsymbol{X}_1, \ldots, \boldsymbol{X}_N$ be an independent identically distributed (iid) sample of observations drawn from a population having multivariate normal distribution with mean vector $\boldsymbol{\mu}$ and covariance matrix $\boldsymbol{\Sigma}$, i.e., $\boldsymbol{X}_i$ are sample observations of random vector $\boldsymbol{X} \sim \mathcal{N}(\boldsymbol{\mu}, \boldsymbol{\Sigma})$. The corresponding likelihood function (up to a constant factor) is

$$\begin{aligned} L(\boldsymbol{\mu}, \boldsymbol{\Sigma}) &= \prod_{i=1}^{N} |\boldsymbol{\Sigma}|^{-1/2} \exp\left\{-\tfrac{1}{2}(\boldsymbol{X}_i - \boldsymbol{\mu})'\boldsymbol{\Sigma}^{-1}(\boldsymbol{X}_i - \boldsymbol{\mu})\right\} \\ &= |\boldsymbol{\Sigma}|^{-N/2} \exp\left\{-\tfrac{1}{2}tr(\boldsymbol{\Sigma}^{-1}\boldsymbol{A})\right\} \exp\left\{-\tfrac{1}{2}N(\bar{\boldsymbol{X}} - \boldsymbol{\mu})'\boldsymbol{\Sigma}^{-1}(\bar{\boldsymbol{X}} - \boldsymbol{\mu})\right\}, \end{aligned} \tag{13.7}$$

where $\bar{\boldsymbol{X}} = N^{-1}\sum_{i=1}^{N}\boldsymbol{X}_i$ and $\boldsymbol{A} = \sum_{i=1}^{N}(\boldsymbol{X}_i - \bar{\boldsymbol{X}})'(\boldsymbol{X}_i - \bar{\boldsymbol{X}})$.

In the saturated model, where there are no constraints on $\boldsymbol{\mu}$ and $\boldsymbol{\Sigma}$ (except for the constraint for $\boldsymbol{\Sigma}$ to be positive definite), the ML estimators of $\boldsymbol{\mu}$ and $\boldsymbol{\Sigma}$ are obtained by maximizing the likelihood function $L(\boldsymbol{\mu}, \boldsymbol{\Sigma})$. It can be immediately seen from (13.7) that the ML estimator of $\boldsymbol{\mu}$ is $\bar{\boldsymbol{X}}$. Therefore in order to obtain the ML estimator of $\boldsymbol{\Sigma}$ we have to maximize $L(\bar{\boldsymbol{X}}, \boldsymbol{\Sigma})$, or equivalently to minimize

$$\begin{aligned} -2\log L(\bar{\boldsymbol{X}}, \boldsymbol{\Sigma}) &= N\log|\boldsymbol{\Sigma}| + tr(\boldsymbol{\Sigma}^{-1}\boldsymbol{A}) \\ &= tr(\boldsymbol{\Sigma}^{-1}\boldsymbol{A}) - N\log|\boldsymbol{\Sigma}^{-1}\boldsymbol{A}| + N\log|\boldsymbol{A}|. \end{aligned} \tag{13.8}$$

Let $\lambda_1, \ldots, \lambda_p$ be eigenvalues of matrix $\boldsymbol{\Sigma}^{-1}\boldsymbol{A}$. Then

$$-2\log L(\bar{\boldsymbol{X}}, \boldsymbol{\Sigma}) = \sum_{i=1}^{p}(\lambda_i - N\log\lambda_i) + N\log|\boldsymbol{A}|.$$

Note that function $f(\lambda) = \lambda - N\log\lambda$ attains its minimum, over $\lambda > 0$, at $\bar{\lambda} = N$. It follows that the ML estimator $\hat{\boldsymbol{\Sigma}}$ satisfies the equation $\hat{\boldsymbol{\Sigma}}\boldsymbol{A} = N\boldsymbol{I}_N$, and hence $\hat{\boldsymbol{\Sigma}} = N^{-1}\boldsymbol{A}$.

This shows that, under the assumption of normally distributed population, the (unconstrained) ML estimators are $\hat{\mu} = \bar{X}$ and

$$\hat{\Sigma} = \frac{1}{N}\sum_{i=1}^{N}(X_i - \bar{X})'(X_i - \bar{X}). \tag{13.9}$$

Recall that an unbiased estimator of the population matrix is obtained if the factor $1/N$ in (13.9) is replaced by $1/(N-1)$, and the obtained estimator is called the sample covariance matrix. Of course, for reasonably large sample size N the difference these two estimators is small and it does not effect the asymptotic theory. Therefore we use the factor $1/(N-1)$ to make $\hat{\Sigma}$ an unbiased estimator of the population covariance matrix, and refer to $\hat{\Sigma}$ as the *sample covariance* matrix.

Consider now a covariance structural model (13.6), where $\Omega \subset \mathbb{S}_{++}^{p\times p}$ is the image set of a corresponding model. The likelihood ratio test statistic for testing the model (13.6) is given by

$$\lambda = \frac{\max\limits_{\mu, \Sigma\in\Omega} L(\mu, \Sigma)}{\max\limits_{\mu, \Sigma\in\mathbb{S}_{++}^{p\times p}} L(\mu, \Sigma)}. \tag{13.10}$$

Note that in this model, μ is viewed as a vector of nuisance parameters. By writing in the saturated ML estimators $\hat{\mu} = \bar{X}$ and $\hat{\Sigma}$, we obtain

$$-2\log\lambda = N\min_{\Sigma\in\Omega} F_{ML}(\hat{\Sigma}, \Sigma), \tag{13.11}$$

where

$$F_{ML}(S, \Sigma) = \log|\Sigma| - \log|S| + tr(S\Sigma^{-1}) - p \tag{13.12}$$

viewed as a function of two symmetric matrix variables S and Σ.

Let $\alpha_1, \ldots, \alpha_p$ be eigenvalues of matrix ΣS^{-1}. Then

$$F_{ML}(S, \Sigma) = \sum_{i=1}^{p}(\log\alpha_i + \alpha_i^{-1}) - p.$$

Since the minimum of function $\log\alpha + \alpha^{-1}$, over $\alpha > 0$, is attained at $\alpha = 1$, we obtain that $F_{ML}(S, \Sigma) \geq 0$ for any positive definite symmetric matrices S and Σ, and $F_{ML}(S, \Sigma) = 0$ if and only if $\Sigma = S$.

Following Browne (1982) we say that a twice continuously differentiable function $F(S, \Sigma)$, of two variables $S, \Sigma \in \mathbb{S}_{+}^{p\times p}$, is a *discrepancy* function if it satisfies the following conditions:

(i) $F(S, \Sigma) \geq 0$ for all $S, \Sigma \in \mathbb{S}_{+}^{p\times p}$,
(ii) $F(S, \Sigma) = 0$ if and only if $S = \Sigma$.

From the above discussion we have that $F_{ML}(\boldsymbol{S}, \boldsymbol{\Sigma})$ is an example of the discrepancy function. Another popular example is the Generalized Least Squares discrepancy function:

$$F_{GLS}(\boldsymbol{S}, \boldsymbol{\Sigma}) = \tfrac{1}{2} tr\left[\boldsymbol{S}^{-1}(\boldsymbol{S} - \boldsymbol{\Sigma})\boldsymbol{S}^{-1}(\boldsymbol{S} - \boldsymbol{\Sigma})\right]. \tag{13.13}$$

Given parametric model (13.1), the statistic $N\hat{F}$, where

$$\hat{F} = \min_{\boldsymbol{\theta} \in \Theta} F(\hat{\boldsymbol{\Sigma}}, \boldsymbol{\Sigma}(\boldsymbol{\theta})), \tag{13.14}$$

is referred to as the *Minimum Discrepancy Function* (MDF) test statistic. The corresponding minimizer $\hat{\boldsymbol{\theta}}$ is referred to as the MDF estimator of the parameter vector $\boldsymbol{\theta}$. Note that equivalently $\hat{F}$ can be defined as

$$\hat{F} = \min_{\boldsymbol{\Sigma} \in \Omega} F(\hat{\boldsymbol{\Sigma}}, \boldsymbol{\Sigma}), \tag{13.15}$$

where Ω is the corresponding image set.

For the diagonal model $\boldsymbol{\Sigma}(\boldsymbol{\theta}) = \text{diag}(\theta_1, \ldots, \theta_p)$, $\boldsymbol{\theta} \in \mathbb{R}^p_{++}$, it is possible to compute the likelihood ratio test in a closed form. We then have that

$$F_{ML}(\hat{\boldsymbol{\Sigma}}, \boldsymbol{\Sigma}(\boldsymbol{\theta})) = \sum_{i=1}^{p} \log \theta_i - \log|\hat{\boldsymbol{\Sigma}}| + \sum_{i=1}^{p} \frac{\hat{\sigma}_{ii}}{\theta_i} - p.$$

By minimizing this with respect to θ_i, we obtain the ML estimator $\hat{\theta}_i = \hat{\sigma}_{ii}$, $i = 1, \ldots, p$. It follows that

$$F_{ML}(\hat{\boldsymbol{\Sigma}}, \boldsymbol{\Sigma}(\hat{\boldsymbol{\theta}})) = \sum_{i=1}^{p} \log \hat{\sigma}_{ii} - \log|\hat{\boldsymbol{\Sigma}}| = -\log|\hat{\boldsymbol{\Sigma}}(\text{diag}\,\hat{\boldsymbol{\Sigma}})^{-1}| = -\log|\boldsymbol{R}|,$$

where $\boldsymbol{R} = (\text{diag}\,\hat{\boldsymbol{\Sigma}})^{-1/2}\hat{\boldsymbol{\Sigma}}(\text{diag}\,\hat{\boldsymbol{\Sigma}})^{-1/2}$ is the sample correlation matrix.

13.3 ALGEBRAIC STRUCTURE OF DISCREPANCY FUNCTIONS

Viewing the discrepancy function $F(\cdot, \cdot)$ as a function of two vector variables $\boldsymbol{s} = vec(\boldsymbol{S})$ and $\boldsymbol{\sigma} = vec(\boldsymbol{\Sigma})$, consider gradient vectors $\partial F(\boldsymbol{s}, \boldsymbol{\sigma})/\partial \boldsymbol{s}$ and $\partial F(\boldsymbol{s}, \boldsymbol{\sigma})/\partial \boldsymbol{\sigma}$ of partial derivatives. Since for given $\boldsymbol{s}$, the function $F(\boldsymbol{s}, \boldsymbol{\sigma})$ attains its minimum (of zero) at $\boldsymbol{\sigma} = \boldsymbol{s}$, we have that $\partial F(\boldsymbol{s}, \boldsymbol{\sigma})/\partial \boldsymbol{\sigma} = \boldsymbol{0}$ at $\boldsymbol{\sigma} = \boldsymbol{s}$, and similarly $\partial F(\boldsymbol{s}, \boldsymbol{\sigma})/\partial \boldsymbol{s} = \boldsymbol{0}$ at $\boldsymbol{s} = \boldsymbol{\sigma}$. Moreover, we have

$$\frac{\partial}{\partial \boldsymbol{s}'}\left[\frac{\partial F(\boldsymbol{s}, \boldsymbol{s})}{\partial \boldsymbol{s}}\right] = \boldsymbol{0}.$$

By using the chain rule of differentiation we can write

$$\frac{\partial}{\partial s'}\left[\frac{\partial F(s,s)}{\partial s}\right]=\frac{\partial^2 F(s,s)}{\partial \sigma'\partial s}+\frac{\partial^2 F(s,s)}{\partial s'\partial s}.$$

It follows that

$$\left.\frac{\partial^2 F(s,\sigma)}{\partial \sigma'\partial s}\right|_{S=\sigma}=-\left.\frac{\partial^2 F(s,\sigma)}{\partial s'\partial s}\right|_{S=\sigma}, \tag{13.16}$$

and similarly

$$\left.\frac{\partial^2 F(s,\sigma)}{\partial \sigma'\partial s}\right|_{S=\sigma}=-\left.\frac{\partial^2 F(s,\sigma)}{\partial \sigma'\partial \sigma}\right|_{S=\sigma}. \tag{13.17}$$

For a given $\sigma_0 = vec(\Sigma_0)$ consider the $p^2 \times p^2$ matrix

$$V=\frac{1}{2}\left.\frac{\partial^2 F(s,\sigma)}{\partial s'\partial s}\right|_{S=\sigma_0,\sigma=\sigma_0} \tag{13.18}$$

of second order partial derivatives. By (13.16) and (13.17) we can write the following second order Taylor expansion of function $F(s, \sigma)$ at the point $(s, \sigma) = (\sigma_0, \sigma_0)$:

$$\begin{aligned}F(s,\sigma)&=F(\sigma_0,\sigma_0)+\left[\frac{\partial F(s,s)}{\partial s}\right]'(s-\sigma_0)+\left[\frac{\partial F(s,s)}{\partial \sigma}\right]'(\sigma-\sigma_0)\\&\quad+(s-\sigma_0)'V(s-\sigma_0)-2(s-\sigma_0)'V(\sigma-\sigma_0)+(\sigma-\sigma_0)'V(\sigma-\sigma_0)\\&\quad+o(\|s-\sigma_0\|^2)+o(\|\sigma-\sigma_0\|^2).\end{aligned}$$

Since the first three terms in the above expansion vanish we obtain the following second order approximation of the discrepancy function at the point (σ_0, σ_0):

$$F(s,\sigma)=(s-\sigma)'V(s-\sigma)+o(\|s-\sigma_0\|^2)+o(\|\sigma-\sigma_0\|^2). \tag{13.19}$$

That is, in a neighborhood of the point (σ_0, σ_0) the discrepancy function can be approximated by the quadratic function $(s - \sigma)'V(s - \sigma)$. This is a basis for deriving asymptotics of the MDF estimators.

Let us compute the weight matrix V for the ML discrepancy function. For small perturbations $d\Sigma$ of Σ we have the following linear (first order) approximations

$$\log|\Sigma+d\Sigma|=\log|\Sigma|+\log|I_p+\Sigma^{-1}d\Sigma|=\log|\Sigma|+tr(\Sigma^{-1}d\Sigma)+o(\|d\Sigma\|)$$

and

$$(\Sigma+d\Sigma)^{-1}=\Sigma-\Sigma^{-1}(d\Sigma)\Sigma^{-1}+o(\|d\Sigma\|).$$

It follows that the differential of F_{ML} with respect to Σ can be written as

$$dF_{ML}=tr\left[(\Sigma^{-1}-\Sigma^{-1}S\Sigma^{-1})d\Sigma\right]=tr\left[\Sigma^{-1}(\Sigma-S)\Sigma^{-1}d\Sigma\right]. \tag{13.20}$$

By using Formula (13.4) we can write (13.20) in the following vector form

$$\frac{\partial F_{ML}}{\partial \boldsymbol{\sigma}} = \boldsymbol{\Sigma}^{-1} \otimes \boldsymbol{\Sigma}^{-1}(\boldsymbol{\sigma} - \boldsymbol{s}), \tag{13.21}$$

and hence

$$\boldsymbol{V} = \frac{1}{2} \frac{\partial^2 F_{ML}(\boldsymbol{s}, \boldsymbol{\sigma})}{\partial \boldsymbol{\sigma}' \partial \boldsymbol{\sigma}} \bigg|_{\boldsymbol{s}=\boldsymbol{\sigma}_0, \boldsymbol{\sigma}=\boldsymbol{\sigma}_0} = \tfrac{1}{2} \boldsymbol{\Sigma}_0^{-1} \otimes \boldsymbol{\Sigma}_0^{-1}. \tag{13.22}$$

We obtain the following second order approximation of the ML discrepancy function

$$F_{ML}(\boldsymbol{s}, \boldsymbol{\sigma}) \approx \tfrac{1}{2}(\boldsymbol{s} - \boldsymbol{\sigma})'(\boldsymbol{\Sigma}_0^{-1} \otimes \boldsymbol{\Sigma}_0^{-1})(\boldsymbol{s} - \boldsymbol{\sigma}). \tag{13.23}$$

Recall that by Formula (13.5) we also have that

$$(\boldsymbol{s} - \boldsymbol{\sigma})'(\boldsymbol{\Sigma}_0^{-1} \otimes \boldsymbol{\Sigma}_0^{-1})(\boldsymbol{s} - \boldsymbol{\sigma}) = tr\left[\boldsymbol{\Sigma}_0^{-1}(\boldsymbol{S} - \boldsymbol{\Sigma})\boldsymbol{\Sigma}_0^{-1}(\boldsymbol{S} - \boldsymbol{\Sigma})\right]. \tag{13.24}$$

As far as the generalized least squares (GLS) discrepancy function is concerned it has the same second order approximation at $(\boldsymbol{s}, \boldsymbol{\sigma}) = (\boldsymbol{\sigma}_0, \boldsymbol{\sigma}_0)$.

Consider a covariance structural model of the form (13.1). Let $\boldsymbol{\Sigma}_0$ be the true (population) covariance matrix. Suppose that there exists a point $\boldsymbol{\theta}_0 \in \Theta$ such that $\boldsymbol{\Sigma}_0 = \boldsymbol{\Sigma}(\boldsymbol{\theta}_0)$, i.e., the model is correctly specified and $\boldsymbol{\theta}_0$ is the corresponding population value of the parameter vector. It is said that the model is identified at $\boldsymbol{\theta}_0$ if $\boldsymbol{\Sigma}(\boldsymbol{\theta}_0) = \boldsymbol{\Sigma}(\boldsymbol{\theta}^*)$ and $\boldsymbol{\theta}^* \in \Theta$ imply that $\boldsymbol{\theta}_0 = \boldsymbol{\theta}^*$, i.e., the population value of $\boldsymbol{\theta}$ is defined uniquely. If this holds for every $\boldsymbol{\theta}^*$ in a neighborhood of $\boldsymbol{\theta}_0$, it is said that the model is locally identified. A sufficient condition for local identification is that the $p^2 \times q$ Jacobian matrix $\boldsymbol{\Delta} = \partial \boldsymbol{\sigma}(\boldsymbol{\theta}_0)/\partial \boldsymbol{\theta}$ has full column rank q. This follows from the following first order approximation of $\boldsymbol{\sigma}(\boldsymbol{\theta})$ at the point $\boldsymbol{\theta}_0$:

$$\boldsymbol{\sigma}(\boldsymbol{\theta}) = \boldsymbol{\sigma}(\boldsymbol{\theta}_0) + \boldsymbol{\Delta}(\boldsymbol{\theta} - \boldsymbol{\theta}_0) + o(\|\boldsymbol{\theta} - \boldsymbol{\theta}_0\|). \tag{13.25}$$

Consider the function

$$\phi(\boldsymbol{S}) = \min_{\boldsymbol{\theta} \in \Theta} F(\boldsymbol{S}, \boldsymbol{\Sigma}(\boldsymbol{\theta})) \tag{13.26}$$

of symmetric matrix variable $\boldsymbol{S}$, and let $\hat{\boldsymbol{\theta}}(\boldsymbol{S})$ be a corresponding minimizer (if it exists). Note that, by the definition, the sample fit $\hat{F}$ of the considered model is given by $\hat{F} = \phi(\hat{\boldsymbol{\Sigma}})$, and that $\hat{\boldsymbol{\theta}}(\hat{\boldsymbol{\Sigma}})$ is the corresponding MDF estimator. Fit of the population matrix $\boldsymbol{\Sigma}_0$ to the model is given by $F_0 = \phi(\boldsymbol{\Sigma}_0)$. If the model is correctly specified, then $F_0 = 0$ and $\hat{\boldsymbol{\theta}}(\boldsymbol{\Sigma}_0) = \boldsymbol{\theta}_0$. Moreover, the minimizer $\hat{\boldsymbol{\theta}}(\boldsymbol{\Sigma}_0)$ is unique if the model is identified at $\boldsymbol{\theta}_0$. Then, under mild regularity conditions (in particular if the set Θ is bounded), any minimizer of the right hand side of (13.26) converges to $\boldsymbol{\theta}_0$ as $\boldsymbol{S} \to \boldsymbol{\Sigma}_0$, i.e., $\hat{\boldsymbol{\theta}}(\boldsymbol{S})$ is continuous at $\boldsymbol{S} = \boldsymbol{\Sigma}_0$ (cf., Shapiro, 1984; Kano, 1986). Uniqueness of the minimizer $\hat{\boldsymbol{\theta}}(\boldsymbol{S})$ for $\boldsymbol{S}$ sufficiently close to $\boldsymbol{\Sigma}_0$ is ensured by the following result.

Proposition 13.3.1: *Let* $\boldsymbol{\Sigma}_0 = \boldsymbol{\Sigma}(\boldsymbol{\theta}_0)$ *for some* $\boldsymbol{\theta}_0 \in \Theta$. *Suppose that:* (i) $\boldsymbol{\theta}_0$ *is an interior point of* Θ, (ii) *the model is identified at* $\boldsymbol{\theta}_0$, (iii) $\hat{\boldsymbol{\theta}}(\boldsymbol{S})$ *is continuous at* $\boldsymbol{S} = \boldsymbol{\Sigma}_0$, (iv) *matrix* $\boldsymbol{V}$ *defined in* (13.18) *is nonsingular,* (v) *the Jacobian matrix* $\boldsymbol{\Delta} = \boldsymbol{\Delta}(\boldsymbol{\theta}_0)$ *has full column rank. Then for all* $\boldsymbol{S}$ *in a neighborhood of* $\boldsymbol{\Sigma}_0$, *the right hand side of* (13.26) *has a unique minimizer* $\hat{\boldsymbol{\theta}} = \hat{\boldsymbol{\theta}}(\boldsymbol{S})$ *and this minimizer is uniquely defined by the following equations (optimality conditions)*

$$\frac{\partial F(\boldsymbol{s}, \boldsymbol{\sigma}(\boldsymbol{\theta}))}{\partial \boldsymbol{\theta}} = \mathbf{0}. \tag{13.27}$$

Proof: Since $F(\cdot, \boldsymbol{\sigma}_0)$ attains its minimum at $\boldsymbol{s} = \boldsymbol{\sigma}_0$, it follows that the matrix $\boldsymbol{V}$ is positive semidefinite, and hence is positive definite by the assumption that it is nonsingular. Moreover,

$$\left.\frac{\partial^2 F(\boldsymbol{\sigma}_0, \boldsymbol{\sigma}(\boldsymbol{\theta}))}{\partial \boldsymbol{\theta} \partial \boldsymbol{\theta}'}\right|_{\boldsymbol{\theta}=\boldsymbol{\theta}_0} = \boldsymbol{\Delta}' \boldsymbol{V} \boldsymbol{\Delta}, \tag{13.28}$$

and since $\boldsymbol{\Delta}$ has full column rank, it follows that matrix $\boldsymbol{\Delta}'\boldsymbol{V}\boldsymbol{\Delta}$ is also positive definite. It follows by the Implicit Function Theorem that there are neighborhoods $\mathcal{V}_s$ and $\mathcal{V}_\theta$ of $\boldsymbol{\sigma}_0$ and $\boldsymbol{\theta}_0$, respectively, such that for all $\boldsymbol{s} \in \mathcal{V}_s$ equations (13.27) have unique solution $\tilde{\boldsymbol{\theta}} = \tilde{\boldsymbol{\theta}}(\boldsymbol{s})$ in the neighborhood $\mathcal{V}_\theta$. By the continuity assumption we have that for all $\boldsymbol{s}$ sufficiently close to $\boldsymbol{\sigma}_0$, every minimizer $\hat{\boldsymbol{\theta}} = \hat{\boldsymbol{\theta}}(\boldsymbol{s})$ lies in the neighborhood $\mathcal{V}_\theta$. We also have that $\hat{\boldsymbol{\theta}}$ satisfies the first order optimality conditions (13.27), and hence it follows that $\hat{\boldsymbol{\theta}}$ coincides with $\tilde{\boldsymbol{\theta}}$ and is unique. This completes the proof. ∎

The above result shows that under the specified assumptions, in particular under the identifiability assumption, for the sample covariance matrix $\hat{\boldsymbol{\Sigma}}$ sufficiently close to the model, the optimality conditions (13.27) uniquely define the corresponding MDF estimator. Of course, this is a qualitative type result since the notion of "sufficiently close" is difficult to quantify.

13.4 STATISTICAL INFERENCE OF THE MDF APPROACH

By employing second order (quadratic) approximation (13.19) of the discrepancy function and the linear approximation (13.25) of the model we can approximate the corresponding MDF test statistic $N\hat{F}$ as

$$N\hat{F} \approx \min_{\boldsymbol{\zeta} \in \mathbb{R}^q} (\boldsymbol{Z}_N - \boldsymbol{\Delta}\boldsymbol{\zeta})' \boldsymbol{V} (\boldsymbol{Z}_N - \boldsymbol{\Delta}\boldsymbol{\zeta}), \tag{13.29}$$

where $\boldsymbol{Z}_N = N^{1/2}(\hat{\boldsymbol{\sigma}} - \boldsymbol{\sigma}_0)$ and $\boldsymbol{\zeta} = N^{1/2}(\boldsymbol{\theta} - \boldsymbol{\theta}_0)$. Note that the above approximation (13.29) is based on the assumption that locally, in a neighborhood of $\boldsymbol{\theta}_0$, there are no restrictions on the parameter vector, i.e., that $\boldsymbol{\theta}_0$ lies in the interior of the parameter space Θ. The right hand side of (13.29) can be viewed as a generalized least squares problem with the weight matrix $\boldsymbol{V}$ (e.g., Seber, 1977, Section 3.6). Assuming that the matrix $\boldsymbol{V}$ is nonsingular and $\boldsymbol{\Delta}$ has full column rank q, the corresponding minimizer is given by

$$\boldsymbol{\zeta}^* = (\boldsymbol{\Delta}'\boldsymbol{V}\boldsymbol{\Delta})^{-1}\boldsymbol{\Delta}'\boldsymbol{V}\boldsymbol{Z}_N \tag{13.30}$$

and the minimum is equal to $\boldsymbol{Z}_N'\boldsymbol{U}\boldsymbol{Z}_N$, where $\boldsymbol{U} = \boldsymbol{V} - \boldsymbol{V}\boldsymbol{\Delta}(\boldsymbol{\Delta}'\boldsymbol{V}\boldsymbol{\Delta})^{-1}\boldsymbol{\Delta}'\boldsymbol{V}$. Alternatively matrix $\boldsymbol{U}$ can be written as (e.g., Rao, 1973, page 77)

$$\boldsymbol{U} = \boldsymbol{\Delta}_c(\boldsymbol{\Delta}_c'\boldsymbol{V}^{-1}\boldsymbol{\Delta}_c)^{-1}\boldsymbol{\Delta}_c', \tag{13.31}$$

where $\boldsymbol{\Delta}_c$ is an orthogonal complement of matrix $\boldsymbol{\Delta}$, i.e., $\boldsymbol{\Delta}_c$ is a $p^2 \times (p(p+1)/2 - q)$ matrix of full column rank such that $\boldsymbol{\Delta}_c'\boldsymbol{\Delta} = \boldsymbol{0}$. Note that each column of the Jacobian matrix $\boldsymbol{\Delta}$, and its orthogonal complement $\boldsymbol{\Delta}_c$, has only $p(p+1)/2$ nonduplicated elements in the same way as the $p^2 \times 1$ vector $\boldsymbol{\sigma} = vec(\boldsymbol{\Sigma})$. This is why $\boldsymbol{\Delta}_c$ has $p(p+1)/2 - q$, rather than $p^2 - q$, columns.

Next by the Central Limit Theorem we have that the random vector $\boldsymbol{Z}_N = N^{1/2}(\hat{\boldsymbol{\sigma}} - \boldsymbol{\sigma}_0)$ converged in distribution to multivariate normal with mean vector zero and covariance matrix $\boldsymbol{\Gamma}$, i.e., $\boldsymbol{Z}_N \Rightarrow \boldsymbol{Z}$ with $\boldsymbol{Z} \sim \mathcal{N}(\boldsymbol{0}, \boldsymbol{\Gamma})$. It follows that

$$N\hat{F} \Rightarrow \boldsymbol{Z}'\boldsymbol{U}\boldsymbol{Z} \tag{13.32}$$

and

$$N^{1/2}(\hat{\boldsymbol{\theta}} - \boldsymbol{\theta}_0) \Rightarrow (\boldsymbol{\Delta}'\boldsymbol{V}\boldsymbol{\Delta})^{-1}\boldsymbol{\Delta}'\boldsymbol{V}\boldsymbol{\Gamma}\boldsymbol{V}\boldsymbol{\Delta}(\boldsymbol{\Delta}'\boldsymbol{V}\boldsymbol{\Delta})^{-1}. \tag{13.33}$$

Note that vector $\boldsymbol{Z}_N$ has $p(p+1)/2$ nonduplicated elements and hence rank of matrix $\boldsymbol{\Gamma}$ cannot be bigger than $p(p+1)/2$, and for a nondegenerate distribution rank of $\boldsymbol{\Gamma}$ is equal to $p(p+1)/2$. So we assume that

$$\text{rank}(\boldsymbol{\Gamma}) = p(p+1)/2. \tag{13.34}$$

A typical element of the $p^2 \times p^2$ covariance matrix $\boldsymbol{\Gamma}$ is given by

$$\boldsymbol{\Gamma}_{ij,kl} = \gamma_{ijkl} - \sigma_{ij}\sigma_{kl}, \; i, j, k, l = 1, \ldots, p,$$

where $\gamma_{ijkl} = \mathbb{E}[(X_i - \mu_i)(X_j - \mu_j)(X_k - \mu_k)(X_l - \mu_l)]$ are fourth-order central moments of the components of the random vector $\boldsymbol{X}$, and $\sigma_{ij} = \mathbb{E}[(X_i - \mu_i)(X_j - \mu_j)]$, i.e., σ_{ij} is the ij-component of the covariance matrix $\boldsymbol{\Sigma}_0$. In particular, if the population distribution is multivariate normal, i.e., $\boldsymbol{X} \sim \mathcal{N}(\boldsymbol{\mu}, \boldsymbol{\Sigma}_0)$, then

$$\boldsymbol{\Gamma}_{ij,kl} = \sigma_{ik}\sigma_{jl} + \sigma_{il}\sigma_{jk}, \; i, j, k, l = 1, \ldots, p.$$

In a matrix form this can be written as

$$\boldsymbol{\Gamma}_{\mathcal{N}} = 2\boldsymbol{M}_p(\boldsymbol{\Sigma}_0 \otimes \boldsymbol{\Sigma}_0)\boldsymbol{M}_p \tag{13.35}$$

with $\boldsymbol{M}_p$ being an $p^2 \times p^2$ symmetric matrix of rank $p(p+1)/2$ with elements in row ij and column kl given by (cf., Browne, 1974)

$$\boldsymbol{M}_p(ij, kl) = \tfrac{1}{2}(\delta_{ik}\delta_{jl} + \delta_{il}\delta_{jk}).$$

Here $\delta_{ik} = 1$ if $i = k$, and $\delta_{ik} = 0$ if $i \neq k$. The notation $\boldsymbol{\Gamma}_{\mathcal{N}}$ emphasizes that this asymptotic covariance matrix of $N^{1/2}(\boldsymbol{s} - \boldsymbol{\sigma}_0)$ is computed under the assumption of a normally distributed population. Note that matrix $\boldsymbol{\Gamma}_{\mathcal{N}}$ has rank $p(p+1)/2$, provided that the covariance matrix $\boldsymbol{\Sigma}_0$ is nonsingular. Note also that matrix $\boldsymbol{M}_p$ has the following properties:

(i) $\boldsymbol{M}_p^2 = \boldsymbol{M}_p$,
(ii) $\boldsymbol{M}_p(\boldsymbol{\Sigma}_0 \otimes \boldsymbol{\Sigma}_0) = (\boldsymbol{\Sigma}_0 \otimes \boldsymbol{\Sigma}_0)\boldsymbol{M}_p$,
(iii) $\boldsymbol{M}_p vec(\boldsymbol{A}) = vec(\boldsymbol{A})$ for any $p \times p$ symmetric matrix $\boldsymbol{A}$, in particular $\boldsymbol{M}_p\boldsymbol{\Delta} = \boldsymbol{\Delta}$ and $\boldsymbol{M}_p\boldsymbol{\Delta}_c = \boldsymbol{\Delta}_c$.

The quadratic form $\boldsymbol{Z}'\boldsymbol{U}\boldsymbol{Z}$ has (central) chi-square distribution if and only if the following condition holds (note that matrices $\boldsymbol{U}$ and $\boldsymbol{\Gamma}$ have the same structure of nonduplicated elements and it is assumed that $\boldsymbol{\Gamma}$ has rank $p(p+1)/2$)

$$\boldsymbol{U}\boldsymbol{\Gamma}\boldsymbol{U}\boldsymbol{\Gamma} = \boldsymbol{U}\boldsymbol{\Gamma}, \tag{13.36}$$

in which case the number of degrees of freedom $df = tr(\boldsymbol{U}\boldsymbol{\Gamma})$ (e.g., Rao and Mitra, 1971, page 71). In particular suppose that $\boldsymbol{V} = \frac{1}{2}\boldsymbol{\Sigma}_0^{-1} \otimes \boldsymbol{\Sigma}_0^{-1}$ and $\boldsymbol{\Gamma} = \boldsymbol{\Gamma}_{\mathcal{N}}$. Then by using properties (i)–(iii) of matrix $\boldsymbol{M}_p$, it is straightforward to verify condition (13.36), and

$$tr(\boldsymbol{U}\boldsymbol{\Gamma}) = tr\left[(\boldsymbol{\Delta}_c'(\boldsymbol{\Sigma}_0 \otimes \boldsymbol{\Sigma}_0)\boldsymbol{\Delta}_c)^{-1}\boldsymbol{\Delta}_c'(\boldsymbol{\Sigma}_0 \otimes \boldsymbol{\Sigma}_0)\boldsymbol{\Delta}_c\right] = tr(\boldsymbol{I}_\nu) = \nu,$$

where $\nu = p(p+1)/2 - q$. That is, we have the following result.

- For the F_{ML} and F_{GLS} discrepancy functions the corresponding MDF test statistics have asymptotically (central) chi-square distribution with $df = p(p+1) - q$ degrees of freedom provided the following basic assumptions hold:

  (A1) the model is correctly specified, i.e., $\boldsymbol{\Sigma}_0 = \boldsymbol{\Sigma}(\boldsymbol{\theta}_0)$ for some $\boldsymbol{\theta}_0 \in \Theta$,
  (A2) the model is identified at $\boldsymbol{\theta}_0$ and the corresponding Jacobian matrix $\boldsymbol{\Delta} = \boldsymbol{\Delta}(\boldsymbol{\theta}_0)$ has full column rank q,

(A3) the population is normally distributed,
(A4) θ_0 is an interior point of the parameter space Θ.

Next we discuss what happens if the above conditions (A1)–(A4) do not hold.

Suppose that the model is *not* correctly specified. Let $\mathbf{\Sigma}^*$ be a minimizer of $F(\mathbf{\Sigma}_0, \mathbf{\Sigma})$ over $\mathbf{\Sigma} \in \Omega$. Of course, for a not correctly specified model $\mathbf{\Sigma}_0 \neq \mathbf{\Sigma}^*$, and hence $F_0 = F(\mathbf{\Sigma}_0, \mathbf{\Sigma}^*) > 0$. By the Law of Large Numbers the sample covariance matrix $\hat{\mathbf{\Sigma}}$ tends with probability one (w.p.1) to $\mathbf{\Sigma}_0$ as $N \to \infty$. By continuity of the function $\phi(\cdot)$ it follows that the sample fit $\hat{F} = \phi(\hat{\mathbf{\Sigma}})$ tends w.p.1 to F_0 with increase of the sample size N. This implies that the MDF test statistic $N\hat{F}$ tends to infinity as $N \to \infty$.

We can use the following first order approximation of the function $\phi(\boldsymbol{s})$ at the point $\boldsymbol{s} = \boldsymbol{\sigma}_0$:

$$\phi(\hat{\boldsymbol{\sigma}}) = \phi(\boldsymbol{\sigma}_0) + \boldsymbol{\gamma}'(\hat{\boldsymbol{\sigma}} - \boldsymbol{\sigma}_0) + o(\|\hat{\boldsymbol{\sigma}} - \boldsymbol{\sigma}_0\|), \tag{13.37}$$

where $\boldsymbol{\gamma} = \partial\phi(\boldsymbol{\sigma}_0)/\partial\boldsymbol{s}$. Assuming that the minimizer $\boldsymbol{\sigma}^*$ is unique, we have by the Danskin Theorem (Danskin, 1967) that

$$\boldsymbol{\gamma} = \left.\frac{\partial F(\boldsymbol{s}, \boldsymbol{\sigma}^*)}{\partial \boldsymbol{s}}\right|_{\boldsymbol{s}=\boldsymbol{\sigma}_0}. \tag{13.38}$$

It follows that

$$N^{1/2}(\hat{F} - F_0) = \boldsymbol{\gamma}'\left[N^{1/2}(\hat{\boldsymbol{\sigma}} - \boldsymbol{\sigma}_0)\right] + o_p(1). \tag{13.39}$$

Since $N^{1/2}(\hat{\boldsymbol{\sigma}} - \boldsymbol{\sigma}_0) \Rightarrow \mathcal{N}(\mathbf{0}, \mathbf{\Gamma})$, this implies the following asymptotics (cf., Shapiro, 1983, Theorem 5.4(b)).

- Let $\mathbf{\Sigma}^*$ be the unique minimizer of $F(\mathbf{\Sigma}_0, \mathbf{\Sigma})$ over $\mathbf{\Sigma} \in \Omega$. Then $N^{1/2}(\hat{F} - F_0)$ converges in distribution to normal $\mathcal{N}(0, \boldsymbol{\gamma}'\mathbf{\Gamma}\boldsymbol{\gamma})$, where the vector $\boldsymbol{\gamma}$ is given in (13.38).

Although correct from a mathematical point of view the above asymptotics typically do not give a good approximation of the MDF test statistic (cf., Chun & Shapiro, 2009). Consider again a minimizer $\mathbf{\Sigma}^*$ of $F(\mathbf{\Sigma}_0, \mathbf{\Sigma})$ over $\mathbf{\Sigma} \in \Omega$, and let $\boldsymbol{\theta}^* \in \Theta$ be the corresponding parameter vector such that $\mathbf{\Sigma}^* = \mathbf{\Sigma}(\boldsymbol{\theta}^*)$. By using the respective quadratic expansion at the point $\boldsymbol{\theta}^*$ we obtain approximation of the form (13.29) with $\boldsymbol{Z}_N = N^{1/2}(\hat{\boldsymbol{\sigma}} - \boldsymbol{\sigma}^*)$. Since $\hat{\boldsymbol{\sigma}}$ is an unbiased estimator of $\boldsymbol{\sigma}_0$, we have that $\mathbb{E}[\boldsymbol{Z}_N] = N^{1/2}(\boldsymbol{\sigma}_0 - \boldsymbol{\sigma}^*)$, and because of $\boldsymbol{\sigma}_0 \neq \boldsymbol{\sigma}^*$, it follows that $\mathbb{E}[\boldsymbol{Z}_N]$ will grow indefinitely with increase of the sample size N. Nevertheless, we can use the quadratic approximation $\boldsymbol{Z}'\boldsymbol{U}\boldsymbol{Z}$ with $\boldsymbol{Z} \sim \mathcal{N}(\boldsymbol{\mu}, \mathbf{\Gamma})$ and $\boldsymbol{\mu} = N^{1/2}(\boldsymbol{\sigma}_0 - \boldsymbol{\sigma}^*)$. The quadratic form

$\boldsymbol{Z}'\boldsymbol{U}\boldsymbol{Z}$ has noncentral chi-square distribution under the same condition (13.36) with the same degrees of freedom and with the noncentrality parameter

$$\delta = \boldsymbol{\mu}'\boldsymbol{U}\boldsymbol{\mu} = N(\boldsymbol{\sigma}_0 - \boldsymbol{\sigma}^*)'\boldsymbol{U}(\boldsymbol{\sigma}_0 - \boldsymbol{\sigma}^*) \approx NF_0.$$

This leads to the following approximate result (cf., Shapiro, 1983, Theorem 5.5).

- For the F_{ML} and F_{GLS} discrepancy functions the corresponding MDF test statistics have approximately noncentral chi-square distribution with $df = p(p+1) - q$ degrees of freedom and noncentrality parameter $\delta = NF_0$, provided the assumptions (A2)–(A4) hold.

Simulation experiments indicate that the noncentral chi-square gives a much better than normal approximation of the distribution of the MDF test statistics for not "too badly" misspecified models.

Now let us look at the assumption (A2). The assumption that the Jacobian matrix $\boldsymbol{\Delta} = \partial\boldsymbol{\sigma}(\boldsymbol{\theta}_0)/\partial\boldsymbol{\theta}$ has full column rank can be relaxed by replacing it with the so-called "constant rank" condition that $\partial\boldsymbol{\sigma}(\boldsymbol{\theta})/\partial\boldsymbol{\theta}$ has the same rank for all $\boldsymbol{\theta}$ in a neighborhood of the considered point $\boldsymbol{\theta}_0$. It is possible to show that this "constant rank" condition in a sense holds generically, and the asymptotics of the MDF test statistics are still chi-squared under basically the same conditions with the number of degrees of freedom $df = p(p+1)/2 - r$, where r is the rank of the Jacobian matrix $\partial\boldsymbol{\sigma}(\boldsymbol{\theta})/\partial\boldsymbol{\theta}$ (cf., Shapiro, 1986).

The assumption that the population has a normal distribution (assumption (A3)) is more difficult to deal with. In general the quadratic form $\boldsymbol{Z}'\boldsymbol{U}\boldsymbol{Z}$, with $\boldsymbol{Z} \sim \mathcal{N}(\boldsymbol{0}, \boldsymbol{\Gamma})$, has the same distribution as $\sum_{i=1}^{d} \lambda_i W_i$, where $\lambda_1, \ldots, \lambda_d$ are nonzero eigenvalues of the matrix $\boldsymbol{U}\boldsymbol{\Gamma}$, and $W_1, \ldots, W_d$ are independent random variables each having chi-square distribution with one degree of freedom. Condition (13.36) ensures that the eigenvalues $\lambda_i = 1$, $i = 1, \ldots, d$. For the ML and GLS discrepancy functions and certain types of models, in particular for the factor analysis model, condition (13.36) can be verified for a considerably larger class of distributions than normal (cf., Browne and Shapiro, 1988). This is a basis of the so-called asymptotic robustness theory of covariance structures.

13.5 CONSTRAINED STATISTICAL INFERENCE

Consider now a situation when the true value $\boldsymbol{\theta}_0$ of the parameter vector lies on the boundary of the set Θ. This may happen when the set Θ is defined by inequality constraints. In that case the set Θ can be approximated near $\boldsymbol{\theta}_0$ by a tangent cone $\mathcal{T}_\Theta(\boldsymbol{\theta}_0)$. For a general (closed) set Θ there are several possible concepts of tangent cones. If the set Θ is *convex*, then the tangent cone is defined as

$$\mathcal{T}_\Theta(\boldsymbol{\theta}_0) = \{\boldsymbol{\zeta} \in \mathbb{R}^q : \operatorname{dist}(\boldsymbol{\theta}_0 + t\boldsymbol{\zeta}, \Theta) = o(t),\ t \geq 0\}.$$

The polar of the tangent cone $\mathcal{T}_\Theta(\boldsymbol{\theta}_0)$ is called the normal cone to the set Θ at the point $\boldsymbol{\theta}_0$ and denoted $\mathcal{N}_\Theta(\boldsymbol{\theta}_0)$. By definition, the cones $\mathcal{T}_\Theta(\boldsymbol{\theta}_0)$ and $\mathcal{N}_\Theta(\boldsymbol{\theta}_0)$ are empty sets if $\boldsymbol{\theta}_0 \notin \Theta$.

Suppose that the parameter set Θ is defined by inequality constraints

$$\Theta = \{\boldsymbol{\theta} \in \mathbb{R}^q : h_i(\boldsymbol{\theta}) \leq 0,\ i = 1, \ldots, m\}, \tag{13.40}$$

where $h_i : \mathbb{R}^q \to \mathbb{R}, i = 1, \ldots, m$, are continuously differentiable functions. By $\boldsymbol{g}_i(\boldsymbol{\theta}) = \partial h_i(\boldsymbol{\theta})/\partial\boldsymbol{\theta}, i = 1, \ldots, m$, we denote gradients of the constraint functions and by

$$\mathcal{I}(\boldsymbol{\theta}) = \{i : h_i(\boldsymbol{\theta}) = 0,\ i = 1, \ldots, m\}$$

the index set of active at $\boldsymbol{\theta}$ constraints. Under a constraint qualification the tangent and normal cones at a point $\boldsymbol{\theta} \in \Theta$ can be written as

$$\mathcal{T}_\Theta(\boldsymbol{\theta}) = \left\{\boldsymbol{\zeta} \in \mathbb{R}^q : \boldsymbol{\zeta}'\boldsymbol{g}_i(\boldsymbol{\theta}) \leq 0,\ i \in \mathcal{I}(\boldsymbol{\theta})\right\}, \tag{13.41}$$

and

$$\mathcal{N}_\Theta(\boldsymbol{\theta}) = \left\{\boldsymbol{\zeta} = \textstyle\sum_{i\in\mathcal{I}(\boldsymbol{\theta})} \lambda_i \boldsymbol{g}_i(\boldsymbol{\theta}) : \lambda_i \geq 0,\ i \in \mathcal{I}(\boldsymbol{\theta})\right\}. \tag{13.42}$$

One such constraint qualification, at $\boldsymbol{\theta} \in \Theta$, is the following so-called Mangasarian–Fromovitz constraint qualification (MFCQ):

There exists $\boldsymbol{\zeta} \in \mathbb{R}^q$ such that $\boldsymbol{\zeta}'\boldsymbol{g}_i(\boldsymbol{\theta}) < 0, i \in \mathcal{I}(\boldsymbol{\theta})$.

The MFCQ holds, in particular, if gradient vectors $\boldsymbol{g}_i(\boldsymbol{\theta}), i \in \mathcal{I}(\boldsymbol{\theta})$, are linearly independent.

For the optimization problem in the right hand side of (13.26), with possibly boundary solutions, the first order necessary optimality conditions (13.27) should be replaced by

$$-\frac{\partial F(\boldsymbol{s}, \boldsymbol{\sigma}(\boldsymbol{\theta}))}{\partial\boldsymbol{\theta}} \in \mathcal{N}_\Theta(\boldsymbol{\theta}). \tag{13.43}$$

By (13.42) the optimality conditions (13.43) at a point $\boldsymbol{\theta} \in \Theta$ can be written in the following form of Lagrange multipliers: There exist multipliers $\lambda_i \geq 0, i = 1, \ldots, m$, such that

$$\frac{\partial L(\boldsymbol{s}, \boldsymbol{\theta}, \boldsymbol{\lambda})}{\partial\boldsymbol{\theta}} = \mathbf{0} \text{ and } \lambda_i h_i(\boldsymbol{\theta}) = 0,\ i = 1, \ldots, m, \tag{13.44}$$

where $L(\boldsymbol{s}, \boldsymbol{\theta}, \boldsymbol{\lambda}) = F(\boldsymbol{s}, \boldsymbol{\sigma}(\boldsymbol{\theta})) + \sum_{i=1}^m \lambda_i h_i(\boldsymbol{\theta})$. Note that if $\boldsymbol{\theta}$ is an interior point of Θ, then $\mathcal{T}_\Theta(\boldsymbol{\theta}) = \mathbb{R}^q$ and $\mathcal{N}_\Theta(\boldsymbol{\theta}) = \{\mathbf{0}\}$, and in that case conditions (13.43) coincide with conditions (13.27). Under the assumptions (ii)–(v) of Proposition 13.3.1 and the

MFCQ, it is also possible to show that for all $\boldsymbol{S}$ in a neighborhood of $\boldsymbol{\Sigma}_0$ the minimizer $\hat{\theta} = \hat{\theta}(\boldsymbol{S})$ is uniquely defined by the optimality conditions (13.44).

Let us discuss now statistical inference of MD estimators under inequality constraints. It is said that the set Θ is approximated at $\boldsymbol{\theta}_0 \in \Theta$ by a cone C, in the sense of Chernoff (1954), if the following two conditions hold

$$\text{dist}(\boldsymbol{\theta}_0 + \boldsymbol{\zeta}, \Theta) = o(\|\boldsymbol{\zeta}\|) \text{ for } \boldsymbol{\zeta} \in C, \tag{13.45}$$

$$\text{dist}(\boldsymbol{\zeta}, C) = o(\|\boldsymbol{\zeta}\|) \text{ for } \boldsymbol{\theta}_0 + \boldsymbol{\zeta} \in \Theta. \tag{13.46}$$

The tangent cone $\mathcal{T}_\Theta(\boldsymbol{\theta}_0)$ gives such approximation if the set Θ is convex, and in the case of set Θ defined by inequality constraints if the MFCQ holds at $\boldsymbol{\theta}_0$.

Assuming that the conditions (A1) and (A2) hold and the set Θ is approximated at $\boldsymbol{\theta}_0$ by its tangent cone $C = \mathcal{T}_\Theta(\boldsymbol{\theta}_0)$, we can use the following approximation of the MDF test statistic (compare with (13.29))

$$N\hat{F} \approx \min_{\boldsymbol{\zeta} \in C} (\boldsymbol{Z}_N - \boldsymbol{\Delta\zeta})' \boldsymbol{V} (\boldsymbol{Z}_N - \boldsymbol{\Delta\zeta}), \tag{13.47}$$

where $\boldsymbol{Z}_N = N^{1/2}(\hat{\boldsymbol{\sigma}} - \boldsymbol{\sigma}_0)$ (cf., Shapiro, 1985). Since $\boldsymbol{Z}_N \Rightarrow \mathcal{N}(\boldsymbol{0}, \boldsymbol{\Gamma})$, it follows that

$$N\hat{F} \Rightarrow \min_{\boldsymbol{\zeta} \in C} (\boldsymbol{Z} - \boldsymbol{\Delta\zeta})' \boldsymbol{V} (\boldsymbol{Z} - \boldsymbol{\Delta\zeta}), \tag{13.48}$$

where $\boldsymbol{Z} \sim \mathcal{N}(\boldsymbol{0}, \boldsymbol{\Gamma})$. The right hand side of (13.47) can be written as (cf., Shapiro, 1985, Theorem 2.1)

$$\min_{\boldsymbol{\zeta} \in C} (\boldsymbol{Z}_N - \boldsymbol{\Delta\zeta})' \boldsymbol{V} (\boldsymbol{Z}_N - \boldsymbol{\Delta\zeta}) = \min_{\boldsymbol{\zeta} \in \mathbb{R}^q} (\boldsymbol{Z}_N - \boldsymbol{\Delta\zeta})' \boldsymbol{V} (\boldsymbol{Z}_N - \boldsymbol{\Delta\zeta}) + \min_{\boldsymbol{\zeta} \in C} (\boldsymbol{Y}_N - \boldsymbol{\zeta})' \boldsymbol{W} (\boldsymbol{Y}_N - \boldsymbol{\zeta}), \tag{13.49}$$

where $\boldsymbol{W} = \boldsymbol{\Delta}' \boldsymbol{V} \boldsymbol{\Delta}$ and $\boldsymbol{Y}_N = (\boldsymbol{\Delta}' \boldsymbol{V} \boldsymbol{\Delta})^{-1} \boldsymbol{\Delta}' \boldsymbol{V} \boldsymbol{Z}_N$ (in a sense Equation (13.49) is the Pythagoras Theorem). Thus

$$N\hat{F} \Rightarrow \boldsymbol{Z}' \boldsymbol{U} \boldsymbol{Z} + \min_{\boldsymbol{\zeta} \in C} (\boldsymbol{Y} - \boldsymbol{\zeta})' \boldsymbol{W} (\boldsymbol{Y} - \boldsymbol{\zeta}), \tag{13.50}$$

where matrix $\boldsymbol{U}$ is given in (13.31) and $\boldsymbol{Y} = (\boldsymbol{\Delta}' \boldsymbol{V} \boldsymbol{\Delta})^{-1} \boldsymbol{\Delta}' \boldsymbol{V} \boldsymbol{Z}$.

The covariance matrix of $\boldsymbol{Y}$ and $\boldsymbol{\Delta}_c' \boldsymbol{Z}$ is $(\boldsymbol{\Delta}' \boldsymbol{V} \boldsymbol{\Delta})^{-1} \boldsymbol{\Delta}' \boldsymbol{V} \boldsymbol{\Gamma} \boldsymbol{\Delta}_c$. For the ML and GLS discrepancy functions and for a normally distributed population this covariance matrix is (see (13.22) and (13.35) and recall properties of the matrix $\boldsymbol{M}_p$)

$$2(\boldsymbol{\Delta}'(\boldsymbol{\Sigma}_0^{-1} \otimes \boldsymbol{\Sigma}_0^{-1})\boldsymbol{\Delta})^{-1} \boldsymbol{\Delta}'(\boldsymbol{\Sigma}_0^{-1} \otimes \boldsymbol{\Sigma}_0^{-1})(\boldsymbol{\Sigma}_0 \otimes \boldsymbol{\Sigma}_0) \boldsymbol{M}_p \boldsymbol{\Delta}_c = \boldsymbol{0},$$

and hence the two terms in the right hand side of (13.50) are independent. Moreover, the covariance matrix of $\boldsymbol{Y}$ is

$$\begin{aligned} &2(\boldsymbol{\Delta}'(\boldsymbol{\Sigma}_0^{-1} \otimes \boldsymbol{\Sigma}_0^{-1})\boldsymbol{\Delta})^{-1} \boldsymbol{\Delta}'(\boldsymbol{\Sigma}_0^{-1} \otimes \boldsymbol{\Sigma}_0^{-1})(\boldsymbol{\Sigma}_0 \otimes \boldsymbol{\Sigma}_0) \boldsymbol{M}_p (\boldsymbol{\Sigma}_0^{-1} \otimes \boldsymbol{\Sigma}_0^{-1}) \boldsymbol{\Delta} (\boldsymbol{\Delta}'(\boldsymbol{\Sigma}_0^{-1} \otimes \boldsymbol{\Sigma}_0^{-1})\boldsymbol{\Delta})^{-1} \\ &\quad = 2(\boldsymbol{\Delta}'(\boldsymbol{\Sigma}_0^{-1} \otimes \boldsymbol{\Sigma}_0^{-1})\boldsymbol{\Delta})^{-1} = (\boldsymbol{\Delta}' \boldsymbol{V} \boldsymbol{\Delta})^{-1} = \boldsymbol{W}^{-1}. \end{aligned}$$

It follows that the distribution of the second term in the right hand side of (13.50) is a mixture of chi-square distributions (cf., Bartholomew, 1961; Kudô, 1963; Shapiro, 1985).

- For the ML and GLS discrepancy functions, asymptotic distributions of the corresponding MDF test statistics are given by a mixture of chi-square distributions, provided that the assumptions (A1)–(A3) hold and the set Θ is approximated at $\boldsymbol{\theta}_0$ by its tangent cone. That is, for any $c \in \mathbb{R}$

$$\lim_{N\to\infty} \text{Prob}\left\{N\hat{F} \geq c\right\} = \sum_{i=0}^{q} w_i \text{Prob}\left\{\chi^2_{\nu+i} \geq c\right\}, \tag{13.51}$$

where χ^2_k denotes a chi-square random variable with k degrees of freedom, $\nu = p(p+1)/2 - q$ and w_i are nonnegative weights summing up to one.
Moreover, consider the MDF difference test statistic $N\hat{F} - N\tilde{F}$, where

$$\tilde{F} = \min_{\boldsymbol{\theta}\in\mathbb{R}^q} F(\hat{\boldsymbol{\Sigma}}, \boldsymbol{\Sigma}(\boldsymbol{\theta})) \tag{13.52}$$

is the corresponding unconstrained fit of the model. It follows that $N\hat{F}$ and $N\hat{F} - N\tilde{F}$ are asymptotically independent and

$$\lim_{N\to\infty} \text{Prob}\left\{N\hat{F} - N\tilde{F} \geq c\right\} = \sum_{i=0}^{q} w_i \text{Prob}\left\{\chi^2_i \geq c\right\}. \tag{13.53}$$

For a discussion of computing the weights $w_i = w_i(q, \boldsymbol{W}, C)$, $i = 0, \ldots, q$, we may refer to Shapiro (1988) and Silvapulle and Sen (2005). In particular, if only one constraint is active at $\boldsymbol{\theta}_0$, and hence the corresponding tangent cone is a half space, then $w_0 = w_1 = \frac{1}{2}$.

In order to compute the difference test statistic $N\hat{F} - N\tilde{F}$ one needs to solve the corresponding constrained and unconstrained problems, and hence to compute the respective MDF estimators $\hat{\boldsymbol{\theta}}$ and $\tilde{\boldsymbol{\theta}}$. Consider now the following Lagrange Multipliers (also called Score Function) test statistic $N\hat{\boldsymbol{\xi}}'\hat{\boldsymbol{W}}^{-1}\hat{\boldsymbol{\xi}}$, where

$$\hat{\boldsymbol{\xi}} = \frac{1}{2}\frac{\partial F(\hat{\boldsymbol{\sigma}}, \boldsymbol{\sigma}(\boldsymbol{\theta}))}{\partial \boldsymbol{\theta}}\bigg|_{\boldsymbol{\theta}=\hat{\boldsymbol{\theta}}} \tag{13.54}$$

and $\hat{\boldsymbol{W}}$ is a consistent estimator of the matrix $\boldsymbol{W} = \boldsymbol{\Delta}'\boldsymbol{V}\boldsymbol{\Delta}$, e.g., for the ML and GLS discrepancy functions one can use $\hat{\boldsymbol{W}} = \frac{1}{2}\hat{\boldsymbol{\Delta}}'\left(\hat{\boldsymbol{\Sigma}}^{-1} \otimes \hat{\boldsymbol{\Sigma}}^{-1}\right)\hat{\boldsymbol{\Delta}}$ with $\hat{\boldsymbol{\Delta}} = \boldsymbol{\Delta}(\hat{\boldsymbol{\theta}})$. Note that by the first order optimality conditions (13.44) we have that $\hat{\xi} = -\frac{1}{2}\sum_{i\in\mathcal{I}(\hat{\theta})}\hat{\lambda}_i \boldsymbol{g}_i(\hat{\boldsymbol{\theta}})$, where $\hat{\lambda}_i$ are respective Lagrange multipliers at the point $\hat{\boldsymbol{\theta}}$. We proceed now to show that the Lagrange Multipliers and the difference test statistics asymptotically are equivalent (cf., Gourieroux, Holly & Monfort, 1982).

- Under the assumptions (A1) and (A2) and that $\hat{\boldsymbol{\theta}}$ and $\tilde{\boldsymbol{\theta}}$ are consistent estimators of $\boldsymbol{\theta}_0$, it holds that

$$N\hat{\xi}'\hat{\boldsymbol{W}}^{-1}\hat{\xi} = N\hat{F} - N\tilde{F} + o_p(1). \tag{13.55}$$

In order to show (13.55) we proceed as follows. Note that $N^{1/2}(\hat{\boldsymbol{\sigma}} - \boldsymbol{\sigma}_0)$ *is bounded in probability, i.e.,* $\hat{\boldsymbol{\sigma}} - \boldsymbol{\sigma}_0 = O_p(N^{-1/2})$. *The same holds for* $\hat{\boldsymbol{\theta}} - \boldsymbol{\theta}_0 = O_p(N^{-1/2})$ *and* $\tilde{\boldsymbol{\theta}} - \boldsymbol{\theta}_0 = O_p(N^{-1/2})$. *Since*

$$\frac{\partial F(\hat{\boldsymbol{\sigma}}, \boldsymbol{\sigma})}{\partial \boldsymbol{\sigma}}\bigg|_{\boldsymbol{\sigma}=\boldsymbol{\sigma}(\hat{\boldsymbol{\theta}})} = 2\hat{\boldsymbol{V}}(\hat{\boldsymbol{\sigma}} - \boldsymbol{\sigma}(\hat{\boldsymbol{\theta}})) + o(\|\hat{\boldsymbol{\sigma}} - \boldsymbol{\sigma}(\hat{\boldsymbol{\theta}})\|),$$

by the chain rule of differentiation we can write $\hat{\xi}' = (\hat{\boldsymbol{\sigma}} - \boldsymbol{\sigma}(\hat{\boldsymbol{\theta}}))'\hat{\boldsymbol{V}}\hat{\boldsymbol{\Delta}} + o(\|\hat{\boldsymbol{\sigma}} - \boldsymbol{\sigma}(\hat{\boldsymbol{\theta}})\|)$, *and hence*

$$\hat{\xi}' = (\hat{\boldsymbol{\sigma}} - \boldsymbol{\sigma}(\hat{\boldsymbol{\theta}}))'\boldsymbol{V}\boldsymbol{\Delta} + o_p(N^{-1/2}).$$

Also since $\tilde{\boldsymbol{\theta}}$ *is an unconstrained minimizer of* $F(\hat{\boldsymbol{\sigma}}, \boldsymbol{\sigma}(\boldsymbol{\theta}))$ *we have that* $\partial F(\hat{\boldsymbol{\sigma}}, \boldsymbol{\sigma}(\tilde{\boldsymbol{\theta}}))/\partial\boldsymbol{\theta} = \boldsymbol{0}$, *and hence*

$$(\hat{\boldsymbol{\sigma}} - \boldsymbol{\sigma}(\tilde{\boldsymbol{\theta}}))'\boldsymbol{V}\boldsymbol{\Delta} = o_p(N^{-1/2}).$$

It follows that

$$\hat{\xi}' = (\boldsymbol{\sigma}(\tilde{\boldsymbol{\theta}}) - \boldsymbol{\sigma}(\hat{\boldsymbol{\theta}}))'\boldsymbol{V}\boldsymbol{\Delta} + o_p(N^{-1/2}).$$

Furthermore,

$$\boldsymbol{\sigma}(\tilde{\boldsymbol{\theta}}) - \boldsymbol{\sigma}(\hat{\boldsymbol{\theta}}) = \boldsymbol{\Delta}(\tilde{\boldsymbol{\theta}} - \hat{\boldsymbol{\theta}}) + o_p(N^{-1/2}),$$

and hence

$$\hat{\xi} = \boldsymbol{\Delta}'\boldsymbol{V}\boldsymbol{\Delta}(\tilde{\boldsymbol{\theta}} - \hat{\boldsymbol{\theta}}) + o_p(N^{-1/2}).$$

We obtain that

$$N\hat{\xi}'\hat{\boldsymbol{W}}^{-1}\hat{\xi} = N(\tilde{\boldsymbol{\theta}} - \hat{\boldsymbol{\theta}})'\boldsymbol{W}(\tilde{\boldsymbol{\theta}} - \hat{\boldsymbol{\theta}}) + o_p(1). \tag{13.56}$$

Now (see (13.30))

$$N^{1/2}\tilde{\boldsymbol{\theta}} = (\boldsymbol{\Delta}'\boldsymbol{V}\boldsymbol{\Delta})^{-1}\boldsymbol{\Delta}'\boldsymbol{V}\boldsymbol{Z}_N + o_p(1) = \boldsymbol{Y}_N + o_p(1).$$

Therefore (13.56) together with (13.49) imply (13.55).

REFERENCES

Bartholomew, D. J. (1961). Ordered tests in the analysis of variance, *Biometrika, 48*, 325–332.

Browne, M. W. (1974). Generalized least squares estimators in the analysis of covariance structures, *South African Statistical Journal, 8*, 1–24.

Browne, M. W. (1982). Covariance structures, in D. M. Hawkins (Ed.), *Topics in applied multivariate analysis*, Cambridge: Cambridge University Press.

Browne, M. W. (1984). Asymptotically distribution-free methods in the analysis of covariance structures, *British Journal of Mathematical and Statistical Psychology, 37*, 62–83.

Browne, M. W. (1987). Robustness in statistical inference in factor analysis and related models, *Biometrika, 74*, 375–384.

Browne, M. W., & Shapiro, A. (1988). Robustness of normal theory methods in the analysis of linear latent variate models, *British Journal of Mathematical and Statistical Psychology, 41*, 193–208.

Chernoff, H. (1954). On the distribution of the likelihood ratio, *The Annals of Mathematical Statistics, 25*, 573–578.

Chun, S. Y., & Shapiro, A. (2009). Normal versus noncentral chi-square asymptotics of misspecified models, *Multivariate Behavioral Research, 44*, 803–827.

Danskin, J. M. (1967). *The theory of max-min and its applications to weapon allocation problems*, Econometrics and Operations Research Vol. 5, Berlin and New York: Springer-Verlag.

Gourieroux, C., Holly, A. & Monfort, A. (1982). Likelihood ratio test, Wald test and Kuhn–Tucker test in linear models with inequality constraints on the regression parameters, *Econometrica, 50*, 63–80.

Kano, Y. (1986). Conditions on consistency of estimators in covariance structure model, *Journal of the Japan Statistical Society, 16*, 75–80.

Kudô, A. (1963). A multivariate analogue of the one-sided test, *Biometrika, 50*, 403–418.

Rao, C. R. (1973). *Linear statistical inference and its applications*, New York: Wiley.

Rao, C. R., & Mitra, S. K. (1971). *Generalized inverse of matrices and its applications*, New York: Wiley.

Seber, G. A. F. (1977). *Linear regression analysis*, New York: Wiley.

Shapiro, A. (1983). Asymptotic distribution theory in the analysis of covariance structures (a unified approach), *South African Statistical Journal, 17*, 33–81.

Shapiro, A. (1984). A note on consistency of estimators in the analysis of moment structures, *British Journal of Mathematical and Statistical Psychology, 37*, 84–88.

Shapiro, A. (1985). Asymptotic distribution of test statistics in the analysis of moment structures under inequality constraints, *Biometrika, 72*, 133–144.

Shapiro, A. (1986). Asymptotic theory of overparametrized structural models, *Journal of the American Statistical Association, 81*, 142–149.

Shapiro, A. (1988). Towards a unified theory of inequality constrained testing in multivariate analysis, *International Statistical Review, 56*, 49–62.

Silvapulle, M. J. & Sen, P. K. (2005). *Constrained statistical inference: Order, inequality, and shape constraints*, New York: Wiley.

Chapter 14

Fisher, Gosset, and Alternative Hypothesis Significance Testing (AHST): Using the Bootstrap to Test Scientific Hypotheses about the Multiple Correlation

Joseph Lee Rodgers and William Howard Beasley

14.1 INTRODUCTION

This chapter, originally presented as part of a conference celebrating the career of Michael Browne, has both an historical and a methodological goal. The historical goal is to speculate about how Fisher and Gosset would have developed hypothesis testing if modern computational methods had been available to them. We propose that Fisher's hypothesis testing system, which has with revision become Null Hypothesis Significance Testing (NHST), might instead have been developed as Alternative Hypothesis Significance Testing (AHST). This suggestion is motivated by the potential to develop empirical sampling distributions using resampling methods (specifically, the bootstrap) to evaluate settings in which features of the model that a researcher wishes to evaluate are reflected in the bootstrap sampling frame. The methodological goal involves extending past AHST theory by building and evaluating sampling distributions for the multiple correlation coefficient, R, within a multiple regression model in the context of AHST.

14.2 BACKGROUND

Recent statistical practice in many applied research settings has already shifted from testing hypotheses about the null, the absence of the researcher's scientific hypothesis, to focusing on the alternative, where the researcher's scientific hypothesis is often defined. Rodgers (2010a) documented this epistemological shift away from the traditional null hypothesis, and referred to the transition as the "modeling revolution." The form of the null hypothesis, and the status of the researcher's hypothesis, are logically quite different when the goal is to define and evaluate models (structural equation modeling, multilevel modeling, categorical data analysis, etc.) compared to the longstanding and popular null hypothesis significance testing (NHST) approach that is usually presented in classic textbook treatments of statistics. We propose that alternative hypothesis significance testing (AHST) may be a better entry point than NHST for students of statistics, given modern statistical tools, because of its relationship to the broader modeling paradigm, and especially because of its match to the way researchers actually think about developing and testing their hypotheses.

NHST answers a certain question: Given the null hypothesis, which is conceptualized in relation to chance outcomes, how probable is the data pattern that was actually observed? If that is the question that the researcher wants to ask, the NHST paradigm is a well-developed and logical way to provide an answer. However, the resulting answer is neither an especially strong answer in relation to many research hypotheses, nor does NHST actually answer many of the questions that it is used to address. It is not an especially strong answer because it only answers the question in relation to the null, the hypothesis of chance effect. In other (relevant) words, the traditional NHST approach answers a question about the status of the world if the researcher's hypothesis is wrong, a curious orientation for the past many generations of researchers to embrace. If the researcher would like stronger statements about the research hypothesis, captured in the alternative to the Fisherian null—which is typically the researcher's hypothesis-of-interest—NHST is not designed to provide that answer. Rather, the NHST conclusion is to either reject the null (which typically implies the legitimacy of many outcomes in addition to any specific alternatives posited by the researcher) or to support the null (which deflates the researcher's confidence in their own working hypothesis). There are several senses in which NHST is often used to address a question other than the one it was designed to address. For example, if the researcher actually wants to know the likelihood of the hypothesis given the data, instead of the other way around, NHST is simply the wrong method to provide an answer. Rather, a Bayesian perspective provides the natural answer to this question (see Carlin & Louis, 2009; Pruzek, 1997; Rindskopf, 1997; Trafimow, 2003; among many others, who develop a cogent and powerful argument for Bayesian methodology). Or, if the researcher has a specific parametric model that s/he wishes to evaluate, to challenge the model by asking whether it improves on a model of pure chance is hardly an appropriate or meaningful scientific challenge. As Rodgers (2010b) has argued, NHST is a useful tool, but is hardly capable—or even designed or intended—to provide an orienting epistemological system at the overarching level that it is taught in behavior sciences educational programs, and still often practiced in applied research settings.

Emerging from the AHST and modeling perspective is the importance of generating sampling distributions explicitly tied to the researcher's scientific hypothesis (which was computationally impossible for many of Fisher's problems, though both he and Gosset clearly wished for such computational power, and likely would have created their methods differently had they been available at that time). In particular, the power of the bootstrap provides the computational mechanism that would have supported the AHST formulation to be pragmatic, as well as advantageous, compared to the NHST paradigm. In the following sections, we develop these ideas in detail by treating both the historical/philosophical basis of NHST, and of AHST, and we demonstrate the ability of the bootstrap to support a viable AHST agenda.

14.3 FISHER, GOSSET, AND THE DEVELOPMENT OF NHST

Both Fisher and Gosset used a rudimentary resampling approach to test hypotheses. Gosset wrote the measures collected from 3,000 prisoners onto slips of cardboard and repeatedly drew them out of a hat to build an empirical sampling distribution, as a starting point to understanding the t distribution: "Before I had succeeded in solving my problem analytically, I had endeavored to do so empirically" (Student, 1908—Gosset used the pen name Student). Fisher felt that the legitimacy of the randomization philosophy at the basis of his analysis of variance procedure depended on the extent to which it matched the re-randomization test procedure that he developed: "[The] conclusions have no justification beyond the fact they could have been arrived at by this very elementary [re-randomization] method" (Fisher, 1936). He actually practiced and applied resampling by using some of Darwin's data on cross-fertilized versus self-fertilized corn, and by building an empirical sampling distribution under the null by re-assigning scores across the category boundary, which was arbitrary under the null hypothesis, H_0: $\mu_D = 0$.

To further develop Fisher and Gosset's obvious interest, it appears from the Fisher and Gosset quotes above that they developed the theoretical t and theoretical F distributions as mathematical shortcuts, as expedients, because they didn't have the computational power to regularly test hypotheses using empirical sampling distributions. Further, the mathematical procedures they developed could only be applied to distributions whose parametric properties could be fully specified, under a set of mathematical assumptions. Because the null was typically the only such distribution whose features were known with sufficient specificity, it is no surprise that hypothesis testing developed focused on the null hypothesis, as NHST.

The very basis of the logic of NHST emerges from the simple fact that a great deal is typically known about the distribution of outcomes under the null hypothesis. For example, to test hypotheses about means, the population mean, variance, and shape of the sampling distribution of the mean can be known quite precisely under the null in fairly broad circumstances, based on the Central Limit Theorem. As implied earlier, whether this information is valuable depends on the question the research wishes to address. However, for many years, questions have typically been developed to match the NHST paradigm. In other words, because we *can* know what the distribution of means looks like under the null, there are whole applied research arenas in which the expectation is that all questions will be asked so that this information *will* be put to use. Many methodologists have critiqued and challenged this practice, though it has had such momentum and popularity that such challenges are often ignored.

The modern and current version of NHST is a blending of the falsification philosophy developed by Fisher, with a focus on the null hypothesis and the p-value, and the decision-making philosophy espoused by Neyman and Pearson, with a focus on α, the alternative hypothesis, and power. Fisher had a strong scientific orientation in his work, and emphasized the importance of replication. Neyman and Pearson were concerned with how to draw conclusions from a single study, in quality control settings. Neither side was friendly to the validity or the goals of the other, and considerable contention resulted (see Serlin, 2010, for an engaging account of some of the exchanges). Ironically, the modern practice certainly would not satisfy either side of the argument; Gigerenzer (1993, p. 314) called modern NHST "an incoherent mismash of some of Fisher's ideas on the one hand, and some of the ideas of Neyman and J. S. Pearson on the other hand."

NHST has been soundly criticized, for a number of different reasons, virtually since its development. Many scholars have traced the history of NHST (see, e.g., Gigerenzer, 1993; Nickerson, 2000; and Rodgers, 2010a, for reviews). Of particular importance is that during the last half of the 1990s, the criticism reached such a volume that the American Psychological Association (APA) appointed a task force of methodologists to evaluate whether to outlaw NHST. Relevant statements from critics included the following: NHST "never makes a positive contribution" (Schmidt & Hunter, 1997, p. 57); NHST "has not only failed to support and advance psychology as a science but also has seriously impeded it" (Cohen, 1994, p. 997); NHST is "surely the most boneheadedly misguided procedure ever institutionalized in the rote training of science students" (Rozeboom, 1997, p. 335). The APA task force stopped short of recommending banishment of NHST, but did strongly recommend the expansion of statistical practice beyond the standard NHST paradigm (Wilkinson, 1999).

14.4 DEVELOPMENT OF AHST AND RELATED METHODS: HOW THINGS MIGHT HAVE BEEN DIFFERENT

Before, during, and after the late-20th-century turmoil over NHST, a modeling paradigm was naturally emerging. Researchers, using structural equation modeling (SEM), multilevel modeling, and other similar methods, reference the null and alternative hypothesis, but the organizing system is substantially more complex than traditional NHST. Widaman and Thompson (2003) addressed the expansion of complexity (also see Browne & Arminger, 1995). They focus on incremental fit indices used in SEM, and reference "a continuum of nested models." At one extreme is the saturated model(s) (actually there are multiple saturated models); a model is saturated if there are as many estimated parameters as there are independent data points, in which case there are no degrees of freedom left to test the fit of the model to the data, and the overall fit of the model must be perfect. Such a model perfectly reproduces the observed covariances and means.

In simpler linear model settings such as ANOVA and regression, saturated models can be specified (when, for example, the number of independent variables in a regression equation is one fewer than the number of subject points being fit). In this case, a simple bivariate regression model can find a perfect line that passes through two data points (or a perfect plane that passes through three data points, etc.). The fit for such models is perfect, though for mathematical and not for any interesting substantive reasons. In linear model settings, such saturated models are seldom referenced or discussed.

Within the structural equation modeling paradigm, Widaman and Thompson (2003, p. 19) discussed a "continuum of nested models." At one extreme is the actual null model, the equivalent of the chance model in NHST. There is some controversy about defining what is the appropriate chance (null) model of variances, covariances, and means in a SEM setting. In principle, the null is a model that does not contain the important structure that is in the researcher's hypothesis, a model that instead replaces the researcher's hypothesized structure with random pattern. Such a model will be simple, in whatever sense parsimony is measured (e.g., relatively few parameters), and will show one or more fit statistics that reflect random data patterns (see Browne & Cudeck, 1993, for a discussion of how to construct fit measures to compare models). There may in fact be multiple ways to conceptualize the null model in relation to a specific researcher's scientific hypothesis. At the other extreme is the saturated model. The saturated model fits perfectly by completely using up all the information (degrees of freedom) in the data. In between those is the researcher's scientific hypothesis (or multiple hypotheses), the statement or statements not only about substantive processes, but using a mathematical structure that leaves (preferably many) degrees of freedom available after estimating the model, so that the model can be evaluated using those remaining degrees of freedom. One of the important logical shifts that occurs in moving to the modeling perspective is that the substantive model is typically compared to the saturated model, rather than to the null model, the reverse direction of comparison compared to the focus in the standard NHST paradigm. Within the SEM (and broader modeling) paradigm, instead of the researcher's scientific hypothesis being compared to the absence of any meaningful model, as in NHST, rather it is compared to the model that can fit the data perfectly, the saturated model. If the substantive model fits the data at a level that is statistically equivalent to that of the saturated model, then support is inferred for the researcher's scientific hypothesis, and the "null hypothesis" of a poorly fitting model is rejected (note that the null model itself is not involved in this comparison, except implicitly). As Widaman and Thompson point out, the chi-square test of fit of a given substantive model is exactly equivalent to the model comparison between that substantive model and the perfectly fitting saturated model. In this case, rejecting the null implies rejecting the researcher's model. Obviously, in a traditional NHST setting, the researcher hopes to reject the null; in a modeling setting, the researcher hopes to not reject (i.e., to support) the null.

A number of scholars have commented on this reversal of focus between NHST and modeling methods. Abelson (1997) wrote, "Most tests of null hypotheses are rather feckless and potentially misleading. However, an additional brand of sensible significance tests arises in assessing the goodness-of-fit of substantive models to data." McDonald (1997) wrote, "After the introduction of ...structural models ..., it soon became apparent that the structural modeler has, in some sense, the opposite intention to the experimentalist. The latter hopes to "reject" a restrictive hypothesis of the absence of certain causal effects in favor of their presence—rejection permits publication. ...The former wishes to "accept" a restrictive model of the absence of certain causal effects—acceptance permits publication." MacCallum, Browne, and Sugawara (1996) wrote that their "approach allows for testing null hypotheses of not-good fit, reversing the role of the null hypothesis in conventional tests of model fit, so that a significant result provides strong support for good fit".

Within each of these quotes, the reference alludes to shifting the focus from the null hypothesis of chance/nil effect, to the researcher's scientific hypothesis, a hypothesis that contains the researcher's theory or working hypothesis. To be clear, the authors we quote in the previous paragraph did not necessarily define this reversal in relation to resampling theory. Rather, our proposal is that the extraordinary focus on NHST quite possibly would never have emerged if Fisher and Gosset had had modern computational ability at their disposal. Such a suggestion justifies our past and current attention to studying how resampling theory can support a shift to a focus on the researcher's scientific hypothesis.

What if Fisher and Gosset had not been so limited by the existing computational methods (where drawing slips of cardboard out of a hat was a legitimate method of re-randomization for world-class statisticians)? Suppose sampling distributions for specified or observed alternative distributions could be as easily defined as those for the null? There is no doubt that both Fisher and Gosset had great appreciation for resampling in principle. Whether they would have applied it to an AHST paradigm is purely speculative. But it seems at least possible, perhaps likely, given their goals and appreciation for such methods, that both Gosset and Fisher would have quickly moved away from testing hypotheses about the null, and rather tested hypotheses about the alternative (or several alternatives). Though the analysis of variance is the dominant statistical tool within the field of psychology, it is relevant to the current historical speculation that Fisher was not nearly as enamored with his method as were the practicing researchers over the past 90 years; rather, he called the ANOVA "a convenient method of arranging the arithmetic" (Fisher, 1934, p. 52).

Beyond the obvious interest (at least to historians of statistics) in this question, it is not easily resolved, because for almost a century there has been a remarkable focus and preoccupation with NHST. But now we have methods and sufficient computational power to do exactly what Fisher and Gosset could not do, to develop empirical sampling distributions in "statistically dense" settings. These empirical sampling distributions have fewer assumptions—virtually none of the statistical assumptions, and only partial assumptions of logic—compared to the more complex and less tractable parametric statistical tests that Fisher, Gosset, and hundreds of others have developed within an NHST context.

A primary reason NHST developed as it did is because statisticians and researchers can formulate specific mathematical and statistical information about the null hypothesis. Because the null is, by definition, a point hypothesis associated with chance outcomes, it has a fixed hypothesized value for the statistic of interest (often, though not always, zero). And as long as psychology, education, and other behavioral sciences are still developing, many alternatives are conceptually directional (or even nondirectional) but not point hypotheses. In other words, for an immature science, it makes sense to formulate a null such as H_0: $\rho = 0$ and test it against an alternative such as H_1: $\rho > 0$. Obviously, the alternative in this case does not support the development of a sampling distribution built around a specified correlation, because the correlation under the alternative is vague and undefined. Program evaluation and testing treatment effects in experimental settings are standard paradigms that are logically emergent from the NHST paradigm (though statisticians have been aware for a very long time that these methods can be easily re-formulated into a model comparison framework that easily attaches to the AHST perspective; e.g., Cramer, 1972). Within domains of behavioral science that are still developing, the NHST logic supports comparing a specific null against a weakly specified alternative.

Within the context of model comparisons, in settings in which a researcher wishes to evaluate stronger statements than just whether a chance model is supported or not, the weakly specified alternative disappears and the researcher has two specified models that are being compared. For example, to compare a three-variable multiple regression model to a two-variable nested multiple regression model is a comparison between two specific statements about how the world may work (or, to restate through a slightly different but equivalent perspective, about what processes generated the sample data). To be sure, there is an NHST hypothesis test that can be used to compare these models, in which the null hypothesis to compare the full model, subscripted f, and the reduced model, subscripted r, is H_0: $R_f^2 - R_r^2 = 0$, tested against H_1: $R_f^2 - R_r^2 > 0$; but this is a mechanistic part of the model comparison approach, and not an overarching epistemological orientation. In other words, the modeling revolution does not in any sense eliminate NHST, it rather moves it from comprising the philosophical foundation for statistical practice to an important procedural detail (see Rodgers, 2010b).

It is, in fact, almost unnatural to force an NHST interpretation (at least one involving a true nil null hypothesis) on a comparison between models. In this setting, the null hypothesis is that the reduced model is a better model than the full model (where "better" refers to the tradeoff between fit and parsimony, resolved statistically). The alternative hypothesis is that the improved fit of the full model is justified even in relation to its additional complexity. This type of null and alternative is rather more useful, specific, and scientifically informative than the traditional null hypothesis of nil/chance effect, compared against a vague and unspecified alternative. In fact, the whole epistemological system shifts. The shift is from a simple method with a vague alternative to one in which formal mathematical models compete against one another to explain empirical data.

14.5 THE BOOTSTRAP AS A BASIS FOR AHST: BACKGROUND

The basis of the AHST philosophy that might have supplanted the development of NHST if resampling methods had been tractable is that a researcher's scientific hypothesis—the alternative hypothesis, in traditional parlance—becomes an explicit component of the data analysis method. In traditional NHST settings, no specified alternative hypothesis even exists, because there is no requirement of the researcher that such be defined. But in AHST, the researcher is expected to specify a substantive scientific model (or preferably, competing nested models, in which case the alternative hypothesis is logically the more complex model). The data are used to evaluate the efficacy of the specified model, or to choose from the competing models. Using the bootstrap (or other resampling method such as the jackknife or randomization test; see Rodgers, 1999, for an organizational framework) provides a valuable approach with substantial flexibility to support the AHST approach.

The use of resampling to build a sampling distribution in relation to the researcher's scientific hypothesis is already reflected in the modeling literature. Bollen and Stine (1992) used a square root transformation to adjust the bootstrap in relation to SEM fit indices. More recently, parametric bootstraps have become popular. These use a distributional model of the data, rather than the observed data themselves, to produce bootstrap samples. For example, Zhang and Browne (2010) used a parametric bootstrap in a dynamic factor analysis of time series data, in which the model was partially imposed within the analysis (i.e., using AHST-type logic) with a moving block bootstrap across the time series. Cudeck and Browne (1992) developed a model to generate a covariance matrix with a specified lack of fit. This method was developed to generate data in Monte Carlo research, but it also illustrates how a researcher's model can be imposed within the analytic method itself.

Beasley et al. (2007) presented a resampling methodology to build an empirical sampling distribution around either an observed nonzero null correlation, or around a specified alternative correlation. The former was referred to as the OI (Observed Imposed) bootstrap procedure, because the empirical sampling distribution is built around the actual observed sample correlation. The latter was referred to as the HI (Hypothesis Imposed) bootstrap method, because the empirical sampling distribution is built around a hypothesized correlation. In the latter case, the hypothesized correlation was viewed as a nonzero null hypothesis. For example, if a behavior geneticist used past data to posit a null heritability of $h^2 = .5$ (tested using a twin correlation with H_0: $\rho = .5$) among an identical-twin-raised-apart sample, testing this null against a directional alternative positing that $\rho > .5$ in some population is a useful test.

The OI, however, is more conceptually flexible. The researcher might well hypothesize a null of H_0: $\rho = .5$, but within the OI method, the empirical sampling distribution is constructed *around the observed correlation.* The hypothesis testing approach involves building a 95% confidence interval (CI), and observing whether the correlation under the null is within the CI. But of critical importance is that the sampling distribution is constructed around the observed correlation; the construction of the sampling distribution does not reflect the null hypothesis. This is the approach, we believe, that Fisher and Gosset might have developed if modern computing had been available to them (and is similar to the construction of a standard parametric

CI, with resampling substituting for the parametric mathematics underlying the usual CI).

Both the OI and HI methods are based on a bootstrapping approach proposed by Lee and Rodgers (1998), which they called a univariate sampling bootstrap. Instead of sampling multivariate observations to construct an empirical sampling distribution (see Diaconis & Efron, 1983, for a description), they resampled data points univariately, which under a null hypothesis of independence will produce the required operating characteristics. Simulation results showed that their method improved on other bootstrapping approaches in small sample and nonnormal settings.

Beasley et al. (2007) evaluated the operating characteristics of the HI and OI methods in relation to a number of distributions (both normal and multiple nonnormal forms), sample sizes, bootstrap methods, and different sized nonzero null correlations. Both methods were found to be superior to parametric and bootstrap competitors in small sample and nonnormal settings. The HI and OI methods were both effective, and there was not a clear "winner," though the OI method appeared to be slightly better in terms of Type I Error control and power.

The conceptual contribution illustrated by these and other previous research articles is that an empirical sampling distribution can be constructed around any posited population correlation whatsoever. Whether it is specified as a null correlation of zero, a nonzero null, or an alternative, doesn't affect the simple production process based on the bootstrap. Further, the univariate sampling method has been shown to improve the performance of the bootstrap for evaluating a wide range of questions about the correlation coefficient. In the following section, we expand and evaluate the performance of the bootstrap of the multiple correlation, R, within a small multiple regression framework.

14.6 EVALUATING *R* IN MULTIPLE REGRESSION USING THE BOOTSTRAP

The above discussion motivates interest in the bootstrap (because it produces empirical sampling distributions about any specified value), and especially in the bootstrap using univariate sampling (because univariate sampling has been shown to have excellent operating characteristics in several important settings). Lee and Rodgers (1998) and Beasley et al. (2007) evaluated the bootstrap using univariate sampling in relation to the correlation coefficient. Our goal in this chapter is to begin to expand this treatment to broader statistical settings. We present results that evaluate the bootstrap using univariate sampling, compared to a number of other popular bootstrap approaches, in a multiple regression setting with up to three independent variables. The statistic that we bootstrap is the multiple correlation, R. The results help expand the idea of using the bootstrap in an AHST conceptualization.

14.7 GENERATION OF THE SAMPLING FRAME

14.7.1 Nonparametric Bootstrap Procedures

Suppose an observed sample contains $N = 60$ scores of a criterion variable and $k = 3$ predictors. The prototypical bootstrap (Efron, 1979) draws 60 intact cases (ordered four-tuple) from the observed sample to form one bootstrap sample of 60 cases; this is repeated many times, say $B = 10{,}000$. The "plug-in statistic" (which is R in this case) is calculated for each of the 10,000 samples, to produce a bootstrap distribution of 10,000 R^* values. Because cases are drawn with replacement, there are N^N possible bootstrap samples.[1] During this sampling process, the same three predictor values (X_1, X_2, X_3) are always matched with the same criterion (Y), that is, the original multivariate observation is respected by the sampling process. We refer to this procedure as the nonparametric multivariate sampling bootstrap (MV).

The *sampling frame* is the pool of scores from which resampled observations are repeatedly drawn to produce the 10,000 bootstrap samples. The current study examines the performance of five different sampling frame constructions. The sampling frame of the MV is simply the observed sample, which are the N rows of $(k + 1)$ values. Its sampling frame is considered *nonparametric*, because it contains no additional structure or assumptions beyond what is present in the observed sample.

In contrast to the MV, the HI and OI procedures in Beasley et al. (2007) draw each of an observation's $k + 1$ values (one Y and kX values) separately, each based on its own independent sampling process. This process is illustrated in Figure 14.1. To create the sampling frame, the values in the $k + 1$ variables are completely crossed with each other, producing a sampling frame that has N^{k+1} rows of $(k + 1)$ values (Figure 14.1, Stage 2b). The hypothesized (or observed) correlation is then imposed on the sampling frame using a transformation operator that reflects the desired correlation (Figure 14.1, Stage 2c). Beasley et al. "diagonalized" the sampling frame to a hypothesized or observed value by multiplying it by the appropriate Cholesky-decomposed correlation matrix. For reasons explained in the appendix of Beasley et al., our attempts to generalize this specific procedure beyond $k = 1$ (using spectral and other methods of diagonalization) have not produced satisfactory performance in simulations. Thus we have pursued two additional approaches to construct the OI's sampling frame, which are motivated, described, and presented in this chapter.

The first we call the *Sample and Iterate* (SI) technique (Ruscio & Kaczetow, 2008), which can be used to produce a sampling frame as well. First, points are drawn from the variables independently. This initial 'intermediate sample' resembles Stage

[1] See Beasley and Rodgers (2009) for a more complete introduction to the bootstrap, and their footnote 4 for the combinatorial equation of samples, irrespective of order.

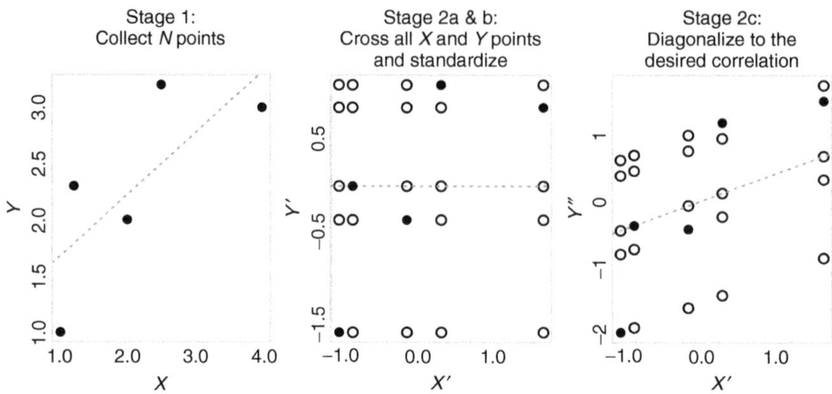

Figure 14.1 The first three stages of the HI sampling frame construction.

2b in Figure 14.1 because (a) its marginal distributions match the observed sample, and (b) the variables are uncorrelated (up to sampling error). Next the points are iteratively adjusted until they possess the user-specified correlation matrix, which is similar to the process in Figure 14.1, Stage 2c. Thus, the SI is a natural candidate to replace the Cholesky diagonalization used in previous HI and OI procedures (Beasley et al., 2007).

After the sampling frame is generated by the SI, it is sampled as before to produce 10,000 bootstrap samples of size N. During development of this specific use of the SI method, we found the best performance when N^2 cases were generated for the sampling frame[2] (though only N cases are sampled for each bootstrap sample). The second additional approach to defining the sampling frame is relatively new, and we describe it in detail in the following section.

14.7.2 Parametric Bootstrap Procedures

The parametric bootstrap was an early variant of the prototypical (nonparametric) bootstrap (Efron, 1979, Remark K). In a regression context, the parametric bootstrap is used when the analyst assumes known marginal distributions for each variable, and then generates scores to match the observed correlation/covariance matrix and $k + 1$ marginal distributions. In the simplest case, the variables are assumed to be multivariate normal, and variates are generated to match the observed means, variances, and covariances (Efron & Tibshirani, 1993, Section 6.5). Because this parametric bootstrap

[2] Generating N^3 points did not further improve performance in the tested conditions.

matches the first two conventional moments of each marginal, we call it the "parametric conventional 2" (PC2) method. Similarly, we implement a PC6 method that matches the first six conventional moments. (This is a new bootstrap method introduced in this chapter, using a sophisticated technique presented in Headrick, 2010.[3])

The Cholesky and SI produce sampling frames that aren't entirely nonparametric, because the analyst specifies a few (possibly implicit) assumptions, such as the strength of the linear relationship. We conceptualize the Cholesky and SI sampling frames closer to nonparametric than parametric, because their scores originate from the observed sample before they are diagonalized. Also, the use of their sampling frames resembles Efron's prototypical nonparametric bootstrap: the same static pool of scores is used for each row of each bootstrap sample. It is theoretically possible (although highly unlikely) to sample an observation $N \times B$ times. In contrast, a point in a parametric sampling frame is ephemeral—once it is generated, it is drawn immediately for the current bootstrap sample and is never repeated in any other rows or bootstrap samples.

14.8 DESIGN

The simulation included four crossed factors: (1) statistical procedure, (2) population distribution, (3) population correlation, and (4) sample size. The sample sizes were $N = 30, 60, 100, 200$.

14.8.1 Statistical Procedures

One analytic and four bootstrap procedures were compared:

1) *Analytic*: A closed-form analytic procedure that generalizes Fisher's r-to-z transformation (Konishi, 1981, Equation 3.3).
2) *MV*: the prototypical bootstrap previously described.
3) *SI*: The SI technique that generates a sampling frame of N^2 points, and N points are drawn for each bootstrap sample.
4) *PC2*: A parametric bootstrap which matches the first two conventional moments of each marginal.
5) *PC6*: A parametric bootstrap which matches the first six conventional moments of each marginal.

[3] Instead of using conventional moments, the univariate L-moment technique in Headrick (2011) uses Legendre polynomials. However, a multivariate L-moment technique has not been completed at the time of this study.

A Bias-Corrected CI (Efron, 1982) was used by all the bootstraps. It is transformation-respecting, so the bootstrap results should be equivalent regardless whether the plug-in statistic is R or R^2. An analytic test of R^2 (Cohen, Cohen, West, & Aiken, 2003, Equation 3.6.3) yielded comparable results to the Konishi (1981) procedure for R.

14.8.2 Parent Population Distributions

Population data were generated using Headrick's (2010) technique[4] from seven different parent populations. The parent populations were defined with one distributional structure for the dependent variable, and another multivariate distributional structure for the three independent variables. The particular populations were chosen because they were the most challenging in Beasley et al. (2007). Three marginal distributions (Normal (Gaussian), $\chi^2_{df=1}$ and $\chi^2_{df=3}$ formed the six population families examined:

1) Normal Y, Normal X
2) $\chi^2_{df=1}Y$, $\chi^2_{df=1}X$
3) $\chi^2_{df=3}Y$, $\chi^2_{df=3}X$
4) $\chi^2_{df=1}Y$, Normal X
5) $\chi^2_{df=3}Y$, Normal X
6) Normal Y, $\chi^2_{df=1}X$
7) Normal Y, $\chi^2_{df=3}X$

Population 7 produced similar results to Population 5, and is not discussed further.

14.8.3 Population Correlations

Two population matrices of different correlations were used; both had equal values of $R^2 = .27$ (Table 14.1). In Figure 14.2, we present bivariate graphs showing the results of the data generating process. We chose not to generate samples from a nil correlation matrix (all intercorrelations $= 0$), because (a) R and R^2 are biased near the boundaries of 0 and 1, and (b) we think the analytic method is most vulnerable to normality violations with a nonnil hypothesis.

[4] Note that this does not give the PC6 an unfair advantage, because the bootstrap procedure doesn't use the population moments, but rather the sample moments. So it is influenced by sampling variability, just as the other five procedures are.

Table 14.1 The two examined population correlation matrices. The R^2 of each is .27

	One factor population					Two or three factor population			
	Y	X_1	X_2	X_3		Y	X_1	X_2	X_3
Y	1	.4	.4	.4	Y	1	.4	.4	0
X_1		1	.4	.4	X_1		1	.2	0
X_2			1	.4	X_2			1	0
X_3				1	X_3				1

14.9 RESULTS

The results are presented graphically in Figures 14.3 and 14.4. The procedures used $\alpha = .05$, and so a gray band extends from $\alpha_{Obs} = .025$ to $\alpha_{Obs} = .075$, which is the range within which we'd like to see empirical Type I error rates fall. The six rows represent the different population families, and the two columns represent the different population correlation structures. Within each of the twelve panels, the four sample sizes extend across the (logged) horizontal axis. At each sample size, the height of the five procedures represents the proportion of times they rejected the hypothesis $R = +\sqrt{.27} = .52$. The ideal simulation outcome for a procedure would be plotted at .05 for all four samples sizes, in all twelve panels.

The analytic procedure was consistently liberal when the Xs and Ys were skewed (both $\chi^2_{df=1}$ and $\chi^2_{df=3}$). However, it controlled α well with other populations, especially when the multivariate normality assumption was justified. This behavior is consistent with our previous bivariate simulation studies (Beasley et al., 2007).

The MV performed fairly well with the examined populations, which is consistent with our previous bivariate simulation studies when the population correlation was $\rho^2 = .16$ and $\rho^2 = .36$ (Beasley et al., 2007, Beasley, 2010). However when $\rho^2 = .64$ in these simulations, the procedure was very liberal when one variable was normal and the other was skewed ($\chi^2_{df=1}$ or $\chi^2_{df=3}$). The MV was robust in the current study (recall $\rho^2 = .27$), but if stronger correlations are encountered in applied settings, we suspect its performance might degrade quicker than the univariate sampling bootstraps. This is an issue we plan to address in future research.

The SI bootstrap did well in the left columns of Figures 14.3 and 14.4, where all variables were equally correlated. However it became more liberal in the right column, where the relationships had different strengths, especially when the marginal distribution of the criterion did match the predictors. The SI generates scores with a mechanism similar to a factor analysis (see Ruscio & Kaczetow, 2008, Steps 4–7);

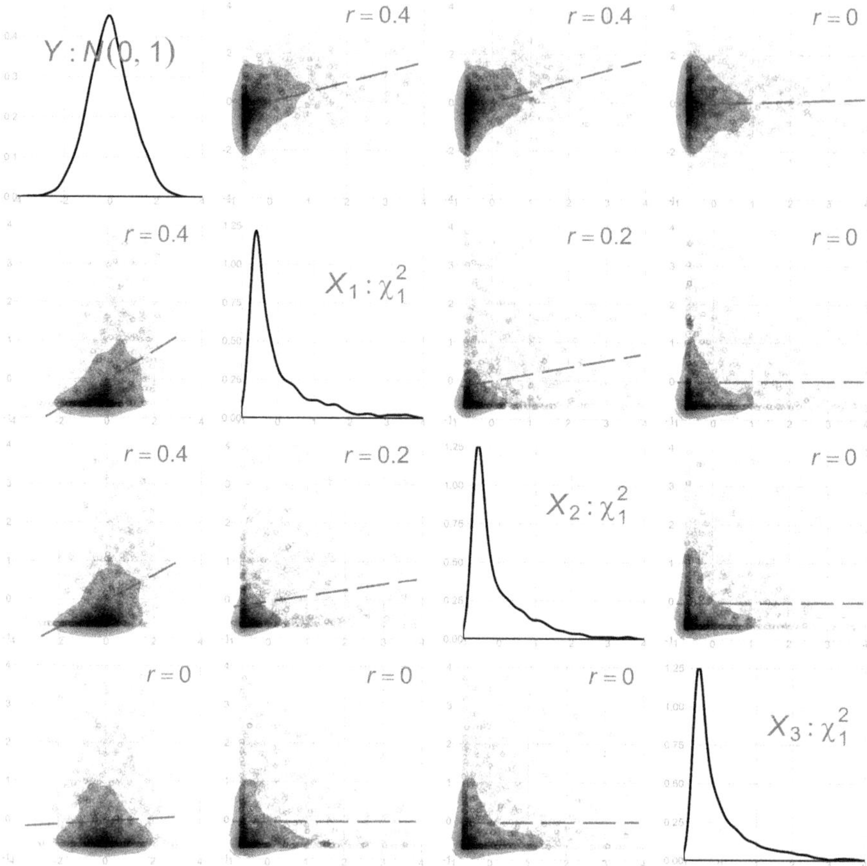

Figure 14.2 Example outcomes from the bivariate data generation process.

the current results might suggest the SI is better suited to bootstrapping populations with simpler factor structures than complicated ones. However, this is our first time we have used the technique, and we hope to learn it better and tailor it to future HI and OI procedures.

The results of the PC2 closely followed the analytic procedure; this is expected because both assume multivariate normality. The PC6 did well with the multivariate normal population, and the populations that contain $\chi^2_{df=3}$, especially when $N \geq 60$. With the $\chi^2_{df=1}$ distributions, PC6 frequently escaped the gray band, but never failed as extremely as the analytic and PC2.

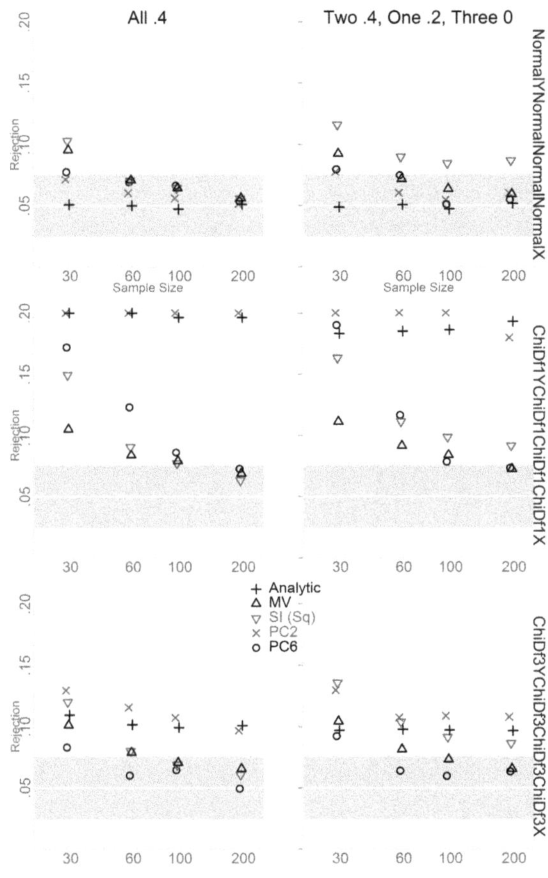

Figure 14.3 Results for Populations 1–3.

14.10 DISCUSSION

The goal of this chapter has been both quite broad and very specific. The broad goal has been to discuss testing hypotheses about the researcher's scientific hypothesis, what is traditionally called the "alternative hypothesis," rather than testing hypotheses about the nonexistence of the researcher's hypothesis, what is traditionally called the "null hypothesis." Although we do not inveigh against NHST—it has served an admirable purpose through the years, even as it has been often and consistently misused—the previous sentence stands on its own. To "test hypotheses about the nonexistence of the researcher's hypothesis" cannot be viewed as an ultimately successful scientific

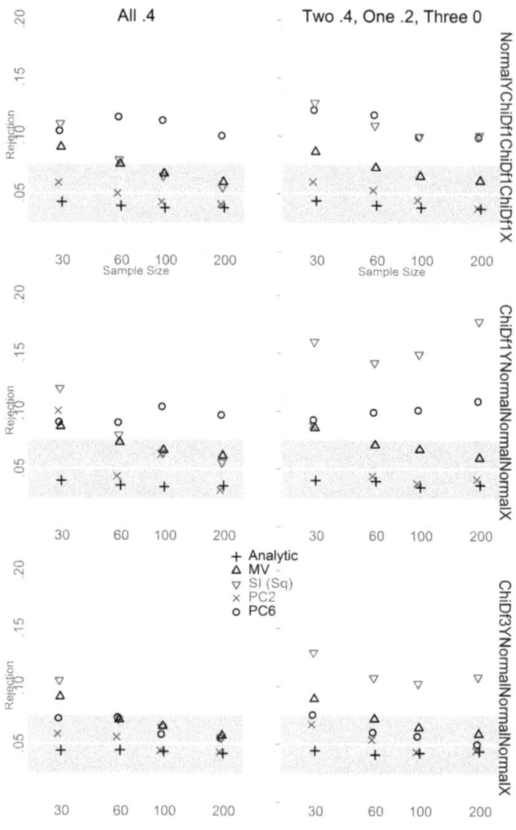

Figure 14.4 Results for Populations 4–6.

paradigm (though it often serves a useful procedural purpose in an immature but developing science). The specific goal has been to advance one specific effort by applying the bootstrap to test alternative hypotheses. These two efforts are united within the same paper through the historical suggestion that Fisher and Gosset might well have preferred the method presented here—or several others that are related conceptually—to the one that they helped develop, the one that has become the epistemological basis for statistical practice during the past 70 or more years.

The conceptual foundation of a shift to AHST is the ability of a researcher to embed information about their hypothesis/theory/explanatory system into a set of data from which can be generated a sampling distribution. In the work illustrated here, we begin with observed correlational data. We generate a sampling frame of data that reflects independence among the variables, and then impose on that sampling frame either a hypothesized or an observed correlation. We then use a bootstrap to generate an empirical sampling distribution, using that transformed sampling frame. In other applications, a researcher could impose distributional knowledge on the data, or even an explicit model to explain the data, and then test hypotheses by using the resulting empirical sampling distribution. Unlike NHST, the data have been informed by the researcher's hypothesis, which is why using a sampling distribution derived from the data is an application of AHST.

We note that this orientation is not new, although our particular approach to implementing it using the univariate sampling bootstrap is new. This kind of thinking underlies structural equation modeling, multi-level modeling, and many other modern statistical analysis systems. For example, the researcher's model is very explicitly stated in SEM, through path diagrams representing a structural model and a measurement model, and then the predictions of the model are compared to either covariances derived from raw data or to the raw data themselves. This is a fundamental shift from NHST, to AHST, one that has been occasionally noted but seldom discussed in any detail.

14.10.1 Recommendations

The MV method worked relatively well in the current study, although we have reason to doubt its ultimate utility, based on past research in which the population marginals were different (Beasley et al., 2007). In the current work, the PC6 method is the only one that is both robust to nonnormality, and also allows the sampling frame flexibility to shift to reflect the researcher's hypothesis (or, alternatively, a nonzero null). We are currently developing further simulation work on this topic, which will eventually include the Legendre L-Moment method within the study as well.

14.10.2 Limitations and Future Directions

Our early work has focused on the correlation coefficient (Lee & Rodgers, 1998; Beasley et al., 2007), and then expanded in the current chapter to a multiple regression/correlation setting. This is a relatively limited setting, although the requirements to evaluate the various bootstrap methods and to build understanding in relation to these has been fairly demanding, even in these simple settings. In the future, we expect to expand the evaluation of this approach to estimation of individual parameters, and to the comparison of the R^2 values from nested linear models (e.g., Appelbaum & Cramer, 1974; Maxwell & Delaney, 2004). This last paradigm is perhaps the one conceptually related the closest to the AHST perspective that we develop and discuss in here, and is not incidentally one that is closest to supporting and facilitating researchers to test their own hypotheses.

14.11 CONCLUSION

In 1984, Browne developed the Asymptotic Distribution Free (ADF) estimator, which incorporates the skew and kurtosis of the data (instead of only the mean and variance). By considering higher moments, the ADF could be more robust than its contemporary methods when the data are not multivariate normal. Similarly, the PC6 method, which we tentatively recommend as the method of choice emerging from the simulation study presented in this chapter, considers more moments of the population distribution than the PC2 considers. Our hope is that similar emerging approaches can capture information hidden beyond the first two moments of the data. The PC6 recreates both the correlational structure and the first six moments of the marginal distributions. The computational demands are increased, but the intellectual demands on the analyst are decreased. The human burden is lightened because a robust tool can be applied confidently in more situations than a nonrobust tool—there is less need to resort to a specialized tool because the generalized tool isn't computationally feasible.

To return to the original theme of our chapter, Gosset and Fisher considered resampling methods as foundational features of their emerging statistical paradigm. However, neither pulling data from a hat nor their hand-cranked mechanical calculators were up to the general computational challenge. Instead, these two statisticians substituted parametric mathematical procedures for the preferred resampling methods, and as a result NHST was born. Now, a century later, NHST can be moved into the background, and the kinds of bootstrapping methods discussed, illustrated, and evaluated in the current chapter can be implemented to provide the statistical support for methods like SEM and Multi-level Modeling as legitimate AHST procedures, capable of addressing much more complex, interesting, sophisticated, and direct research questions in the constantly maturing and developing behavioral science arena.

REFERENCES

Abelson, R. P. (1997). A retrospective on the significance test ban of 1999 (if there were no significance tests, they would be invented). In L. L. Harlow, S. A. Mulaik, & J. H. Steiger (Eds.) *What if there were no significance tests?* (pp. 117–141). Mahwah, NJ: Erlbaum.

Appelbaum, M.I. & Cramer, E. M. (1974). Some problems in the nonorthogonal analysis of variance. *Psychological Bulletin, 81*, 335–343.

Beasley, W.B. (2010). *Univariate sampling bootstrap procedures using prior information* (Unpublished doctoral dissertation). University of Oklahoma, Norman.

Beasley, W. H., DeShea, L., Toothaker, L. E., Mendoza, J. L., Bard, D. E., & Rodgers, J. L. (2007). Bootstrapping to test for nonzero population correlation coefficients using univariate sampling. *Psychological Methods, 12*, 414–433.

Beasley, W. H., & Rodgers, J. L. (2009). Resampling methods. In R. E. Millsap & A. Maydeu-Olivares (Eds.) *Quantitative methods in psychology* (pp. 362–403). Thousand Oaks, CA: Sage.

Bollen, K. A., & Stine, R. A. (1992). Bootstrapping goodness-of-fit measures in structural equation models. *Sociological Methods Research, 21*, 205–229.

Browne, M. W. (1984). Asymptotically distribution-free methods for the analysis of covariance structures. *British Journal of Mathematical and Statistical Psychology, 37*, 62–83.

Browne, M. W., & Arminger, G. (1995). Specification and estimation of mean- and covariance-structure models. In G. Arminger, C. C. Clogg, & M. E. Sobel (Eds.), *Handbook of statistical modeling for the social and behavioral sciences* (pp. 185–249). New York: Plenum Press.

Browne, M. W., & Cudeck, R. (1993). Alternative ways of assessing model fit. In K. A. Bollen & J. S. Long (Eds.), *Testing structural equation models* (pp. 136–162). Newbury Park, CA: Sage.

Carlin, B. P., & Louis, T. A. (2009). *Bayesian methods for data analysis.* Boca Raton, FL: Chapman & Hall.

Cohen, J. (1994). The earth is round ($p < .05$). *American Psychologist, 49*, 997–1003.

Cohen, J., Cohen, P., West, S.G., & Aiken, L.S. (2003). *Applied multiple regression/correlation analysis for the behavioral sciences* (3rd ed.). Mahwah, NJ: Erlbaum.

Cramer, E. M. (1972). Significance tests and tests of models in multiple regression. *The American Statistician, 26*, 26–30.

Cudeck, R., & Browne, M. W. (1992) Constructing a covariance matrix that yields a specified minimizer and a specified minimum discrepancy function value. *Psychometrika, 57*, 357–369.

Diaconis, P., & Efron, B. (1983). Computer-intensive methods in statistics. *Scientific American, 248*, 116–130.

Efron, B. (1979). Bootstrap methods: Another look at the jackknife. *Annals of Statistics, 7*, 1–26.

Efron, B. (1982). *The jackknife, the bootstrap, and other resampling plans.* Philadelphia: Society for Industrial and Applied Mathematics.

Efron, B., & Tibshirani, R. J. (1993). *An introduction to the bootstrap.* Boca Raton: Chapman & Hall/CRC.

Fisher, R. A., (1934), Discussion of "Statistics in agricultural research" by J. Wishart, *Journal of the Royal Statistical Society, Supplement, 1*, 26–61.

Fisher, R. A. (1936). "The coefficient of racial likeness" and the future of craniometry. *Journal of the Royal Anthropological Institute of Great Britain and Ireland, 66*, 57–63.

Gigerenzer, G. (1993). The superego, the ego, and the id in statistical reasoning. In G. Keren & C. Lewis (Eds.), *A handbook for data analysis in the behavioral sciences: Vol 1. Methodological issues* (pp. 311–339). Hillsdale, NJ: Erlbaum.

Headrick, T. C. (2010). *Statistical simulation: Power method polynomials and other transformations*. Chapman & Hall: Boca Raton.

Headrick, T. C. (2011). A characterization of power method transformations through l-moments. *Journal of Probability and Statistics, Article ID 497463*, 1-22. doi:10.1155/2011/497463.

Konishi, S. (1981). Normalizing transformations of some statistics in multivariate analysis. *Biometrika, 68*, 647–651.

Lee, W., & Rodgers, J. L. (1998). Bootstrapping correlation coefficients using univariate and bivariate sampling. *Psychological Methods, 3*, 91–103.

MacCallum, R. C., Browne, M. W., & Sugawara, H. M. (1996). Power analysis and determination of sample size for covariance structure modeling. *Psychological Methods, 1*, 130–149.

Maxwell, S. E., & Delaney, H. D. (2004). Designing experiments and analyzing data: A model comparison perspective (2nd ed.). Mahwah, NJ: Erlbaum.

McDonald, R. P. (1997). Goodness of approximation in the linear model. In L. L. Harlow, S. A. Mulaik, & J. H. Steiger (Eds.) *What if there were no significance tests?* (pp. 199–219). Mahwah, NJ: Erlbaum.

Nickerson, R. S. (2000). Null hypothesis significance testing: A review of an old and continuing controversy. *Psychological Methods, 5*, 241–301.

Pruzek, R. M. (1997). An introduction to Bayesian inference and its applications. In L. L. Harlow, S. A. Mulaik, & J. H. Steiger (Eds.) *What if there were no significance tests?* (pp. 287–318). Mahwah, NJ: Erlbaum.

Rindskopf, D. M. (1997). Testing "small," not null, hypotheses: Classical and Bayesian approaches. In L. L. Harlow, S. A. Mulaik, & J. H. Steiger (Eds.) *What if there were no significance tests?* (pp. 319–332). Mahwah, NJ: Erlbaum.

Rodgers, J. L. (1999). The bootstrap, the jackknife, and the randomization test: A sampling taxonomy. *Multivariate Behavioral Research, 34*, 441–456.

Rodgers, J. L. (2010a). The epistemology of mathematical and statistical modeling: A quiet methodological revolution. *American Psychologist, 65*, 1–12.

Rodgers, J. L. (2010b). Statistical and mathematical modeling versus NHST? There's no competition! *Journal of Modern and Applied Statistical Methods, 9*, 340–347.

Rozeboom, W. W. (1960). The fallacy of the null hypothesis significance test. *Psychological Bulletin, 57*, 416–428.

Rozeboom, W. W. (1997). Good science is abductive, not hypothetico-deductive. In L. L. Harlow, S. A. Mulaik, & J. H. Steiger (Eds.) *What if there were no significance tests?* (pp. 335–392). Mahwah, NJ: Lawrence Erlbaum Associates.

Ruscio, J., & Kaczetow, W. (2008). Simulating multivariate nonnormal data using an iterative algorithm. *Multivariate Behavioral Research, 43*, 355–381. doi: 10.1080/00273170 802285693.

Schmidt, F. L., & Hunter, J. E. (1997). Eight common but false objections to the discontinuation of significance testing in the analysis of research data. In L. L. Harlow, S. A. Mulaik, & J. H. Steiger (Eds.), *What if there were no significance tests?* (pp. 37–64). Mahwah, NJ: Erlbaum.

Serlin, R. C. (2010). Fisher was right. *Journal of Modern Applied Statistical Methods, 9*, 2–8.

Student (1908). The probable error of a mean. *Biometrika, 6*, 1–25.

Trafimow, D. (2003). Hypothesis testing and theory evaluation at the boundaries: Surprising insights from Bayes' theorem. *Psychological Review, 110*, 526–535.

Widaman, K. F. & Thompson, J. S. (2003). On specifying the null model for incremental fit indices in structural equation modeling. *Psychological Methods, 8*, 16–37.

Wilkinson, L. and the Task Force on Statistical Inference, APA Board of Scientific Affairs (1999). Statistical methods in psychology journals: Guidelines and explanations. *American Psychologist, 54*, 595–604.

Zhang, G., & Browne, M. W. (2010) Bootstrap standard error estimates in dynamic factor analysis. *Multivariate Behavioral Research, 45*, 453–482.

Chapter 15

On The Equilibrium Dynamics of Meaning

Steven M. Boker[1] and Mike Martin

15.1 LATENT VARIABLES

Latent variables provide a way to infer properties of constructs that cannot be directly measured. We build statistical models, such as confirmatory factor models, to identify latent constructs and test theories about them. This methodology has been extraordinarily successful and has become extremely popular. Increasingly, the methods and models for identifying and estimating properties of latent variables are becoming reified. That is to say, it is increasingly common for an experiment to be designed around the methodology of latent variable modeling and perhaps around a particular type of model. This is not necessarily a bad thing. One would hope that an expensive data collection initiative would be likely to provide convincing evidence about a thesis or its antithesis. But the well–oiled gears of latent variable modeling

[1] Funding for this work was provided in part by NIH Grant 1R21DA024304–01. Any opinions, findings, and conclusions or recommendations expressed in this material are those of the authors and do not necessarily reflect the views of the National Institutes of Health.

may be so attractive that we do not consider alternate theories that do not fit into this smoothly functioning machine.

It is important at regular intervals to step outside the currently popular framework of thinking and to reconsider basic premises. With that in mind, this chapter will present a challenge to what we see as a limitation of latent variable modeling as it is currently practiced. We do not intend to pull down any of the marvelously functional architecture of confirmatory factor analysis that is at the heart of latent variable modeling. Instead, if we are successful in our intent, we shall broaden the scope of how one might identify a latent variable, and consequentially introduce an expanded definition of how meaning might be attached to latent constructs.

Let us begin with the idea of a latent variable as it is used in psychology and illustrate it with an example. A psychologist administers a questionnaire about, for instance, anxiety to a sample of participants. The questionnaire has items that, in the mind of the psychologist, are related to what she means when she thinks of anxiety. Each item is structured to have numerically mapped answers that, again in the mind of the psychologist, are related by some function to the meaning of that item. Each participant reads each item, perceives a meaning for the item, and then responds according to some function that, in the participant's mind, relates the meaning of the item to some property of him – or herself. Finally, the psychologist submits the matrix of all item responses to latent variable modeling in order to identify a latent construct that, to her, means "anxiety."

In analysis of results from this example, one problem is that the stated numerical mapping of each item to its ordinal numerical responses is unknown. Multivariate ordinal analysis provides a way to linearize this mapping from observed data. Ordinal threshold estimation is an important and underutilized tool in the latent variable modeler's toolbox. However, our paper does not address this aspect of the example. Instead we will focus on something quite a bit slipperier—meaning. What is meaning, how is it assigned, and what role does meaning play in the statistical identification of latent variables?

15.2 MEANING

The premise of this paper is the following: Meaning is at the heart of what we do in latent variable modeling. A latent construct is a way to aggregate and focus meaning into quantifiable constructs. Structural models, and in particular factor models, are a way to use the considerable power of product moment matrices to focus meaning in such a way that it aggregates across participants or within participant across time. The goal is that the meaning the psychologist had in mind is the meaning that emerges in the latent variable as indicated by the participants' responses.

These are deep waters—What is meaning? And how is it attached to words or utterances? Philosophers of language have written at length about these problems, and it may be useful to consider a selected sample of ideas from epistemology.

A good deal of twentieth century philosophy of language explored connections between concepts of truth and meaning. Bertrand Russell's theory of denoting (Russell, 1905) makes a distinction between that which we *apprehend* and that with which we are *acquainted.* We are *acquainted* with facts that are directly perceivable in the environment. In latent variable modeling terms, we are acquainted with manifest variables. That which we *apprehend* follows as a consequence of a combination of facts with which we are acquainted, and a system of logic or mathematics to lead to a conclusion. Russell's definition of apprehension is very much like a latent variable: something we believe to exist and to have meaning, but with which we are not directly acquainted. We estimate properties of the latent variable through facts with which we are acquainted, i.e., manifest variables, and a system of logic, i.e., linear structural relations of product moment matrices.

15.3 CONTEXT

Grice's influential paper on meaning (Grice, 1957) altered the discussion by bringing up a series of examples that pointed to the influence of context on meaning. He argued that the definition of "meaning" requires that there is a *communicator* with an *intention* and an *audience* who is intended to recognize the meaning. To quote Grice, "'A meant$_{NN}$ something by x' is (roughly) equivalent to 'A intended the utterance of x to produce some effect in an audience by means of the recognition of this intention'" (Grice, 1957, p. 385).[2] Grice argues that once one has accepted that the communicator, the intention, and the audience are required, then meaning can shift based on who is speaking, what they intend, and who is listening.

Contextual effects in psychological measurement are often acknowledged, but too frequently left untested. Nesselroade (2006) describes an example where, in an instrument designed to measure state anxiety, some participants interpreted the word "anxious" as "having anxiety" and others interpreted "anxious" to mean "being eager." Items are constructed by instrument designers who intend to convey a specific meaning to an audience consisting of research participants. In the Nesselroade example, the intention of the instrument designer was not interpreted in the same manner by all participants. Disambiguating the affected items might resolve the "anxious" example; however, this is only the visible part of a much larger iceberg.

[2] Grice used the notation "meant$_{NN}$" as a way to distinguish the way that "meant" is used in the quote from the sense of the word in which one "meant (intended) to do something."

The "anxious" item itself is a manifest variable—in Russell's terms it is a fact with which we are acquainted. The latent variable "state anxiety" is intended by the research psychologist to convey meaning to an audience consisting of readers of scientific publications. As a participant answers items, he or she is likely to form an apprehension of intention attributed to the psychologist. This forms a context in which the meaning of latent variables indicated by items in the psychometric instrument affects the meaning of individual items. It is difficult to avoid this type of contextual effect, since the meaning of each individual item must be unambiguous and the item meanings must overlap with the target latent construct meaning. However, the latent variable meaning apprehended by participants may vary considerably from one individual to the next.

Grice argues that, "... for x to have meaning$_{NN}$, the intended effect must be something which in some sense is within the control of the audience" (Grice, 1957, p. 385). Each participant has a life context that can affect the apprehended meaning of latent variables indicated by a psychometric instrument and thus the meaning of each of the items. As latent variable modelers, we hope for factor invariance, but inevitably there will be individual differences in how each item is interpreted. And if participants' life contexts are reliably associated with the latent variables indicated by the instrument (which is often why the constructs were chosen for study in the first place), individual differences in measurement model loadings are likely to vary in reliable ways.

15.4 TRANSLATION

Quine took the position that meaning is inherently underdetermined (Quine, 1970). Consider the problem of meaning as being a problem of translation. Suppose that Alf produces utterances that are intended to be understood by Betty. Betty must find a way to translate what Alf uttered into her own internal language. In order to help us focus on the difficulties involved in this translation, let us suppose a case when the translation is difficult for Betty. In the worst case, Alf is an alien. Betty must find a way to divine meaning from a corpus of Alf's strange clicks and whistles. Quine argued that we are all, to some degree, isolated from each others' minds and so fall somewhere on a continuum of alien–ness to one another. We must go through a process of divining meaning from first principles as we develop the ability to use language; a process that involves building a model of how Alf's alien utterances map onto physical reality as we perceive it. The next step in this argument is to note that no model of reality is entirely determined. That is to say, for any set of physical processes there are always at least two models that account for the data equally well. From this set of premises, Quine argues for the "indeterminacy of translation"—there must always be some ambiguity in the meaning of communication (Quine, 1970).

As latent variable modelers, we are familiar with Quine's argument as it applies to problems such as unique factors, indeterminacy of factor scores, and factor rotation. These problems are intrinsic to the latent variable modeling framework. Consider the inherent indeterminacy in meaning of a latent variable model as it is apprehended both by the intended audience (i.e., readers of scientific articles), and by its unintended audience (i.e., participants providing responses to a psychometric instrument). How can one expect a latent variable to have *any* sort of reliable meaning?

By this point in the chapter, one might conclude that the authors have a particularly bleak outlook on empirical research. One might wonder if there is any way out of this indeterminate and confounding thicket. So let us look on the bright side. While there is more than sufficient misunderstanding in the world, people do, for the most part, communicate successfully with one another. Sometimes the meaning of a selected latent variable model may be translated in surprisingly different ways by two different readers of a scientific article. However, it is likely that most scientists are largely in agreement about the meaning of most models. Clearly, there must be processes—some sort of dynamics—that lead to inter–individual agreement in meaning in the face of its theoretical indeterminacy. This is the fundamental postulate around which our article is organized.

15.5 THESAURUS

Consider Betty's problem as being similar to that of creating a thesaurus.[3] If Betty manages to create an internal thesaurus for Alf's language, some sort of translation will become possible. That is to say, by understanding the relationship between words, one can create a semantic network where each word gives meaning to its near synonyms within the language. Then, with only a few direct translational referents, the semantic network of Alf's language can be mapped onto the semantic network of Betty's language. But in order for a reliable translation to happen, Betty's internal thesaurus for her own language must remain reasonably stable while a thesaurus for Alf's language converges to an equilibrium. If Betty's internal thesaurus did not remain stable while developing a thesaurus for Alf's language, Betty would be learning a translation into some idiosyncratic language that no one but Betty spoke. The processes—the dynamics—that lead to inter–individual agreement in meaning must tend toward intra–individual stability in meaning over time if we are to continue to understand one another while being exposed to novel linguistic information.

To review, we have concluded that there must be processes that lead to inter–individual agreement in meaning and that lead to intra–individual stability of meaning

[3] The authors would like to acknowledge Ryne Estabrook, who suggested pursuing the thesaurus argument.

over time. These two processes may both be the outcome of a single intrinsic dynamic or they may be two separate dynamic processes, but they must, in the long run lead to stability in meaning. Thus, meaning must have stable equilibria.

Grice argued that one must include contextual information in a semantic network. That is to say, Betty would need to take into account the context in which Alf found himself when he produced an utterance, and she would need to find a way to map his context onto similar contexts from her own life. Thus Betty needs a context–dependent "active thesaurus." For a given entry, the words that appear as synonyms and antonyms will actively change based on contextual information.

The number of degrees of freedom in Betty's context–dependent thesaurus may not be as large as it first appears. The process of the formation of context mapping may reduce the number of stable equilibria so that only a few context dependencies exist for each thesaurus entry. The number of contexts per entry, and the words in each context–dependent entry may be a source of inter–individual differences. However, we can expect that there is a large degree of between–person overlap in internal thesauri in order for between–person communication to be reliable. We expect that there is likely to be intra–individual change in the internal thesaurus as a person develops and is exposed to more contextual information. However, the process that stabilizes meaning within a person over time will act to revise these contextual changes in order that internal self–consistency in the latent variable be maintained.

15.6 IMPLICATIONS OF EQUILIBRIUM DYNAMICS OF MEANING

Our thesis is this: Intra–individual context–dependent meaning has an intrinsic dynamic that tends towards stable equilibria. If this thesis holds, we can think of three implications for the construction and identification of latent variable measurement models.

First, one would expect that there are inter–individual differences in the equilibrium values of the meaning of a selected latent construct. Aggregation over a large sample of individuals with stable intra–individual equilibria results in a mean stable equilibrium. Thus, our thesis is not incompatible with decades of work using latent variable and factor models to account for the structure of personality, mood, depression, cognitive abilities, and so forth. By adopting our thesis, we do not abandon previous results.

Second, the interpretation of cross–sectional results becomes more rich if we acknowledge that observed individual differences may have three sources: One, there may be individual differences in the traits or states that were intended to be measured. Two, there may be individual differences in equilibria of the meaning of a latent variable as apprehended by participants in the study. And three, there may be individual differences in equilibria of the meaning of a latent variable as apprehended by readers of scientific articles.

Third, when data are longitudinal, one may identify intra–individual variability in the intended state or trait plus two more potential sources of intra–individual variability in construct meaning as apprehended by participants: One, there may be variability due to the context of the participant at the moment of measurement. And two, there may be longer term intra–individual fluctuations in context–dependent meaning as the intrinsic dynamic process adapts to new contextual and semantic information. Although the process of forming equilibria in the semantic network will tend towards stability, adaptation to new contextual information must occur in order to adapt to our changing environment. It is also evident that in order to learn new vocabulary or newly accepted meanings for old words, we must adapt our thesauri to include this new semantic information. Let us next examine in more detail these two sources of intra–individual variability or adaptation by use of example.

15.6.1 Context–Dependent Meaning

Consider the following example of context–dependent meaning. Suppose Andrew and Betty are native English speakers. Over time, they have each built up a context–dependent thesaurus by communicating with thousands of other English speakers. Yesterday, Andrew told Betty, "I feel especially happy." Betty noted that it was a sunny day, Andrew was walking with a bounce in his step, and he was smiling broadly. Betty consults her context–dependent thesaurus and concludes that Andrew meant that his mood was elevated, and that he felt something similar to "pleased," "joyful," and "cheerful." Today, Andrew again tells Betty, "I feel especially happy." Betty notes that Andrew looks downcast and recalls that he just received a rejection notice from a journal editor. Betty consults her context–dependent thesaurus and concludes that Andrew means to convey a sense of ironic detachment. Betty has concluded that "happy" means something different in Andrew's sentence today than what he meant yesterday. But if you ask Betty to recall what Andrew meant yesterday, she will likely still conclude that he meant he was joyful yesterday. Given this consistency in recalled meanings, we expect that context–dependent meaning can be considered to be an expansion of the thesaurus. One word may have many context–dependent meanings while each of those meanings maintains a stable equilibrium over time.

The implication for psychometric measurement is that a carefully controlled context is critical if one is to use verbal items to focus a particular context–dependent meaning into a latent variable. Of course, context cannot be completely controlled. Each participant brings his or her own context to an experiment. Thus, even after controlling for context, it would be desirable to measure individual context in order to estimate its effect on shifting the meaning of items. One way to do this is to develop items that probe for both the level of the intended trait or state as well as for the meaning of the items themselves. Since items' meanings can shift with context, manipulating priming and sequence effects within an instrument can be used to estimate context–dependent meaning shifts, and thereby provide an estimate of the meaning used in an earlier item.

15.6.2 Adaptation of Equilibria

Over longer periods of time, the context–dependent thesaurus may adapt by revision of existing entries or by addition of new entries. As Betty becomes older, she is exposed to a wider variety of life contexts. A new experience may lead her to create a new context–dependent meaning of a word. A new entry in the context–dependent thesaurus implies a new equilibrium, i.e., a new attractor in the semantic network. Alternatively, a new experience may lead Betty to revise an existing context–dependent meaning. Revising the words in a context–dependent thesaurus entry for a target word could imply a change in the equilibrium for the context–dependent meaning of that target word. But it could also be that the process maintaining stability in the thesaurus would create a cascade of revisions, with many thesaurus entries shifted and reweighted in order that the equilibria themselves maintain stable relationships with each other.

As an example of a new attractor in a semantic network, suppose that when Betty was a teenager she lived by a lake. Her childhood friend Andrew said to her, "I felt weightless today." Betty might consult her context–dependent thesaurus and decide that Andrew meant that he felt like he was floating on the lake on a summer day, contented and relaxed. But a few years later, Betty becomes an astronaut. Now, when Andrew tells her, "I felt weightless today," Betty might decide that Andrew means that he feels disoriented and nauseous. This new attractor in the semantic network might be so strong that Betty might have a hard time assigning the old meaning of contented and relaxed even when speaking with someone she knows has never been in orbit.

To illustrate revision of an existing entry, suppose that at age 25 Betty marries her childhood friend Andrew. Soon after, Andrew says, "I feel contented." Betty recalls contexts in which she has felt contented, many of them shared with her childhood friend, and now husband, Andrew. These contexts include floating on the lake in the summer sun, and other childhood and young adult experiences. Fifty years later, Andrew again tells Betty, "I feel contented." Betty now has a lifetime of experiences including times that she and Andrew spent raising their children and playing with their grandchildren. We might suspect that Betty's overall context–dependent equilibrium point for her interpretation of Andrew's meaning of "contented" may not have shifted much over a lifetime. However, we expect that her thesaurus is likely to be richer than it was when she was 25. That is to say, many of the synonyms for this entry for "contented" are likely to be retained from her youth. Some synonyms may have been dropped, or may count less towards the central meaning. But other words are likely to have been added to the entry as she has more life experience.

Betty's context–dependent semantic network is likely to become more complex as she ages. Old context–dependent entries may include new words as she gains more life experiences. And new entries may be formed: new attractors in her semantic network. One implication of increasing complexity of the context–dependent semantic network is that verbal abilities would be expected to show less age–related decline than other abilities (Horn & Cattell, 1966; Salthouse & Davis, 2006, e.g.,).

Next, let us consider three possible mechanisms by which an entry in a context–dependent semantic network might be revised. First, a word or experience might be added to the pool of synonyms, antonyms, and life experiences in a thesaurus entry. Second, a word or life experience might be removed from the pool. And third, there might be some sort of weighting function that is applied, indicating how central a chosen word or experience in the pool is to the overall meaning. The weighting function would constitute a metric that measures the distance from a chosen word or experience in the pool to the equilibrium of the entry. These three mechanisms could be used to create a process–oriented definition of an entry in a context–dependent thesaurus.

But note that this process–oriented definition bears a striking resemblance to a latent variable as identified by confirmatory factor loadings and indicator variables. While some of the indicators may themselves be latent variables (i.e., words with entries in the thesaurus), some of the indicators are manifest (i.e, life experiences). There is a weighting function, a metric similar to factor loadings that relates each of the indicators to a *subjective* latent variable. We refer to an equilibrium of meaning in a semantic network as a subjective latent variable, since a latent variable and the model that identifies it exist in the mind of the communicator, and another similar latent variable and model exist in the mind of the audience.

15.7 SEMANTIC NETWORKS AND LATENT VARIABLE MODELS

Might it be that we could recast a context–dependent semantic network as one very large latent variable model? If so, we would expect that some objective function would be continuously updated as new life experiences occurred. An objective function that would permit maximum correct transmission of meaning would need to attend to maximizing the likelihood that we can correctly perform interpretations of others intentions, while at the same time maximizing the likelihood that others can correctly interpret what we intend to communicate. Based on the arguments presented so far, we conclude that this objective function would produce revisions in the semantic network that tend towards stable equilibria for context–dependent meanings. Of course, other objective functions are possible. For instance, a poet might wish to use words in ways that induce multiple ambiguous meanings so as to produce an artistic effect. Or a magician might intentionally mislead his audience by manipulating contextual effects.

Some form of weighted aggregation operation appears to be necessary in order to calculate a value for the intended objective function. One form of weighted aggregation function is the linear combination as used in a standard latent variable model. But a more general form of weighted aggregation may be required: In algebra, *rings* constitute a class of operations that includes weighted aggregation. It may be useful to expand the operators under consideration so as to include rings other than linear combinations of variables.

We have spoken of several operators in describing revisions of the semantic network. Inclusion, exclusion, re–weighting, and weighted aggregation over the pool of words in an entry appear to be necessary operations in order to update an entry in the semantic network. Creation of a new attractor or merging two attractors into one might be achieved with the same set of operations, similar to how factors can be more or less correlated while being indicated by a shared set of manifest variables. Or it may be that other operators, such as those from Boolean set theory would be useful.

In order to clarify the notion of how equilibrium dynamics of a subjective latent variable might be observed in psychological data, we next present an example: Quality of Life. Quality of Life is a particularly useful example since there appear to be both inter–individual differences in its meaning as well as intra–individual change in its meaning over the lifespan. Quality of Life is also helpful in pointing out that maintenance of a stable equilibrium of meaning may involve a dynamic process that could include regulation of behavior. That is to say, the process of maintaining stable meaning could itself affect the contexts to which we choose to expose ourselves.

15.8 QUALITY OF LIFE

Consider the meaning of "Quality of Life." Suppose that Quality of Life is a subjective latent variable with meaning that maintains a relatively stable equilibrium across the lifespan. Many resources considered to be substantially related to Quality of Life change, fluctuate, and eventually decline as people age. However, a number of studies have concluded that self–reported level of Quality of Life remains surprisingly age–invariant (Staudinger, Freund, Linden & Maas, 1999; Staudinger, 2000, see e.g.,). Perhaps, this subjective latent variable absorbs much of its meaning from one's interests and activities, in other words, from manifest variables—the facts with which one is acquainted. But, each of us can regulate our behavior and interests, implying that the manifest variable contribution to the meaning of Quality of Life could be revised as part of a regulatory process relative to a stable equilibrium. In other words, the manifest variables indicating the subjective latent variable could be adjusted in order that the subjective latent variable retains a stable meaning.

Suppose that a stable individual–specific equilibrium dynamic for Quality of Life motivates an individual to perform multiple resource–regulating activities. Thus, what remains stable over time may be the equilibrium value and attractor for Quality of Life; and thus the dynamics of the resource–regulating activities. Which resources are functionally related to an individual's Quality of Life may differ within and between individuals. We still expect that the relation between resources and Quality of Life would be systematic and could be described by a model.

One instrument for measuring Quality of Life is the Schedule for the Evaluation of Individual Quality of Life (SeiQOL) (O'Boyle, McGee, Hickey, Joyce, Brown, & O'Malley, 1993). Quality of Life is assessed by asking individuals to (a) name the five most important resources that determine their Quality of Life, (b) rate each area on a scale of 0 to 100, and (c) weight each resource for its relative importance by distributing 100 percent between the five resources. A Quality of Life score is then calculated by multiplying the rating for each resource by the weighting percentage and taking the sum.

If we are to now estimate the dynamics of an intra–individual attractor for Quality of Life, we need to define rules or mechanisms for resource–regulating activities. Longitudinal SeiQOL data have shown instances where individuals (a) shift relative weights, (b) increase or decrease activities so as to increase or decrease a resource, (c) replace one or more resources with others, and (d) adjust the midpoint and/or range of optimal Quality of Life. The midpoint of optimal Quality of Life is one way of estimating a stable equilibrium value. The range of optimal Quality of Life is a way of estimating the strength of the attractor—wide ranges imply a weak attractor while narrow ranges imply a strong attractor. The individual patterns in the SeiQOL data imply that the same equilibrium value and strength of attractor Quality of Life can be achieved with multiple combinations of resource levels. In addition, resource–regulating activities can result in intra–individual change in these combinations.

If an important resource for Quality of Life becomes depleted, we can think of two regulatory strategies that might be employed. First, one might re–weight the remaining resources to maintain the value of the subjective latent variable Quality of Life near its preferred equilibrium value—a good strategy if activities cannot improve the resource as in cases of chronic illnesses. Second, one might increase the weight applied to the low resource. This would lead to a magnified change in the subjective latent variable such that it was far from the stable equilibrium and would thus increase motivation for activities to regain the original resource. In this strategy each improvement in the target resource produces a high weighted change in the subjective latent variable—a useful strategy if resource impairment can be overcome.

If multiple resources cannot be rehabilitated, widening the Quality of Life attractor would allow deviations from the equilibrium value to not be as important, and thus reduce the need for engaging in resource regulating activities. On the other hand, narrowing the Quality of Life attractor would increase the importance of deviations from the equilibrium and could be used to motivate resource–regulating activities.

15.9 CONCLUSIONS

Davidson stated that some form of theory similar to that presented here is required.

> "The psychologist, skeptical of his ability to be certain how a subject is interpreting his instructions must add a theory of verbal interpretation to the theory to be tested. If we think of all choices as revealing a preference that one sentence rather than another be true, the resulting total theory should provide an interpretation of sentences, and at the same time assign beliefs and desires, both of the latter conceived as relating the agent to sentences or utterances. This composite theory would explain all behaviour, verbal and otherwise." (Davidson, 1984, p. 163)

Davidson makes a strong claim. In comparison, our goals are more modest, but not insubstantial. We believe that modeling the dynamics of how individuals maintain stable equilibria of meaning will lead to a broader understanding of a wide variety of latent constructs from fields such as personality, affect, motivation, and cognitive abilities. We believe that such models have the potential to partition variance in new and interesting ways that take into account the contributions of individual differences in participants' apprehension of a psychometric instrument.

Furthermore, we have had a glimpse of a largely unexplored area of inquiry for statistical theorists: algebraic rings that generalize on the notion of association as estimated by a scaled inner product between vectors. If we look carefully at empirical patterns of regulation in longitudinal data, we expect that an outline for a necessary and sufficient set of operations for models of the equilibrium dynamics of meaning will become apparent. While we do not make Davidson's claim, we do believe that such models will move the field closer to integrating interpretation of meaning into models of behavior.

REFERENCES

Davidson, D. (1984). Thought and talk. In D. Davidson (Ed.), *Truth and interpretaion* (pp. 155–170). Oxford: Clarendon Press.

Grice, H. P. (1957). Meaning. *The Philosophical Review, 66*(3), 377–388.

Horn, J. L., & Cattell, R. B. (1966). Age differences in primary mental ability factors. *Journal of Gerontology, 21*, 210–220.

Nesselroade, J. R. (2006). Quantitative modeling in adult development and aging: Reflections and projections. In C. S. Bergeman & S. M. Boker (Eds.), *Quantitative methodology in aging research* (pp. 1–19). Mahwah, NJ: Lawrence Erlbaum Associates.

O'Boyle, C. A., McGee, H. M., Hickey, A., Joyce, C. R. B., Brown, J., & O'Malley, M. (1993). *The schedule for the evaluation of individual quality of life*. Dublin: Department of Psychology, Royal College of Surgeons in Ireland.

Quine, W. V. (1970). On the reasons for indeterminacy of translation. *The Journal of Philosophy, 67*(6), 178–183.

Russell, B. (1905). On denoting. *Mind, 14*(56), 479–493.

Salthouse, T. A., & Davis, H. P. (2006). Organization of cognitive abilities and neuropsychological variables across the lifespan. *Developmental Review, 26*(1), 31–54.

Staudinger, U. M. (2000). Many reasons speak against it, yet many people feel good: The paradox of subjective well-being. *Psychologische Rundschau, 51*(4), 185–197.

Staudinger, U. M., Freund, A., Linden, M., & Maas, I. (1999). Self, personality, and life regulation: Facets of psychological resilience in old age. In P. B. Baltes & K. U. Mayer (Eds.), *The Berlin Aging Study: Aging from 70 to 100* (pp. 302–326). New York: Cambridge University Press.

Chapter 16
Applying Components Analysis to Attitudinal Segmentation

Krishna Tateneni and Megan Schiller[1]

Principal components analysis (PCA) is well known in both the quantitative psychology and marketing research communities. However, there is notable confusion of PCA with factor analysis (FA) that persists both in the literature and in practice. While graduate training plays a major role in this confusion, the situation is exacerbated by the use of computer programs with options for both PCA and FA (e.g., PROC FACTOR in SAS software). Users who run such programs without changing the default options are often unaware that they might have carried out a PCA while intending to do FA (defaults in most major statistical packages make it unlikely that FA will be carried out instead of PCA.)

Browne and Tateneni (2009) examine the similarities and differences between FA and PCA by considering the dominant purpose of each of these methods of data analysis. The dominant purpose of a method of data analysis is determined by the assumptions made in the derivation of the method, not by common usage (or misuse). Rethinking PCA in terms of dominant purpose leads us away from

[1] The authors thank Michael W. Browne for extensive commentary on an earlier draft of this chapter.

the restrictive identification conditions of principal components to thinking more generally about "optimal components." This approach also enables the rotation of principal components, which destroys the identification conditions of PCA, but retains optimality in the sense of predicting the original variables from a small set of "component" variables.

In this chapter, we continue the effort started in Browne and Tateneni (2009) to elucidate the differences between PCA and FA in terms of dominant purpose, which has motivated the development of a dedicated software program for Data-Based Components Analysis (DBCA; Browne and Tateneni, 2011). We also provide an example of the application of DBCA in the context of consumer segmentation. While the example is limited in scope, we will present all aspects of segmentation, from generating a candidate solution to profiling segments and developing a classification tool.

This chapter is intended to serve as a tutorial to those interested in PCA and its application to market segmentation, and a thorough literature review of PCA or market segmentation approaches is not included. For readers interested in a more detailed review of PCA, we suggest Kshirsagar (1972; see Chapter 11, *Principal Components*, and Section 11.2, *Optimality Properties of Principal Components*) and Rao (1964). For those interested in market segmentation, we suggest Wedel and Kamakura (2000). We do express our opinion at various points on both PCA and market segmentation, and although such comments are not intended to be controversial, they are based on our own experience rather than incontrovertible proof.

16.1 FACTOR ANALYSIS VS. PRINCIPAL COMPONENTS

Both FA and PCA may be represented by the same equation, a multivariate multiple regression of p dependent variables on m independent variables:

$$X = ZB' + E. \tag{16.1}$$

where X is an $N \times p$ data matrix, usually with rows representing persons and columns representing variables (N notes the number of persons). For convenience, we assume that the variables are standardized to have zero means and unit variances. Z is an $N \times m$ matrix of independent variables, and B is a $p \times m$ matrix of regression coefficients:

$$B' = \left(Z'Z\right)^{-1} Z'X. \tag{16.2}$$

There are two key differences between FA and PCA. First, the independent variables z in FA are latent variables. That is, the values of z are unobservable at the individual (or case) level. The scales for z are usually identified by setting means to 0 and standard deviations to 1. In contrast, the values of the independent variables z in PCA are known at the individual (or case) level; they are manifest variables called "components," and observations on these variables are called "component scores."

Second, in FA, there is a testable model. This model states that the correlations among the manifest dependent variables x are explained completely by the latent independent variables z. The implication is that the error terms e are uncorrelated. Goodness-of-fit of the FA model is measurable, and FA models may be compared using goodness-of-fit statistics. In PCA, however, there is no model. The error terms are correlated, and errors decrease in magnitude as more components are extracted.

The dominant purpose of FA is to analyze the pattern of interrelationships among the p dependent variables. These interrelationships are represented by the $p \times p$ variance–covariance or correlation matrix of the dependent variables, which is the primary source of data for analysis. In contrast, the dominant purpose of PCA is to condense data involving many variables p into data involving a much smaller number of variables m with a minimal loss of information. Although standard methods of PCA operate on the same $p \times p$ covariance/correlation matrix as FA, the primary source of data for analysis in PCA is the $N \times p$ data matrix. It is from this data matrix that component scores are calculated, so the original data matrix is replaced by a smaller data matrix of order $N \times m$.

16.2 OPTIMAL COMPONENTS VS. PRINCIPAL COMPONENTS

Let $r^2 (x_j.z)$ represent the squared multiple correlation of the jth dependent variable x_j on all m components. We choose the component matrix Z so as to retain as much of the information in the original data matrix X as possible, in the sense of maximizing the average squared multiple correlation with respect to Z:

$$\overline{r^2} = \frac{1}{p} \sum_{j=1}^{p} r^2 (x_j.z) = \frac{1}{Np} \text{tr} \left[X'Z (Z'Z)^{-1} Z'X\right]. \qquad (16.3)$$

The set of m components that maximize $\overline{r^2}$ are "optimal components" in that they predict the original data matrix as accurately as possible in the sense above. The elements of Z are the component scores and those of B are component loadings.

However, the optimal components are not uniquely defined because if Z maximizes $\overline{r^2}$, then so does $\tilde{Z} = ZT^{-1'}$ where $T^{-1'}$ is any nonsingular $m \times m$ matrix. To define T uniquely, identification conditions must be imposed on Z. The singular value decomposition of X (Eckart & Young, 1936) provides a set of optimal components with component scores Z and the associated component loadings B that satisfy the identification conditions:

$$B'B = \text{Diag}\ (B'B) \tag{16.4}$$

$$\text{Cov}\ (z, z') = N^{-1}Z'Z = I \tag{16.5}$$

The optimal components obtained by singular value decomposition of the data matrix are known as "principal components" and they satisfy the above identification conditions. From this viewpoint, "principal components" are a special class of optimal components; while they are easily computed, principal components are not usually convenient for interpretation. Other, more interpretable optimal components may be obtained by rotation.

Orthogonal or oblique rotation of principal components does not alter the optimal properties of the components. The rotated components continue to be optimal in the sense of minimizing the loss of information contained in the original variables, and the new optimal components satisfy other identification conditions associated with the rotation method that is employed. However, rotation destroys the identification conditions satisfied by principal components.

When PCA is viewed from the perspective of its dominant purpose, which is to condense data while minimizing loss of information, rotation is not only possible, but desirable in order to produce optimal components that are interpretable. The original identification conditions of principal components often do not yield interpretable components; they are mathematically convenient, but not practically relevant.

16.3 MARKET SEGMENTATION AND COMPONENT ANALYSIS

Market segmentation is a decision by a marketer to divide the overall market into two or more distinct segments of individuals, such that individuals within the same segment are similar in ways important to the marketer, while individuals in different segments are distinct from each other. Metrics for measuring similarity depend on the specific business objectives to be addressed by segmentation, and may include such variables as interest in using a new product, demographics, or attitudes. Segmentation is often described as a compromise between individually tailored and mass marketing, and is an important component of the lifecycle management of a product or service.

PCA has long been a part of algorithms developed for market segmentation, and there is a natural connection between the "dominant purpose" view of components analysis and segmentation. In the latter, one often starts the analysis with a large set of variables, and data condensation is the first step of any segmentation algorithm. Additionally, the matrix of component loadings together with the matrix of component scores addresses the respondent profiling problem, which is to replace complex response profiles on many different variables with more interpretable profiles on the small set of components. The pattern of component loadings is used to interpret the components, while the component scores are used to assign individuals to segments. The matrix of component scores is the basis for segmenting respondents, with the objective of segmentation being the grouping of respondents with similar patterns of component scores.

In our experience, PCA has been applied as a first step in market segmentation without much regard for the interpretability of the components. We believe that a better understanding of components analysis, and in particular the value of analytic rotation, will lead to better applications. To that end, we present here an example of a specific type of marketing problem, that of a "needs based segmentation." The objective of such a segmentation is to identify segments with different needs. For example, one segment of fast-food-restaurant goers may be seeking to fulfill a need for affordable dining, while a different segment may be seeking to fulfill a need for convenient dining options.

Applying components analysis to a set of ratings by a large number of individuals, where all the ratings are on the same Likert-type scale, seems a natural fit for the method. By defining a few key components, the large set of ratings can be condensed to a few key components that are easy to understand, especially with careful use of analytic rotation. As with traditional segmentation algorithms, the component scores could be input into a clustering algorithm or latent class analysis; however, with closer attention paid to the interpretability of the components, it may be possible to define segments simply by examining means or medians of the component scores, leading to greater clarity in the solution.

16.4 NUTRITIONAL ATTITUDES SURVEY

We surveyed 310 individuals, each of whom had at least one child between the ages of 9 months and 10 years. To qualify for the study, the individual also had to have an active role in purchasing and preparing food for their children. We asked the survey respondents to characterize their level of agreement with 18 attitudinal statements, half of which were about organic foods and half about food preparation. These statements are given in Table 16.1. They were presented in randomized order to respondents, with randomization taking place within each group of nine statements.

We also included some demographic questions so that we could later characterize our segments. These questions included items such as gender, age, marital status, and employment status. Additional questions in the survey addressed food purchasing behaviors (types of stores, frequency of purchase).

Table 16.1 Attitudinal statements included in the study

No.	*Statement*
1	Organic foods make grocery shopping less convenient because I have to shop at specialty stores
2	Organic foods usually taste better than nonorganic foods
3	Organic foods are not convenient because they have a shorter shelf life than nonorganic foods
4	Organic foods make it more difficult to ensure balanced nutrition because they lack added vitamins and minerals
5	Organic foods are healthier than nonorganic foods
6	Organic foods are not cost effective for my family
7	Organic foods are safer for my family than nonorganic foods
8	Organic foods are generally of a higher quality than nonorganic foods
9	Organic foods are better for the environment than nonorganic foods
10	I enjoy preparing meals from scratch for my family
11	I think restaurants offer healthy alternatives that I can feed my child/children
12	I prefer serving ready-to-eat meals because there is less cleaning up to do afterwards
13	Ease of preparation is a critical factor for me when I make food-buying decisions
14	I consider it important to prepare food for my child/children myself so I can be certain of the contents
15	I often buy ready-to-serve meals because it frees up my time to do other things with my family
16	I am opposed to feeding my child/children fast food
17	I often don't have the time to make my child/children dinner
18	I find it easier to get my child/children to eat pre-prepared foods than homemade food

Respondents were asked to rate their level of agreement with each statement using a five-point scale: Strongly disagree, somewhat disagree, neither agree nor disagree, somewhat agree, and strongly agree.

16.4.1 Segmentation Basis

When conducting a segmentation study, it is important to have specific business objectives drive the design of the survey. The business objectives should guide the key dimensions to be addressed, which in turn should provide clear direction on the variables that will constitute the segmentation basis (i.e., the variables which will be included in creating segments.) Often in practice, especially when PCA is involved, there is a tendency to include all possible variables in the analysis in the hope that the segments will be differentiated on a wide variety of variables. In our opinion, investing the initial effort in refining the segmentation basis down to a few key variables results in a clearer and more stable solution that is not susceptible to quirks in the specific

Table 16.2 Subset of six attitudinal statements selected for components analysis

No.	*Statement*
5	Organic foods are healthier than non-organic foods
7	Organic foods are safer for my family than nonorganic foods
8	Organic foods are generally of a higher quality than nonorganic foods
12	I prefer serving ready-to-eat meals because there is less cleaning up to do afterwards
15	I often buy ready-to-serve meals because it frees up my time to do other things with my family
18	I find it easier to get my child/children to eat pre-prepared foods than homemade food

segmentation algorithm being used. We also note that the larger the segmentation basis, the more likely the segmentation algorithm is to capitalize on chance characteristics in the sample.

For this study, we examined the pattern of correlations among the 18 attitudinal statements, and selected a subset of six statements, half dealing with organic foods and half with food preparation. These are given in Table 16.2.

We selected these statements because they are strongly correlated with each other within their two groupings and weakly correlated across grouping. In our study, which is intended to be tutorial in nature, there is no obvious business objective. However, our goal was to segment on the two dimensions of attitudes towards organic foods and food preparation, which guided the development of the 18 attitudinal statements. Further, we would like to segment the respondents on the basis of a few key indicators of the two dimensions.

16.4.2 Optimal Components

We used the program DBCA (Browne & Tateneni, 2011) to conduct the component analysis. DBCA is a program dedicated to components analysis, and it applies a singular value (Eckart–Young) decomposition (Eckart & Young, 1936; also see Weisstein, 2011) directly to the data matrix, without calculating correlations. A wide range of orthogonal and oblique rotation methods are built into the program. Unlike in FA, the rotation in DBCA is double-sided: Both component loadings and component scores are changed by rotation.

Table 16.3 gives the cumulative proportion of variance accounted for by each additional component. The first two components account for over two-thirds of the total variance in the six attitudinal variables. In this analysis, we standardized the data matrix so that the six variables have means of 0 and standard deviations of 1.

Table 16.3 Cumulative proportion of variance accounted for by each additional component

No.	*Cumulative proportion of total variance*
1	40%
2	69%
3	82%
4	89%
5	95%
6	100%

Table 16.4 Variance composition for a two-component solution

Statement	*Common variance*	*Residual variance*
Organic foods are healthier than nonorganic foods	76%	24%
Organic foods are safer for my family than nonorganic foods	79%	21%
Organic foods are generally of a higher quality than nonorganic foods	75%	25%
I prefer serving ready-to-eat meals because there is less cleaning up to do afterwards	68%	32%
I often buy ready-to-serve meals because it frees up my time to do other things with my family	74%	26%
I find it easier to get my child/children to eat pre-prepared foods than homemade food	44%	56%

The variances were fairly uniform, so we chose to have component loadings that could be interpreted as correlations in orthogonally rotated solutions.

Table 16.4 gives the variance composition for a two-component solution. The statements about organic foods were fairly strongly correlated with each other, while the food preparation statements had weaker within-group correlations. This is reflected in Table 16.4, with greater residual variances for two of the food preparation statements.

Table 16.5 gives the principal-components solution. A pattern is apparent, however the components are hard to interpret. To aid the interpretation, we rotated the principal-components solution to a partially specified target, as given in Table 16.5. We found that an oblique rotation makes the components slightly negatively correlated; here we present an orthogonal rotation, which is nearly identical. The difference in interpretability between rotated and unrotated component loadings is obvious.

Table 16.5 Unrotated and rotated component loadings for a two-component solution

Statement	*Unrotated loadings*		*Partial target*		*Rotated loadings*	
Organic foods are healthier than nonorganic foods	.83	.28	–	0	.87	−.08
Organic foods are safer for my family than nonorganic foods	.83	.31	–	0	.89	−.05
Organic foods are generally of a higher quality than nonorganic foods	.80	.34	–	0	.87	−.01
I prefer serving ready-to-eat meals because there is less cleaning up to do afterwards	−.37	.74	0	–	−.04	.82
I often buy ready-to-serve meals because it frees up my time to do other things with my family	−.39	.76	0	–	−.05	.86
I find it easier to get my child/children to eat pre-prepared foods than homemade food	−.33	.57	0	–	−.07	.66

In the rotated solution, the first component is clearly associated with positive attitudes towards organic foods while the second component is clearly associated with positive attitudes towards prepared foods.

16.4.3 Segmentation

We used median splits on the two sets of component scores to create four segments of respondents. We also analyzed the behavioral and demographic data from the survey separately for each segment. Table 16.6a shows that these segments are fairly uniform in size. In Table 16.6a, we give the "Top 2 Box" scores (percent who somewhat agree or strongly agree with each statement) for the 6 attitudinal statements that comprised the segmentation basis. As expected, we see strong differences in these attitudes across the segments.

In Tables 16.6b and 16.6c, we give data on selected demographic and behavioral variables. Differences in these sets of variables are seen by virtue of the correlation that exists between these variables and those selected to be part of the segmentation basis. Careful planning *prior* to launching a quantitative segmentation study can help uncover important variables to include in the survey. Unfortunately, in practice, many segmentation studies simply include a large mass of variables with the hope that several will "stick" and differences will be discovered in the analysis stage of the study.

It is useful to have a verbal summary of the "segment profiles" which summarizes the patterns seen in the data, and highlights similarities and differences among the segments. Table 16.7 shows what such a summary might look like for the segmentation

Table 16.6a Selected attitudinal data by segment (percent who agree or strongly agree)

	Segment 1	*Segment 2*	*Segment 3*	*Segment 4*
Proportion of sample	23%	27%	22%	28%
Organic foods are healthier than nonorganic foods	99%	60%	32%	97%
Organic foods are safer for my family than nonorganic foods	100%	56%	43%	96%
Organic foods are generally of a higher quality than nonorganic foods	97%	45%	25%	97%
I prefer serving ready-to-eat meals because there is less cleaning up to do afterwards	0%	4%	39%	34%
I often buy ready-to-serve meals because it frees up my time to do other things with my family	1%	1%	53%	64%
I find it easier to get my child/children to eat pre-prepared foods than homemade food	3%	12%	44%	44%

Table 16.6b Selected demographic data by segment

	Segment 1	*Segment 2*	*Segment 3*	*Segment 4*
Proportion of sample	23%	27%	22%	28%
Gender				
Female	76%	64%	59%	67%
Male	24%	36%	41%	33%
Age				
40 years and over	45%	39%	40%	26%
Less than 40 years	55%	61%	60%	74%
Location				
Urban	17%	29%	25%	38%
Suburban	79%	62%	62%	60%
Rural	4%	9%	13%	2%
Political leaning				
Conservative	27%	21%	26%	32%
Moderate	28%	41%	46%	34%
Liberal	44%	34%	24%	34%
Prefer not to answer	1%	4%	4%	0%

Table 16.6c Selected behavioral data by segment

	Segment 1	*Segment 2*	*Segment 3*	*Segment 4*
Proportion of sample	23%	27%	22%	28%
Frequency of serving pre-prepared food				
Never	30%	24%	1%	7%
1–3 times a week	63%	69%	76%	71%
4+ times a week	7%	7%	23%	22%
Frequency of organic food purchases				
Never	3%	26%	22%	3%
Occasional	38%	51%	71%	49%
Regular	59%	23%	7%	48%
Grocery shopping locations used regularly				
Large supermarket chains	73%	67%	72%	84%
Natural and organic stores	68%	40%	25%	65%
Wholesale or warehouse	54%	35%	32%	53%
Convenience stores	1%	1%	0%	6%
Specialty food stores	23%	8%	9%	24%
Farmers markets/stands	48%	27%	28%	40%

Table 16.7 Verbal summary comparing each segment to the others

	Organic on-your-own	*You process I prep*	*Conventional convenience*	*Healthy but helpless*
Proportion of sample	23%	27%	22%	28%
Description	Organic purchasers, prepare food themselves	Nonorganic purchasers, prepare food themselves	Nonorganic and pre-prepared food purchasers	Organic and pre-prepared food purchasers
Demographics	More likely to be older, female, liberal, in a suburban setting	More likely to be male, politically and geographically diverse	More likely to be male, in a rural setting	More likely to be younger, female, liberal, in an urban setting
Shopping behavior	More likely to shop at organic stores and farmers markets	Diverse shopping habits	Less likely to shop at organic stores and farmers markets	More likely to shop at organic stores and farmers markets

Table 16.8 A classification tool for the four-segment solution

Variable	*Values*
What is the highest level of education you have completed?	Some high school High school/GED Some college Associates Bachelors Advanced degree
Which of the following best describes the community where you live?	Urban Suburban Rural
How often do you buy organic food for your family?	Regularly Occasionally Never
(If applicable) In the last year my purchasing of organic foods has:	Increased Stayed the same Decreased
In a typical week I feed my child/children frozen food or pre-prepared food:	Never 1–3 times 4–5 times More than 5 times
In a typical week I eat out with my child/children:	Never 1–3 times 4–5 times More than 5 times
Do you shop for groceries at natural and organic grocery stores such as Whole Foods and Trader Joe's?	Yes No

Table 16.9 Predictive accuracy of a naïve Bayes classifier in the sample of 310 respondents

Predicted segment[a]	*Organic on-your-own*	*You process I prep*	*Conventional convenience*	*Healthy but helpless*
Organic on-your-own	61%	23%	2%	15%
You process I prep	14%	55%	16%	14%
Conventional convenience	11%	23%	45%	21%
Healthy but helpless	16%	21%	16%	47%

[a] The rows in the above table show the percent of respondents in each predicted segment that fall into each of the actual segments (i.e., the diagonal entries are the likelihood that the predicted segment is correct.)

in this tutorial example; it is not intended to be detailed or comprehensive. We were also able to give each segment a descriptive name on the basis of the summary. The names are an easy way to refer to the segments, and often they also serve as mnemonics for the segment profiles themselves.

In many segmentation studies, the final step is to develop a "classification" or "typing" tool, to provide guidance on classifying individuals outside the study sample. For example, it may be useful to pre-classify a database of physicians so that pharmaceutical sales representatives can prioritize their calling plans. Another example might involve on-the-fly classification of visitors to a retail website in order to customize the information presented by segment.

We examined the nonattitudinal questions in our survey to see which, if any, were predictive of segment membership. In practice, when high accuracy is a business need, the choice of variables in the segmentation basis should be made carefully to ensure that a suitable classification tool can be developed. In our study, we found that a set of seven different questions was reasonably predictive of segment membership. These are given in Table 16.8.

A naïve Bayes classifier (Zhang, 2004) based on these seven questions is moderately effective at classifying respondents in our sample, with an overall accuracy of 50%, which is about twice the accuracy expected if respondents are classified with no prior information except segment sizes. In Table 16.9, each row gives the percent of respondents in the predicted segment that fall into each of the original segments. Accuracy does vary by segment, but not greatly.

16.5 CONCLUSIONS

In our opinion, traditional approaches to market segmentation rely too much on a "kitchen blender" approach, in which the analyst uses most if not all of the variables in the segmentation basis. We believe it is better to focus on one or two key business issues and select a small segmentation basis to address those issues. Segmentation is not a scientific theory, but a business decision; as such, segmentation methods will always be algorithmic in nature. We believe that components analysis, with intelligent use of rotation methods, is a useful tool in creating segmentation solutions. Although the example we have provided in this chapter is limited in scope, our hope is that we have outlined how components analysis may be used in a market segmentation study.

REFERENCES

Browne, M. W., & Tateneni, K. (2009). Principal components as a method for data condensation. Paper presented at the annual meeting of the Society of Multivariate Experimental Psychology.

Browne, M. W., and Tateneni, K. (2011). DBCA: Data-Based Components Analysis [Document and computer software]. http://quantrm2.psy.ohio-state.edu/browne/software.htm

Eckart, C., & Young, G. (1936). Approximation of one matrix by another of lower rank. *Psychometrika, 1*, 211–218.

Kshirsagar, A. M. (1972). *Multivariate statistical analysis*. New York: Marcel Dekker.

Rao, C. R. (1964). The use and interpretation of Principal Component Analysis in applied research. *Sankhya: The Indian Journal of Statistics, Series A, 26*, 329–358.

Wedel, M., & Kamakura, W. (2000). *Market segmentation—Conceptual and methodological foundations*, 2nd ed. Boston: Kluwer Academic Publishers.

Weisstein, Eric W. (2011). Singular Value Decomposition: From MathWorld—A Wolfram Web Resource. http://mathworld.wolfram.com/SingularValueDecomposition.html (accessed February 12, 2011).

Zhang., H. (2004). The optimality of naïve Bayes. In *Proceedings of the 17th International FLAIRS conference*. California: Association for the Advancement of Artificial Intelligence (AAAI) Press.

Author Index

Subject Index

For Product Safety Concerns and Information please contact our EU representative GPSR@taylorandfrancis.com
Taylor & Francis Verlag GmbH, Kaufingerstraße 24, 80331 München, Germany

www.ingramcontent.com/pod-product-compliance
Ingram Content Group UK Ltd.
Pitfield, Milton Keynes, MK11 3LW, UK
UKHW021300190926
475701UK00009B/165

* 9 7 8 0 4 1 5 6 3 7 7 8 7 *